Nanostructured Thermoelectric Films

Zhiyu Hu · Zhenhua Wu

Nanostructured Thermoelectric Films

上海交通大學出版社
SHANGHAI JIAO TONG UNIVERSITY PRESS

Springer

Zhiyu Hu
Shanghai Jiao Tong University
Shanghai, China

Zhenhua Wu
Shanghai Jiao Tong University
Shanghai, China

ISBN 978-981-15-6520-5 ISBN 978-981-15-6518-2 (eBook)
https://doi.org/10.1007/978-981-15-6518-2

Jointly published with Shanghai Jiao Tong University Press
The print edition is not for sale in China (Mainland). Customers from China (Mainland) please order the print book from: Shanghai Jiao Tong University Press.

This Springer imprint is published by the registered company Springer Nature Singapore Pte Ltd.
The registered company address is: 152 Beach Road, #21-01/04 Gateway East, Singapore 189721, Singapore

The soul to dare, the will to do!

To our families and friends for their selfless love and continued support

Foreword

Energy and environmental protection are major issues related to China's national economic development and national security. In today's world, energy security is a priority area for national security in all countries. As the world's largest energy consumer, how to effectively guarantee national energy security and effectively protect the country's economic and social development has always been the primary issue of China's energy development. The development trend of upcoming energy revolution presents several characteristics: the primary energy structure is in the process of changing from high carbon to low carbon, and new energy and renewable energy will form innovated low-carbon energy structures. In this round of evolution, electricity will become the main body of final energy consumption, and energy technology innovation plays a decisive role in the energy revolution.

In the modern society, with the increasing popularity of electricity-consuming equipment, people's demand for electricity is increasing day by day. At present, China's power supply is mainly supported by burning coal, oil, natural gas, and other petrochemical energy sources, although some improved power generation efficiency technologies with environmental protection control are gradually being put into use. Yet large-scale high-temperature combustion still inevitably brings environmental pollution and releases a large amount of greenhouse gases.

Everything grows by the power of sun in the nature. Due to the continuous irradiation of the sun, the earth will always maintain a certain temperature; coupled with human activities, global warming has become a fact that humans have to face directly. People have been actively looking for technologies to slow or reduce global temperature rise. In fact, from the Industrial Revolution to the present, human consumption of countless petrochemical energy has already caused the temperature of the earth to rise by about 1 °C. What if we take advantage of the huge thermal energy contained in this 1 °C temperature difference, we can obtain a sustainable energy supply forever. According to the Paris Agreement adopted by nearly 200 world leading parties in 2015, the main goal is to control the global average temperature rise within 2 °C in this century and the global temperature rise to within 1.5 °C above the pre-industrial level.

For a long time, mankind has always been searching for an inexhaustible energy supply in nature. Is it possible to obtain a technology that directly converts the ultra-low-quality (temperature difference less than 25 °C) inexhaustibly ambient heat energy of the earth's surface to electrical energy without providing additional energy? Once with such advanced energy technologies, after large-scale implementation in the future, mankind will then obtain truly sustainable and completely environmentally friendly green energy, and completely get rid of dependence on petrochemical energy. There is also a very important reason for using environmental thermal energy to generate electricity. It is possible to fundamentally solve the balance and fairness that exist in the world's energy sources. At present, all energy sources, whether oil, natural gas, hydropower, nuclear power, solar energy, and wind energy, have a fundamental problem of which the energy resources distributions are uneven and unfair. The uneven distribution of these energy sources has caused countless disputes and many wars that have brought countless disasters to the people of all countries in the world.

Only the distribution of environmental thermal energy on the surface of the earth is ubiquitous and completely free for charge. Physics tells us that thermal energy is volumetric energy. The thermal energy contained in a 1 °C (or K) temperature difference of an object of a certain weight without phase change is exactly the same. For example, a piece of 1 kg iron block, the thermal energy contained between 40 °C and 41 °C above zero is exactly the same as the thermal energy between 40 °C and −41 °C below zero. If we can use temperature difference to generate electricity, we can completely solve this most difficult world problem!

Traditional heat engines convert heat energy into electrical energy through mechanical motion. According to Carnot's law, they must work under large (generally greater than 200–300 °C) temperature difference conditions to effectively work and obtain more economical output. At present, the overall power generation efficiency of the diesel/gasoline power generation system is less than 30%, and 70% of the energy is wasted as heat loss. According to the 2018 National Yearbook, the comprehensive power generation efficiency of China's thermal power plants and nuclear power plants is 44.6%, and the outlet temperature of a large amount of cooling water is 30–60 °C. Coupled with other industrial waste heat, the heat loss wasted each year is equivalent to the annual power generation of more than 100 Three Gorges Dams (the annual power generation of the Three Gorges Power Station in 2018, was 101.6 billion kWh, and the investment cost of the Three Gorges Dam in 1993, was nearly 100 billion Yuan Renminbi). In addition, solar photovoltaic power generation only uses part of the energy in the visible spectrum of solar energy, accounting for 2/3 of the thermal energy, which is not only unusable, but also increases the temperature of the photovoltaic panel and reduces the efficiency of power generation. And these ultra-low-quality heat energy with huge content, because the temperature difference is very small, the traditional heat engine cannot be used at all!

There are two main factors restricting the development and application of current thermoelectric materials: one is the low thermoelectric conversion efficiency of the materials, and another is the high cost of thermoelectric devices and the immaturity of large-scale manufacturing technology. In recent years, the research of thermoelectric materials has mainly focused on the middle and high temperature range (such as PbTe, Cu2Se, half-Heusler, SnSe, etc.), and has achieved remarkable results. However, the research on thermoelectric materials in the middle and low-temperature regions is lacking, and the progress is slow. In recent years, the ZT value of Bi_2Te_3, GeTe, and other low-temperature thermoelectric material systems has increased, but the efficiency of the device is still not high, and the economy, reliability, and stability are still poor, which is far from enough to support the future scale and commercialization of thermoelectric power generation technology application.

The performance of thermoelectric materials can be characterized by thermoelectric figure of merit ZT. The theoretical predictions and experimental results at this stage show that the two-dimensional superlattice and nanocomposites can significantly improve the thermoelectric figure of merit. When the material reaches the nanometer scale, the density of electron energy states near the Fermi level changes, which causes dimensional and size limitations and interface scattering effects on phonon transmission, increasing the degree of freedom in regulating heat and electricity transport. Therefore, the low-dimensional material is an effective means to improve thermoelectric performance.

Due to the advancement of modern advanced processing technologies and characterization methods, traditional bulk thermoelectric materials containing nano-structured components have been developed to obtain high ZT values. At present, the research on thermoelectric materials mainly focuses on two methods: one is a bulk thermoelectric material containing nanostructures, and the other is a nanothermoelectric material. At present, the ZT value of the best commercial thermoelectric materials is only about 1, and the conversion efficiency is only about 57%. Therefore, when the ZT value of the thermoelectric material is lower than 1, the conversion efficiency is quite low. When it reaches 2, it can be used for the recovery of waste heat. Only when the ZT value reaches 4 or 5, it has the ability to cool the refrigerator.

The thermoelectric material can be designed by controlling different dimensions, and the low-dimensional thermoelectric material has excellent thermoelectric performance. After the thermoelectric material is reduced in dimension, the density of states near the Fermi level will increase, and the effective mass of carriers and the absolute value of the Seebeck coefficient will increase. On the other hand, the quantum confinement effect of phonons reduces the thermal conductivity. In addition, effects such as quantum confinement increase carrier mobility. Compared with the development of more mature bulk thermoelectric materials and devices, thermoelectric thin films are easily combined with modern micro/nano fabrication process technology to make micro devices, which are suitable for a wider field. This book aims to construct thermoelectric thin films in nanometers, regulate the factors that affect the performance of thermoelectric thin films, and provide reference for the low-dimensional application of thermoelectric thin films.

In this book, we use a variety of physical vapor deposition techniques and micro/nano processing techniques, trying to explore a method that can be applied to large-scale preparation of high-efficiency nano-structured thin film thermoelectric materials. All content in this book is a team effort of the Institute of Nano/Micro Energy, Shanghai Jiao Tong University, Shanghai, China. Authors would like to thank Zeng Zhigang, Yang Penghui, Wu Yigui, Liu Yanling, Cao Yi, Xiao Danping, Tian Zunyi, Shen Binjie, Shen Chao, Lin Cong, Fang Bo, Ye Fengjie, Zhang Xiangpeng, Zhang Haiming, Wang Zhichong, Zhang Ziqiang, Yang Gang, Hu Yangsen, Mu Erzhen, Wu Zhimao, and Liu Yang. All experimental results were obtained by ourselves with the limited scientific research conditions of our laboratory in the past decade. These first-hand experimental results and the data are true and reliable.

Zhiyu Hu
Zhenhua Wu
Shanghai Jiao Tong University
Shanghai, China

Acknowledgements

We are very grateful to many experts and friends who have always cared about and supported our work. We also would like to thank the National Natural Science Foundation of China (51776126) and Shanghai Jiao Tong University for their supports!

Contents

Symbols

Eg	Band gap
Φ	Work function
E_A	Electronic affinity
E_F	Fermi level
E_0	Vacuum level
E_c	Conduction band energy level
E_v	Valence band energy level
v_p	Phonon group velocity
ZT	Figure of merit
α	Seebeck coefficient
σ	Electrical conductivity
κ	Thermal conductivity
κ_e	Electronic thermal conductivity
κ_L	Lattice thermal conductivity
n	Carrier concentration
η	Reduced Fermi Level
γ	Scattering factor
m^*	Effective mass
B	Magnetic induction
U_H	Hall voltage
R_H	Hall coefficient
m_0	Electronic quality
ϕ_b	Barrier height
C	Specific heat capacity
λ	Wavelength
D	Grain size
T_h	Hot-end temperature

T_c	Cold-end temperature
μ	Carrier mobility
k_B	Boltzmann constant
e	Electronic charge
L	Lorentz constant
l	Carrier mean free path
Λ	Phonon mean free path
PF	Power factor

Chapter 1
Research Background and Current Situation

The high-temperature combustion that relies on burning coal, oil, natural gas and other primary petrochemical energy sources inevitably brings environmental pollution, and the release of a large amount of greenhouse gases also causes global temperatures to rise. The massive consumption of energy has also led to an energy crisis. There is an urgent need to obtain new types of sustainable, environmentally friendly green energy to dispose of dependence on petrochemical energy. Thermoelectric conversion directly converts environmental thermal energy into electrical energy, providing a new way.

1.1 Research Background

The energy issue has become one of the hottest issues that people are most concerned about in the 21st century. Energy is an important material foundation for the development of human society and economy. With the continuous improvement of human society and the continuous progress of the economy, energy consumption has become larger and larger, and the energy crisis has also emerged in recent years. The development of human civilization is accompanied by the increase of energy consumption. According to the prediction of the United Nations, in the first 20 years at the beginning of the 21st century, world energy consumption will increase at a rate of 2% per year. According to the forecast of the US Department of Energy, the total energy consumption of the world in 2020 will increase by about 59% compared to 1999, and will increase by 25% by 2040. Since human society entered the industrial era, coal has become the main source of energy. After the Second World War, oil and gas were quickly used as m main energy sources. By the late 20th century, a world energy system with fossil fuels such as coal, oil and natural gas as the main body was gradually formed. It was not until the end of the 20th century and the beginning of the 21st century that a world energy system based on fossil fuels and supplemented

Z. Hu and Z. Wu, *Nanostructured Thermoelectric Films*,
https://doi.org/10.1007/978-981-15-6518-2_1

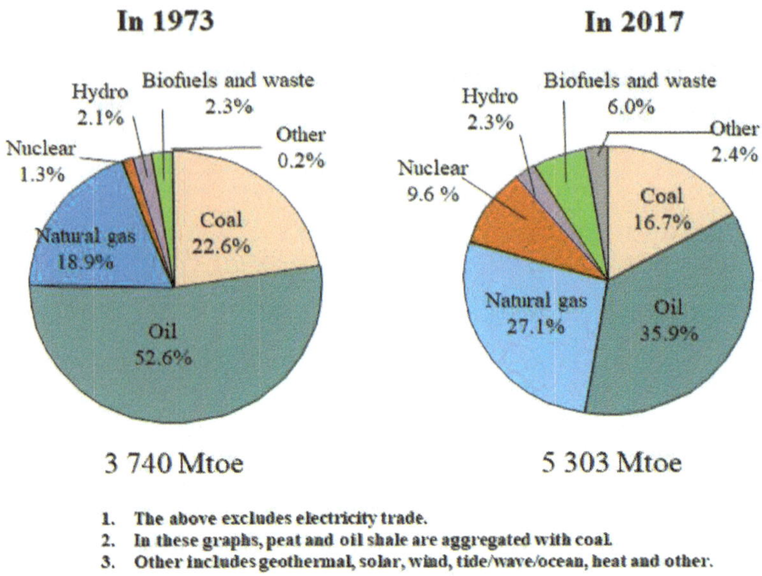

Fig. 1.1 Current world energy consumption (IEA2018 Key World Energy Statistics)

by new energy was formed. According to the International Energy Agency's report, oil, natural gas, and coal are still the main sources of energy in the world (Fig. 1.1).

In recent years, with the continuous development of new energy technologies, the proportion of new energy in the world's energy structure has continued to increase. However, according to research reports by the World Energy Council (WEC) and the International Institute for Applied Systems Analysis (IIASA), it is expected that traditional fossil fuels such as coal, oil and natural gas will still be the main body of world energy by the middle of the 21st century. It is estimated that by 2100, the proportion of renewable energy such as solar energy, wind energy and biomass energy will increase, which may account for about half of the world's energy composition. However, currently, non-renewable energy sources such as coal, oil, and natural gas, which are the main energy sources, are almost exhausted [1]. At the same time, the harmful substances generated during the use of traditional fossil fuels have caused great harm to the environment and caused serious environmental pollution problems. In addition, the current low energy utilization rate is also an important issue in the energy utilization process. The low effective utilization of energy causes a large amount of energy waste. Table 1.1 shows the utilization efficiency of fossil energy in the major energy consuming countries [2]. As can be seen from the table, about two-thirds of energy in China is wasted during the energy utilization process, and half of the world's energy is wasted. The waste of energy is mainly lost in the form of heat, such as the combustion of internal combustion engines, factory chimneys, and heating of electrical components.

Table 1.1 Energy structure and utilization efficiency of major world countries

Countries	Coal %	Oil %	Natural gas %	Energy efficiency %
China	65.59	24.62	2.71	36.81
United States	24.15	39.00	26.20	50.00
Japan	20.67	47.62	13.68	52.51
Germany	25.68	38.62	22.56	52.51
India	55.61	30.05	7.81	40.06
Russia	15.39	19.20	54.61	54.08
Brazil	6.76	48.11	6.93	62.26
World average	25.50	37.45	24.26	50.32

In order to solve the above problems, the search for new environmentally friendly energy sources that can replace traditional fossil fuels has become the main way to solve energy problems. The emergence of thermoelectric technology provides a new way to solve the energy crisis and environmental pollution.

1.2 Thermoelectric Effects and Development

Thermoelectricity refers to the phenomenon that thermal energy and electrical energy can be converted to each other under a temperature gradient or differential voltage. Thermoelectric effects include Seebeck effect, Thomson effect, and Peltier effect [3]. The Seebeck effect refers to the phenomenon of materials producing potential differences in conditions of temperature gradients, which were discovered by German scientist Seebeck in 1821. The Peltier effect is the inverse effect of the Seebeck effect. The Peltier effect was discovered by French scientist Peltier in 1834 which refers to in addition to Joule heat, currents are generated at both ends of the conductor when current is passed through them. The exotherm and endotherm depend on the direction of the incoming current. The Thomson effect refers to the fact that when a current is passed through a conductor with a temperature difference, in addition to the Joule heat generated by the conductor, an endothermic or exothermic reaction can occur at both ends of the input current of the conductor. This effect was discovered by British scientist Thomson in 1856. Since the pyroelectric effect was discovered, its application has developed rapidly, and various thermoelectric devices have been continuously developed. Thermoelectric device is a device that can use the Seebeck effect of thermoelectric materials to directly convert thermal energy into electrical energy [4]. Thermoelectric conversion is a promising green energy utilization method. Thermoelectric devices have the inherent advantages of no noise, pollution, and mechanical vibration [5–9], which makes them widely used in vehicles [10–12], wearable devices [13–19], and solar systems [11, 20–22] and industrial waste heat recovery systems [22–24] and so on. Thermoelectric devices

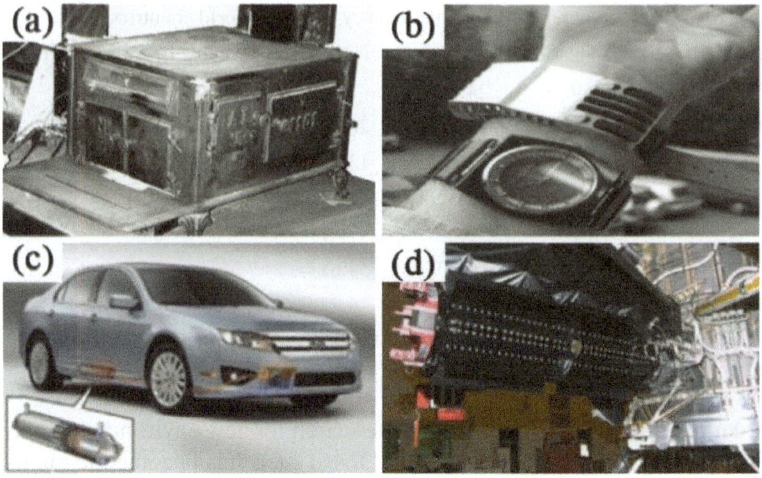

Fig. 1.2 Partial application of thermoelectric devices: **a** Wood-fired furnace heating power generation [153], **b** Human thermal power generation watch [154], **c** Ford automobile exhaust pipe waste heat power generation [155], **d** Voyager radiant heat source [40]

can directly convert human body heat or waste heat into electrical energy, thereby improving energy efficiency and reducing energy costs [25]. Shows some applications of thermoelectric devices. In recent decades, with the development of new thermoelectric materials and new thermoelectric device processing methods, such as screen printing [13, 19, 26, 27], physical vapor deposition [28–31], spark plasma sintering [32–36], hot pressing [37, 38], metal organic chemical vapor deposition [39], etc. The research and application of thermoelectric materials and devices have once again attracted researchers' interest (Fig. 1.2).

1.2.1 Seebeck Effect

In this book we mainly concerned with the application of the Seebeck effect: the conversion of heat to electricity. The Π-like structure shown in Fig. 1.3 is called a P-N thermoelectric pair. When there is a temperature difference between the two ends of the thermoelectric materials, the carriers in the materials will move from the high temperature end to the low temperature end under the driving of the temperature difference, thereby generating a voltage in the closed loop. In practice application, in order to obtain a higher voltage, multiple thermoelectric pairs are often connected in series. Among them, the carriers of the P-type semiconductor material are mainly holes; and the carriers of the N-type semiconductor material are mainly electrons. The Seebeck voltage generated by a material is proportional to the temperature difference across the material. This proportionality factor is called the Seebeck coefficient. The Seebeck coefficient of a P-type semiconductor is positive, and the Seebeck

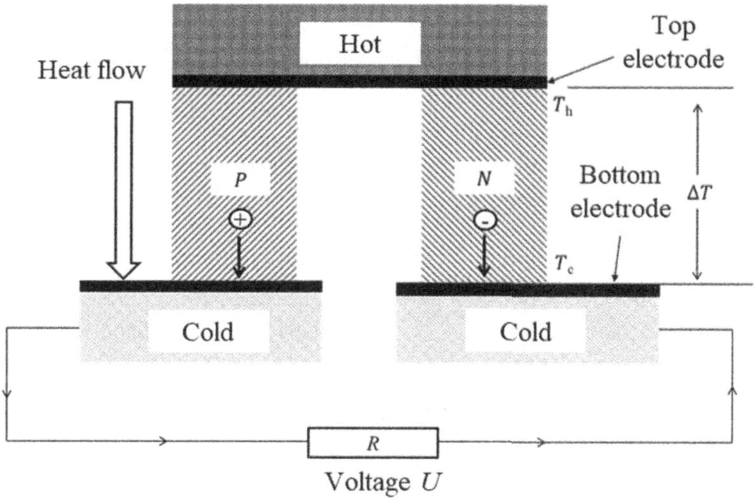

Fig. 1.3 Seebeck effect schematic

coefficient of an N-type semiconductor is negative. The Seebeck coefficient is the Seebeck voltage generated when the temperature difference between the two ends of the thermoelectric material is 1 K. The Seebeck effect principle is shown in Fig. 1.3 and is expressed by Eq. 1.1, where α is the Seebeck coefficient and U is Seebeck voltage, is the temperature difference, T_h is the hot-end temperature, and T_c is the cold-end temperature:

$$U = \alpha \cdot \Delta T = (\alpha_P - \alpha_N) \cdot (T_h - T_c). \tag{1.1}$$

1.2.2 Thermoelectric Material Performance Parameters

Thermoelectric materials play a vital role in thermoelectric devices [40]. The development and application of thermoelectric devices is accompanied by the development and research of thermoelectric materials. The performance of thermoelectric materials can be characterized by the thermoelectric figure of merit, which is related to the physical properties of thermoelectric materials and can be expressed as [41]

$$ZT = \frac{\sigma \cdot \alpha^2}{\kappa} \cdot T, \tag{1.2}$$

where δ is the conductivity of the thermoelectric material, α is the Seebeck coefficient of the thermoelectric material, $\kappa = \kappa_e + \kappa_L$ is the thermal conductivity of the thermoelectric material, κ_e and κ_L are the electronic thermal conductivity and

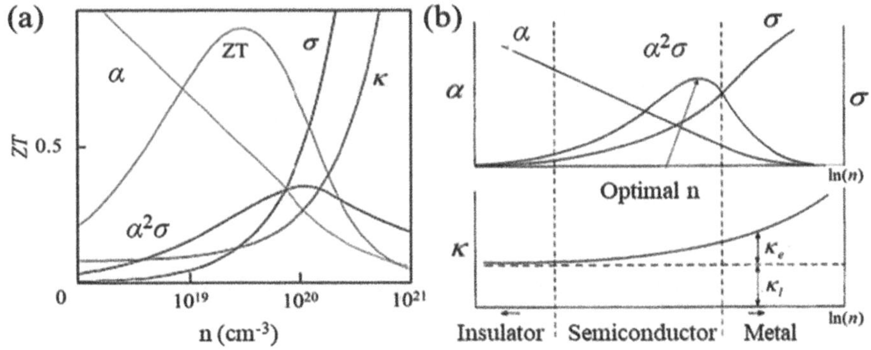

Fig. 1.4 Thermoelectric parameters **a** relationship between related parameters of thermoelectric materials, **b** Seebeck coefficient, thermal conductivity, and electrical conductivity at the optimal carrier concentration 1×10^{19} cm^{-3} [44]

the lattice thermal conductivity, respectively, and are temperature. According to the thermoelectric figure-of-merit Eq. 1.2, it can be seen that high-performance thermoelectric materials require high electrical conductivity, high Seebeck coefficient, and low thermal conductivity. Where represents the electrical properties of thermoelectric materials, also known as power factor [40], and represents the thermal properties of thermoelectric materials. The Seebeck coefficient, thermal conductivity, electrical conductivity, and ZT value of a thermoelectric material are functions related to the material's energy band structure, carrier concentration, and other parameters [42]. The relationship between the ZT value, power factor, thermal conductivity, electrical conductivity, and carriers of thermoelectric materials is shown in Fig. 1.4 [43].

For a long time after the Seebeck effect was discovered, the thermoelectric materials were mainly metals and their alloys. Because metals have very low Seebeck coefficients or output values, early thermoelectric materials were mainly used for thermocouples [44]. Until the 1930s, with the discovery of semiconductor thermoelectric materials and the development of semiconductor thermoelectric material theory [45], the performance of thermoelectric materials has been greatly improved [46]. However, from the 1960s to the 1990s, the development of thermoelectric materials was relatively slow, and no significant progress was made [47]. Until the mid-1990s, theories predicted that nanostructure engineering could greatly improve thermoelectric efficiency. Since then, there has been another wave of research into thermoelectric materials [48] (Fig. 1.5).

With the advancements of modern micro/nano-fabrication process technology and characterization methods, traditional bulk thermoelectric materials containing nanostructured components have been developed, and the bulk thermoelectric materials can obtain high ZT values through nano-process. At present, research on thermoelectric materials is mainly focused on two aspects: one is bulk thermoelectric materials containing nanostructures, and the other is nano-thermoelectric materials (Fig. 1.6).

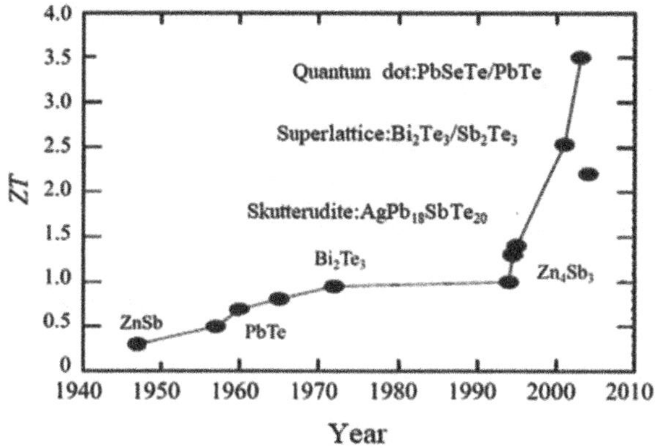

Fig. 1.5 Changes in ZT value of thermoelectric materials in different years [156]

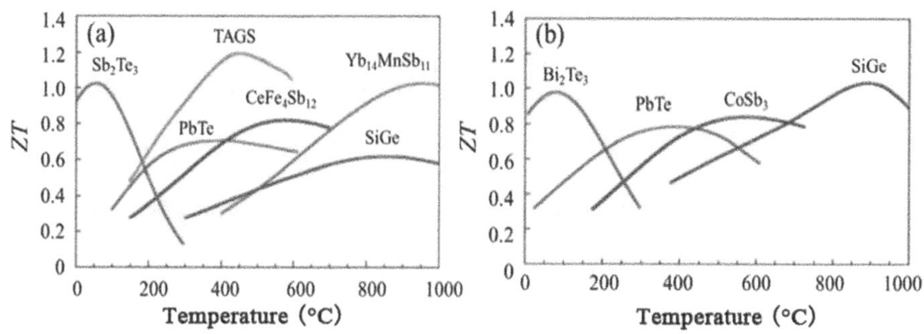

Fig. 1.6 ZT values of typical thermoelectric materials [43]

Increasing the ZT value of a thermoelectric material can improve the thermoelectric conversion efficiency of the material. The relationship between the thermoelectric conversion efficiency η and the ZT value is shown in Eq. (1.3):

$$\eta = \frac{T_h - T_c}{T_h}\left(\frac{\sqrt{Z\overline{T}+1}-1}{\sqrt{Z\overline{T}+1}+T_c/T_h}\right), \tag{1.3}$$

where T_h is the hot-end temperature, and T_c is the cold-end temperature, and the average temperature, that is $\overline{T}=\frac{T_h+T_c}{2}$.

At present, the best commercial thermoelectric materials have a ZT value of only about 1, and their thermoelectric conversion efficiency is only about 3–7%. According to the estimation in Fig. 1.7, its thermoelectric efficiency is far lower than Carnot efficiency. When the ZT value of the thermoelectric material is lower than 1, the thermoelectric conversion efficiency is quite low. When the ZT value reaches 2, it

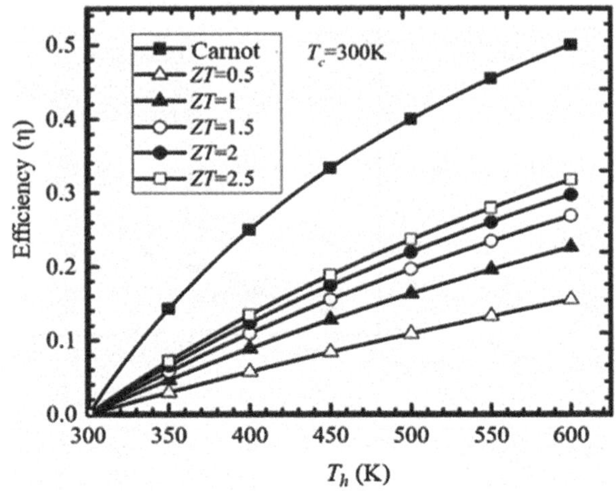

Fig. 1.7 Thermoelectric conversion efficiency and Carnot efficiency with different ZT values

can be used for the recovery of waste heat. Only when the ZT value reaches 4 or 5 can the thermoelectric material have the ability to cool the refrigerator [40].

1.2.3 Progress in Research on Si-Based Thermoelectric Materials

Si is the most widely used semiconductor material and is also a type of thermoelectric material. It is the foundation of modern integrated circuits. It has huge reserves on the earth and is extremely convenient to mine and process. At the same time, the performance of Si is stable, non-toxic and pollution-free. Therefore, it has broad application prospects as a thermoelectric material.

Si is a Group IVA element and has high thermal conductivity, and its unit cell type is a face-centered cube with a diamond structure. Its lattice constant is 0.543 nm, which is a covalent bond type, the number of moles is 28.0855 g mol^{-1}, the theoretical density is 2.33 g cm^{-3}, and the melting point is 1690 K. Among them, the Si power factor of the block is large, and the electrical performance is also superior. However, at room temperature, Si the thermal conductivity of single crystal Si can reach 148 W m^{-1} K^{-1} [49], and its high thermal conductivity makes it difficult to establish that results in poor thermoelectric performance of pure Si. Therefore, in order to improve the application value of Si materials, it is necessary to reduce its thermal conductivity. In order to reduce the thermal conductivity of Si, scientists have proposed various solutions. The more common methods are: compounding Si with other materials [50, 51]; lowering the dimensionality of Si [52–54] to prepare two-dimensional thin films or forming one-dimensional nanowires.

Si can be combined with Ge to form a SiGe alloy, which is an excellent high-temperature thermoelectric material. Since the atomic masses and atomic radii of

Si and Ge are quite different, if they are recombined, there will be strong mass fluctuations and stress-strain scattering, which will have a significant impact on phonon motion [55], resulting in significant decreasing of thermal conductivity that leads the thermoelectric performance is drastically improved. It has been reported that when it is used as a P-type thermoelectric material, the ZT value of the SiGe alloy can reach 1.08 at a high temperature of 1000 K [56]. Because of the excellent thermoelectric properties of SiGe alloys, thermoelectric devices based on SiGe alloys have been widely used in the aerospace field [57].

Mg_2Si formed by the combination of Mg and Si is considered to be a very promising thermoelectric material [58]. The best working range is in the middle temperature zone (400–700 K). Mg_2Si-based thermoelectric materials have the characteristics of large effective mass, high mobility, and small lattice thermal conductivity. Under the increasingly serious energy crisis and environmental crisis, Mg_2Si-based thermoelectric materials are considered to be one of the most potential thermoelectric materials in the middle temperature area because of their superior thermoelectric performance and environmental friendliness. Mg_2Si was first prepared in 1955, and then the scientific community conducted in-depth exploration of Mg_2Si-based thermoelectric materials. In 2006, Fedorov et al. [59–61] prepared a series of Mg2Si-based solid solution thermoelectric materials by smelting, and studied the influence of energy band structure on their performance. In the same year, Nemoto [61] and Tani et al. [58] also prepared Mg_2Si-based thermoelectric materials by sintering and hot pressing, respectively, and doped with other elements to improve their performance.

Due to its photoluminescence properties, porous silicon attracted widespread attention in the 1990s. Subsequent research found that nano-crystallization can effectively reduce the thermal conductivity of Si materials, thereby improving their thermoelectric properties. Gesele et al. [62] found that porous silicon with randomly distributed pores and a porosity of 64%-89% has a thermal conductivity of 0.1 W m^{-1} K^{-1} at room temperature, which is 3 orders of magnitude smaller than bulk silicon. Although it has greatly helped to reduce the thermal conductivity, the electrical properties of this traditional porous structure Si will also be greatly reduced. Yamamoto et al. [63] found that this porous structure with randomly distributed pore shapes leads to a very low conductivity ($\sigma = 0.2$ S cm^{-1}). Therefore, the ZT at room temperature is only 0.03. Song et al. [64] found that if the pores are ordered and micron-sized, the electrical conductivity will be significantly improved, and the thermal conductivity can be reduced to 40 W m^{-1} K^{-1}. Lee et al. [65] predicted by kinetic simulation that the thermal conductivity of ordered nanoscale porous Si can be as low as 0.6 W m^{-1} K^{-1} due to phonon scattering.

The preparation of Si-based films by two-dimensional methods is also an important means to improve the thermoelectric properties of Si-based thermoelectric materials. Huxtable et al. [66] tested $Si/Si_{0.75}Ge_{0.25}$ and $Si_{0.75}Ge_{0.25}/Si_{0.25}Ge_{0.75}$ at different temperatures. Thermal conductivity values of two longitudinal films of $Si/Si_{0.75}Ge_{0.25}$ superlattice. The experimental results show that the thermal conductivity of the surface $Si/Si_{0.75}Ge_{0.25}$ superlattice is related to its periodic thickness, which decreases with the reducing of the periodic thickness value; while $Si_{0.75}Ge_{0.25}/$

$Si_{0.25}Ge_{0.75}$ superlattice thermal conductivity and cycle thickness have no significant dependence. Lee [67] and Borca-Tasciuc [68] and others measured the thermal conductivity of the $Si_{0.75}Ge_{0.25}$ superlattice and show that the thermal conductivity first increases with the cycle thickness, and when the cycle thickness increases, at 10 nm, it starts to decline. T. Koga [69] prepared a Si/Ge superlattice using the "carrier pocket" concept using molecular beam epitaxy and obtained a large ZT value. In this system, the lattice strain at the Si/Ge interface provides another degree of freedom is provided to control the conduction band structure of the superlattice.

In recent decades, scientists have continued to study one-dimensional thermoelectric materials. In 2008, Boukai [52] of the California Institute of Technology prepared 10 nm wide Si nanowires and found that the average free path of phonons in the nanowires was effectively reduced, resulting in a thermal conductivity drop to 0.5% of the bulk. In the same year, HochbAum [70] of the University of California, Berkeley used chemical methods to prepare Si nanowires with a diameter of 20–300 nm and obtained extremely high ZT values. At room temperature, the ZT value of the 50 nm Si nanowires reached 0.6, the reason being that the higher surface roughness caused the thermal conductivity to be two orders of magnitude lower than that of the bulk Si.

1.2.4 Research Progress of Sb_2Te_3 and Bi_2Te_3 Thermoelectric Materials

Antimony telluride (Sb_2Te_3) and bismuth telluride (Bi_2Te_3) have a graphite-like hexahedral layered structure. The compounds and their derivatives are the earliest and one of the most mature thermoelectric materials. The compounds and derivatives of this system have been widely used in the market, mainly used for thermoelectric power generation and energized refrigeration. When introducing the research status of materials, if there is no special explanation, Sb_2Te_3 is p-type and Bi_2Te_3 is n-type. It is divided into two parts: nanocomposite structure (mainly bulk) and thin film. The introduction of bulk nanocomposite structure is intended to provide a reference for the study of the composite structure of thin film.

(1) **Nano-composite structure**

Ko et al. [71] synthesized Pt-Sb_2Te_3 using the solution method, and its power factor was improved compared to pure Sb_2Te_3. H. Q. Yang et al. [72] used organic surfactants or inorganic salts to modify Sb_2Te_3, and analyzed the formation mechanism of Sb_2Te_3 nanosheets. Das et al. [73] synthesized graphite-containing Sb_2Te_3 composites using solid-phase synthesis, which greatly reduced the lattice thermal conductivity (0.8 $W\ m^{-1}\ K^{-1}$) and maintained a high Seebeck coefficient. The room temperature ZT value of the graphite with different contents was adjusted to be 0.18–0.38. The Pb-modified Sb_2Te_3 nanosheets have a ZT value of 0.24–0.37 in the temperature range of 50–200 °C. Liu et al. [74] used the solution method combined with hot pressing to prepare the Te-Sb_2Te_3 composite, and the ZT was as high as 0.29

at 475 K. Schaumann et al. [75] used microwave-assisted thermal decomposition to prepare Sb_2Te_3 in an ionic liquid. The thermoelectric performance of the material was the best when ion exchange was performed in C_4mimI, and the ZT value was 0.59 at 300 K. Mukherjee et al. [76] used solid-phase synthesis to synthesize Cu_2Te-Sb_2Te_3. By adjusting the content of Cu_2Te, the maximum ZT value obtained was 0.6 (300 K). Mehta et al. [77] synthesized Sb_2Te_3 by sulfur doping using microwave dissolution thermal method. The ZT value of the bulk nanocrystals reached 0.95 at 423 K. Mun et al. [78] prepared the Sb_2Te_3 bulk material containing Te by the melt method. Te atoms with high self-diffusion rate are favorable for grain rearrangement at the liquid-solid homogeneous interface, forming a low energy with dense dislocation array. Grain boundaries reduce the thermal conductivity and significantly improve the thermoelectric performance. Yang et al. [79] used solution method to synthesize Sb_2Te_3 nanocomposites with different Te content. When the composition is 2.6 mol% Sb_2Te_3 and 97.4 mol% Te, the ZT value at 623 K is close to 1.0. Zheng et al. [80] synthesized the Sb_2Te_3/PEDOT composite using the solution method. The thermal conductivity is about 0.14 W m^{-1} K^{-1} in the temperature range of 300–523 K, and the maximum ZT value is 1.18 at 523 K. Zhang et al. [81] synthesized $Bi_{0.5}Sb_{1.5}Te_3$ by solution method using substitutional doping (Bi instead of Sb position). At 375 K, the ZT value of the sample was greater than 1.4. The minority carrier blocking suppresses the bipolar thermal conductivity and reduces the lattice thermal conductivity, resulting in an increase in ZT at high temperatures. Lee et al. [82] synthesized Sb_2Te_3-Ag_2Te bulk composites. When the ratio was controlled at 1: 1, a ZT value of 1.5 was obtained at 700 K.

Sinduja et al. [83] successfully synthesized Sb_2Te_3 nanorods at different reaction temperatures using a surfactant-assisted hydrothermal method. The sheet-shaped modified Bi_2Te_3 nanorods have a power factor of 1.32 μW cm^{-1} K^{-2} at 410 K. Sinduja et al. [84] used helium ions to modify Bi_2Te_3 nanopillars, and obtained a maximum power factor of 8.2 μW cm^{-1} K^{-2} at 390 K. Du et al. [85] prepared reduced graphene oxide and Bi_2Te_3 nanocomposite bulk materials by reduction method, and obtained a maximum power factor of 13.4 μW cm^{-1} K^{-2} at 150 °C. Stavila et al. [86] prepared Bi_2Te_3 by wet chemical method, and the maximum ZT value was 0.38 at room temperature. Guo et al. [87] synthesize Bi_2Te_3 nanostructures by liquid phase reaction and regulate the transition from pure Bi_2Te_3 hexagonal nanosheets to Te-Bi_2Te_3 heterostructures to Bi_2Te_3 nanostrings, that is, homogeneous-heterogeneous-homogeneous structure, Bi_2Te_3 The maximum ZT value of the nanostring at 450 K is 0.42. A. Soni et al. [88] doped Se with Bi_2Te_3 to synthesize nanosheet structures using a polyol solution method. The maximum ZT value at room temperature was 0.54, and the corresponding Seebeck coefficient was −259 μV K^{-1}. Zhang et al. [89] used the solution method to prepare Bi_2Te_3-Te nanoparticles and made them into bulk materials by hot pressing. The ZT value was 0.42 at room temperature and 0.55 at 400 K. Li et al. [90] synthesized a Bi_2Te_3 composite containing graphene quantum dots using a solution method, and obtained a maximum ZT value of 0.55 at 425 K. Son et al. [91] used the solution method to synthesize Bi_2Te_3 nanosheets, and the maximum ZT was 0.62 at 400 K. Min et al. [92] used a solution to synthesize Bi_2Te_3 nanocomposite doped with Bi_2Se_3, and used spark plasma sintering (SPS) to make blocks for performance testing. The interface between the nano-particles and the heterostructure in the nanocomposite reduces the thermal conductivity greatly and the carrier filtering effect to increase the power factor. The maximum ZT value of nanocomposites with 10–15% Bi_2Se_3 is 0.7 at 400 K. Cheng et al. [93] used the solution method to synthesize Bi_2Te_3 nanosheet-Te nanocolumn heterostructures, and obtained bulk samples by centrifugation, washing and drying. The maximum ZT value at 320 K was 0.73. Zhang et al. [94] synthesized a Ag-modified Bi_2Te_3 using a chemical synthesis method with a ZT value of 0.77 at 475 K. Ivanov et al. [95] used microwave solvothermal method and SPS to mix Lu and Tm in Bi_2Te_3, the density of states near the Fermi level became sharp, the effective mass of electrons increased, and the electronic thermal conductivity decreased as a result that the ZT value was greatly improved. The maximum ZT value of Lu-doped films is about 0.86 around 430 K. Hong et al. [96] successfully prepared Te/Bi_2Te_3 layered nanostructures using microwave-assisted solvothermal methods using Te nanotubes as templates. Compared with pure Bi_2Te_3 nanosheets, ZT increased from 0.75 to 1. Xu et al. [97] synthesized a hollow nanostructure of $Bi_2Te_{2.5}Se_{0.5}$ using a liquid phase route. After sintering, the pore-containing nanocomposite material sintered in a wide temperature range (388–513 K) has a ZT value greater than 1. Hao et al. [98] used metal M (M = Cd, Cu, and Ag) to composite with Bi_2Te_3 to prepare P-type Bi_2Te_3. The average ZT value at 100–300 °C reached 1.0–1.2. Wu et al. [99] used the Bridgman method to synthesize Cu-doped Bi_2Te_3 alloy, and obtained a P-type alloy with a ZT value of 1.2 (300 K) and an N-type ZT value of 1.09 (363 K). Han et al. [100] synthesized a Cu_2 and Sn co-doped Bi_2Te_3 alloy by a high-temperature solid-phase reaction method. The average ZT value was 1.02 and the maximum ZT value was 1.24 (425 K) in 300–525 K.

(2) **Thin Films**

Li et al. [101] calculated that the ZT value of the five-layer Sb_2Te_3 film at room temperature can be greater than 2, and the ZT value decreases to about 0.5 after the dimension becomes a bulk. Wanarattikan et al. [102] used magnetron sputtering to deposit Sb_2Te_3 films on flexible substrates, controlled different thicknesses, and studied the effects of grain size and thickness on thermoelectric properties. The results show that as the film thickness increases, the grain size increases, and the mean free path of the carriers increases, which leads to a decrease in Seebeck coefficient and an increase in conductivity. When the film thickness is less than 1 μm, the contribution of phonon scattering to thermal conductivity is dominant. When the film thickness is greater than 1 μm, both phonon and carrier scattering are the main factors affecting thermal conductivity. Zhang et al. [103] used MBE to deposit Sb_2Te_3 films of different thicknesses on Si(111). The films were all oriented in high (00 l) orientation. As the film thickness increased (28–121 nm), The mean free path of the carrier increases, and the conductivity increases from 425.7 to 1036 S cm^{-1}. Kim et al. [104] obtained γ-Sb_2Te_3-Sb_2Te_3 composite films by electrodeposition and post-annealing, and the Seebeck coefficient of room temperature was 320 μV K^{-1}. Zhang et al. [105] used MBE to deposit Sb_2Te_3 films on the quartz glass substrate by co-evaporation with Sb and Te, adjusted the Sb content, and observed the Sb-Sb_2Te_3 heterojunction interface by TEM. The Seebeck coefficient reached to 536 μV K^{-1} and this phenomenon was explained by carrier energy filtering. In addition, MBE was used to deposit Sb_2Te_3-Te films on P-type (100) high-resistance Si, and Te (0/1/2/4 nm) layer nanoparticles were periodically inserted into multiple Sb_2Te_3 (5 nm) layers. At room temperature, the Seebeck coefficient of Sb_2Te_3 films containing 1 nm Te nanoparticles increased by about 25%, the power factor increased by about 50%, and the thermal conductivity decreased by about 26% compared to pure Sb_2Te_3. The Te-Sb_2Te_3 nanocrystalline interface and Te nanoparticles explain the reasons for their optimization [106]. Khumtong et al. [107] deposited Sb_2Te_3 film on polyimide by magnetron sputtering with a room temperature power factor of 1.07 μW cm^{-1} K^{-2}. Dun et al. [108] synthesized a Ag-modified Sb_2Te_3 film on a flexible substrate using a solution method, introduced a metal/semiconductor interface, and improved the Seebeck coefficient and conductivity. The highest power factor at room temperature was 3.71 μW cm^{-1} K^{-2}. Shi et al. [109] used magnetron sputtering to prepare Cu-containing Sb_2Te_3 films. Cu doping reduced the Seebeck coefficient and increased conductivity, and the power factor was 3.8 μW cm^{-1} K^{-2}. Yoo et al. [110] used electrodeposition to obtain Sb_2Te_3 films with Te nanoparticles. The maximum room temperature power factor was 7.16 μW cm^{-1} K^{-2}, which was greatly improved compared to Sb_2Te_3 without Te. Kim et al. [111] used electrodeposition to prepare Ag_2Te-Sb_2Te_3 composite films, controlled different Ag_2Te content, and obtained a maximum power factor of 18.7 μW cm^{-1} K^{-2} at 300 K. Bendt et al. [112] prepared Sb_2Te_3 films with high (00 l) orientation on Si (95) and Al (0001) by thermal evaporation. The room temperature power factor of Sb_2Te_3 films grown on Al reached 33 μW cm^{-1} K^{-2}. Shen et al. [113] used thermal co-evaporation to prepare Sb_2Te_3 films with different thicknesses (1/6/10/16 μm) on Si (95). The room-temperature power factor of 1 μm films was as high as 33 μW cm^{-1} K^{-2}. Goncalves et al. [114] deposited Sb_2Te_3 film on polyimide by thermal co-evaporation, and controlled the substrate temperature and Te content. The estimated maximum room temperature ZT value was 0.3. Thankamma et al. [115] used physical vapor deposition to add sulfur to Sb_2Te_3 to obtain $Sb_2Te_{3-x}S_x$ crystals. When x was 0.3, the maximum ZT was 0.54 (room temperature). Zhang et al. [116] prepared Ag_xTe_y- Sb_2Te_3 heterostructure film using solution method. Due to the interface barrier, cold carriers scatter stronger than hot carriers, and the power factor is increased by more than 50% at 150 °C (~2 μW cm^{-1} K^{-2}). In addition, a chemical method was used to introduce the oxide layer [117], and the Ag/oxide/Sb_2Te_3-Te metal-semiconductor heterostructure was constructed, and the ZT value was 1.0 at 460 K. Trung et al. [118] used electrochemical deposition of Sb_2Te_3 (50 μm) and Bi_2Te_3 (600 μm) films. The maximum power factors of the films at room temperature were 11.2 and 15 μW cm^{-1} K^{-2}. Lee et al. [119] used plasma enhanced chemical vapor deposition to deposit Sb_2Te_3 and Bi_2Te_3 films on graphene and silicon oxide substrates. The maximum power factor of the films deposited on graphene at room temperature was 35.9 and 35.2 μW cm^{-1} K^{-2}. EMF Vieira et al. [120] used co-evaporation to deposit Sb_2Te_3 and Bi_2Te_3 films on borosilicate glass and polyimide. The room temperature power factors of the films on polyimide were 12 and 27 μW cm^{-1} K^{-2}. At 373 K, the power factor is the largest, 23 and 58 μW cm^{-1} K^{-2}, respectively. C. Dun et al. [121] synthesized Ag/Cu modified Sb_2Te_3 and Bi_2Te_3 nanosheet films by solution reaction method. Due to the energy filtering interface, the conductivity was significantly increased, and the two types of films were used for power generation.

Suh et al. [122] bombarded the Bi_2Te_3 film with helium ions, adjusted the inherent defects in the film, broke the coupling of conductivity and Seebeck coefficient, and greatly enhanced the thermoelectric properties of the film. Liu et al. [123] prepared a Bi_2Te_3 film on a flexible substrate by magnetron sputtering, and annealed it at different temperatures. Bi_2Te_3 nanosheets appeared on the surface, which can be attributed to thermal stress. Annealing is an effective means for increasing the electrical conductivity, and the lamellar structure can reduce the thermal conductivity. Compared with that before annealing, the ZT value of Bi_2Te_3 film was increased by more than 4 times. Talebi et al. [124] prepared a P-type Bi_2Te_3 thin film by electrophoretic deposition, sintered at 693 K, and the in-plane Seebeck coefficient was 239 μV K^{-1} at 500 K. Lin et al. [125] prepared a Bi_2Te_3 thin film on a silicon oxide substrate using a thermal evaporation method. The maximum power factor obtained by annealing control was 6.05 μW cm^{-1} K^{-2}. Zeng et al. [126] used co-sputtering to prepare Bi_2Te_3 films with different thicknesses on glass substrates, and studied the variation of the conductivity of annealed films with different thicknesses (70–350 nm) with temperature. Electrical conductivity increases with increasing film thickness and grain size, and activation energy decreases with increasing film thickness. The power factor increases with the thickness of the film and is generally less than 8 μW cm^{-1} K^{-2}. Wang et al. [127] used MBE to deposit Bi_2Te_3 films on high-resistance Si(111), and compared the performance of the two-step method (seed layer deposition) and the one-step method. After the pre-deposition of the seed layer, the surface of the film was crossed hexagonal single crystal nanosheets, and its power factor (8.13 μW cm^{-1} K^{-2}) increased by 4.5 times compared to that without the seed layer. Shang et al. [128] used magnetron sputtering to prepare high (00 l) oriented Bi_2Te_3 and Te composite films on single crystal MgO, and the power factor reached 25 μW cm^{-1} K^{-2} at room temperature. Zhang et al. [129] used magnetron co-sputtering to grow (00 l) crystal-oriented Bi_2Te_3 thin films on quartz glass (substrate temperature 350 °C), with in-plane power factor is 33.7 μW cm^{-1} K^{-2} at room temperature, out-of-plane thermal conductivity is 0.86 W cm^{-1} K^{-1}. Jin et al. [130] synthesized a (00l) oriented Bi_2Te_3 flexible film on a single-walled carbon nanotube (SWCNT) support. The maximum ZT value of the material at room temperature was 0.89. Kim et al. [131] introduced a conductive polymer (polypyrrole) for hybridization in N-type Bi_2Te_3, which increased the Seebeck coefficient. Due to the phonon scattering effect, the thermal conductivity decreased, and the ZT significantly increased (0.98 at 25 °C and up to 1.21 at 100 °C). Choi et al. [132] used the MBE to deposit a Te-rich Bi/Te multilayer structure on a silicon oxide substrate and annealed it to construct a Te-Bi_2Te_3 film. At 375 K, the ZT value reached 2.27.

In summary, various methods for enhancing the performance of thermoelectric materials mainly increase the Seebeck coefficient (including energy filtering effect) under the premise of ensuring electrical conductivity, so that the power factor is increased. In addition, the introduction of interfacial scattered phonons reduces thermal conductivity. For thin films, the morphology, thickness, annealing temperature, substrate, and the introduction of heterostructures all affect their thermoelectric properties.

1.3 Periodic Nano Multi-layer Film

1.3.1 Overview of Periodic Nano-Multi-layer Films

In order to improve the performance of the material, two different materials can be superimposed, and the different properties of the material and the interface characteristics between the materials can be used to optimize the performance of the material. The periodic nano multilayer film is a functional nano film constructed based on this idea. As shown in Fig. 1.8, its basic definition is a cyclically changing composite film structure formed by alternately superimposing a variety of different components. The two groups of elements are arranged in an alternating cycle. Several adjacent components form a cycle, which is called the modulation cycle. It is denoted as $\Delta = h_A + h_B$ where h_A and h_B are respectively the thicknesses of a single layer constituting one component of a multilayer periodic film. The ratio of the component thickness is called the modulation ratio $n = h_A/h_B$. In periodic nano-multilayer films, the most important feature is that there are multiple phase interfaces between nano-scale components, which results in nanoscale effects and interface effects that are different from bulk materials [133].

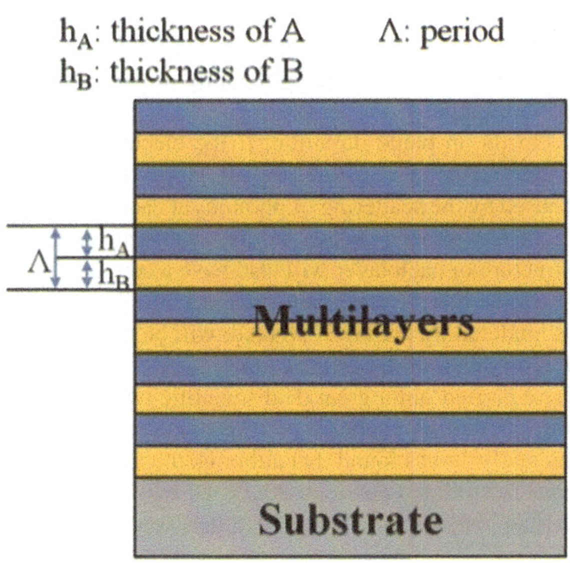

Fig. 1.8 Schematic of multilayer structure

The periodic nano-multilayer structure has important applications in many fields because of its special properties. One of the important uses is to make quantum well lasers [134, 135], because it has the advantages of stable performance, low temperature influence, and low threshold, which will meet the needs of the future communication industry. Another important application of periodic nano-multilayer films is to manufacture optical stable devices. Such optical non-linear elements take advantage of the quantum confinement effect of lasers and have outstanding advantages such as fast response and low working environment requirements. For example, GaAs/AlAs multilayer structures have been engineered [135]. Periodic nano-multilayers are also widely used in large extreme ultraviolet lithography machines and soft X-ray equipment due to their special optical properties [136, 137]. In the range of soft X-rays with a wavelength of 4.5–30 nm, periodic nano-multilayers have proven to be an effective reflective coating for normal incidence. Such periodic nano-multilayers are generally composed of alternating layers of high atomic number and low atomic number materials, with the thickness of each layer in the range of 1–10 nm. The coating superimposes the quarter wave, which enhances the reflectivity of the spectral band [137].

1.3.2 Thermoelectric Properties of Periodic Nano-Multilayer Films

There are many ways to prepare periodic nano-multilayer films. In order to obtain high-quality multilayer films (precise and uniform film thickness, precise chemical composition and stoichiometry), several methods are generally employed to prepare them, such as molecular beams epitaxial (MBE), chemical vapor deposition (CVD), and magnetron sputtering. However, when studying the thermoelectric properties of multilayer thermoelectric thin films, the longitudinal (in-plane) and facing (on-plane) thermoelectric properties of the films are quite different, forcing researchers to study them separately (Fig. 1.9).

In the in-plane direction, if the electron transport is not affected, the thermal conductivity can be significantly improved by the decrease in thermal conductivity caused by the scattering of phonons at each interface. However, the transport parameters in the plane may also be affected by the layered structure, and the quantum well structure of each layer will also have a certain effect on the transport [138].

In the direction perpendicular to the plane, the conduction of thermoelectricity is hindered by the layer, so there will be different phenomena. In order to study the longitudinal thermoelectric conversion of periodic nano-multilayer films, scientists have carried out a great deal scientific research. In 1987, Yao [139] first measured the longitudinal thermal conductivity of GaAs/AlAs superlattice multilayers experimentally. It was found that although the thermal conductivity of the film was greater than the thermal conductivity of the bulk, it was better than their weighted average but the overall value is small. Chen et al. [140] tested the thermal conductivity of

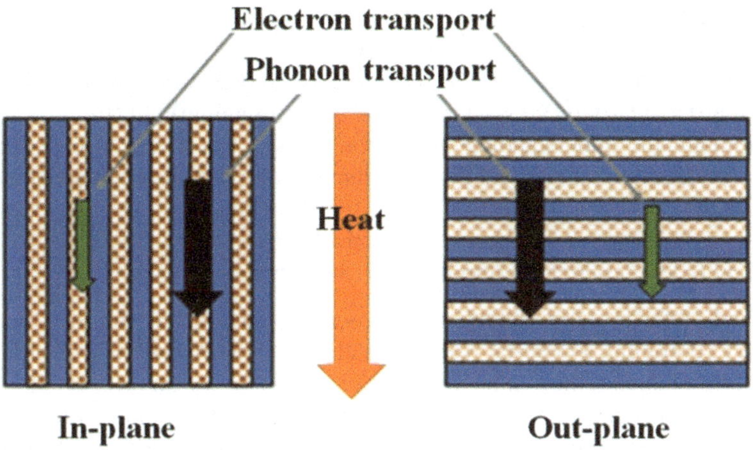

Fig. 1.9 Different transmission directions have different transport characteristics [157]

GaAs/AlAs multilayer thin film structures in the transverse and longitudinal directions using a new test scheme designed by themselves, and found that the thermal conductivity of the films is 7 times smaller than their bulk thermal conductivity. Lee [67] et al. Experimentally tested the longitudinal thermal conductivity of Si/Ge superlattice multilayer films and found that the thermal conductivity obtained was basically equivalent to that of SiGe alloy.

Over the years, people have obtained superlattice multilayer thermal conductivity values of various material systems, such as Bi_2Te_3/Sb_2Te_3, Si/Ge, InAs/AlSb [141], InP/InGaAs [142], SbTe-based superlattices. These experiments show that the thermal conductivity of these superlattice thin film systems is smaller than their bulk values, both in the longitudinal and transverse directions. Yao's research results on the thermal conductivity of GaAs/AlAs parallel to the film and Capinski's [143] research results on its longitudinal thermal conductivity show that the thermal conductivity of the superlattice decreases with increasing cycle thickness. That indicates the thermal resistance at the interface increases with periodic thickness.

So far, the thermoelectric performance of multilayer systems has not been greatly improved. Harman [144] and others used MBE to deposit N-type PbTe/Te multilayer system, and tuned the thickness and carrier concentration of each layer. The results show that the Seebeck coefficient and power factor are significantly improved compared with similar bulk materials, but there is no significant improvement in thermal conductivity. Se-doped (N-type) PbSeTe/PbTe superlattices were prepared using MBE. They found that self-assembled tapered quantum dots can further improve the performance by epitaxial strain, and obtained a ZT value of 1.5 at room temperature [145, 146].

Summarize the research of multilayer thermoelectricity, one may notice the following facts:

(1) Most of the research focuses on a few material systems, especially V–VI compounds (antimony and bismuth telluride), IV–VI binary compounds (lead selenide, telluride), III–V semiconductors (arsenides of aluminum, gallium, and indium) and Group IV systems (silicon and germanium) [147];
(2) Although the understanding of phonon transport and scattering is far from complete, most efficiencies improvement comes from the decrease of thermal conductivity;
(3) There is less research on in-plane thermoelectricity;
(4) There is less research on improving ZT by increasing power factor.

Recently, Mahan [148] and others proposed a special idea to improve the power factor. By combining a material with a high electron concentration (such as a metal) with a semiconductor, the conduction electron distribution with respect to the asymmetry of the Fermi level is introduced, resulting in significant improvement in power factor. At the same time, high interfacial density can also effectively reduce thermal conductivity, leading to an increase in the overall ZT value.

This view point was also introduced into the study of multilayer structures consisting of rock salt structure nitride phases (ScN and GaN) and metal transition nitrides (ZrN and ZrWN). Metal transition nitrides have a resistivity of 15-50 Ω cm similar to that of metals. At the same time, these materials also have very high thermal and chemical stability. The melting point is usually above 2773 K, which has high oxidation resistance at high temperatures. These two compound systems are not easy to fuse, and the layered structure is not easily affected by high temperature. Therefore, it has great application value. Rawat et al. [149] used reactive metal sputtering to deposit ScN/(Zr,W)N metal-semiconductor superlattice at a substrate temperature of 1123 K. At the same time, they evaluated the room temperature thermal conductivity of the ScN/(Zr, W)N superlattice and showed that ScN/ZrN can obtain the smallest thermal conductivity in a period of 3–7 nm, and its value is 5 W/m K, which is much lower than the thermal conductivity of the constituent materials (the total thermal conductivity of ZrN is measured at 47 W m^{-1} K^{-1}, and the contribution of the calculated lattice is 18.7 W m^{-1} K^{-1}). By alloying W-N to reduce the lattice mismatch of ScN, the thermal conductivity was further reduced to 2.2 W m^{-1} K^{-1} [150].At the same time, Zebarjadi et al. [151] studied the ScN(6 nm)/ZrN(4 nm) superlattice and tested the Seebeck coefficient at room temperature to be 840 μV K^{-1}.

Therefore, multilayers composed of semiconductors and metals have demonstrated great thermoelectric potential.

1.3.3 Thermal Stability of Periodic Nano-Multilayer Films

Since the thermoelectric film refrigeration system can obtain a high cooling power density, it has great application value in miniaturized electronic chips. When studying the properties of thermoelectric films, not only their scientific value, but also their engineering value. In engineering applications, problems such as contact resistance, substrate thermal conductivity, integrated thermoelectric materials, and classic device vertical height problems [138, 152], all of which are of great concern in engineering. In addition, the long-term thermal stability, interdiffusion, and coarsening of nanocrystals are also a focus of attention in applications.

1.4 Chapter Summary

This chapter introduces the effects and progress of thermoelectric conversion on energy issues, and provides the basis and reference for the work. Because our earth is powered by the sun every day, the environmental thermal energy is abounded and completely cost-free. The temperature differences wildly exist in every corner of the world at all-time disregarding the weather conditions, in-door/out-door, in the sky or underground/ocean. The current technology of converting thermal energy to electrical power requires a complex, heavy and high temperature mechanical system normally contents a steam turban engine and electrical power generator. In the future, its widespread implantations might greatly reduce the dependence on petrochemical energy, which is of great significance to ecological protection. The solar thermal energy can be used as the main energy source to support the development of human society. The economical scale of the new clean energy industry could be in the trillion dollars in future.

References

1. Sun X, Sun Y (2004) 10 trends in world energy development in the 21st century. Sci Technol Rev 5:51–53
2. Dai Z (2004) Comparison of energy structure and utilization efficiency of major world countries shanghai electric power 17(6):545
3. Min G, Rowe DM (2007) Conversion efficiency of thermoelectric combustion systems. IEEE Trans Energy Convers 22(2):528–534
4. Cornett J, Chen B, Haidar S, Berney H, McGuinness P, Lane B, Gao Y, He Y, Sun N, Dunham M (2017) Fabrication and characterization of Bi_2Te_3-based chip-scale thermoelectric energy harvesting devices. J Electron Mater 46(5):2844–2846
5. Jang B, Han S, Kim J-Y (2011) Optimal design for micro-thermoelectric generators using finite element analysis. Microelectron Eng 88(5):775–778
6. Xie J, Lee C, Feng H (2010) Design, fabrication, and characterization of CMOS MEMS-based thermoelectric power generators. J Microelectromech Syst 19(2):317–324
7. Wang Z, Leonov V, Fiorini P, Van Hoof C (2009) Realization of a wearable miniaturized thermoelectric generator for human body applications. Sens Actuators, A 156(1):95–102
8. Lim JR, Whitacre JF, Fleurial JP, Huang CK, Ryan MA, Myung NV (2005) Fabrication method for thermoelectric nanodevices. Adv Mater 17(12):1488–1492
9. Matsubara I, Funahashi R, Takeuchi T, Sodeoka S, Shimizu T, Ueno K (2001) Fabrication of an all-oxide thermoelectric power generator. Appl Phys Lett 78(23):3627–3629
10. Bell LE (2008) Cooling, heating, generating power, and recovering waste heat with thermoelectric systems. Science 321(5895):1457–1461

11. Li P, Cai L, Zhai P, Tang X, Zhang Q, Niino M (2010) Design of a concentration solar thermoelectric generator. J Electron Mater 39(9):1522–1530
12. Orr B, Akbarzadeh A, Lappas P (2017) An exhaust heat recovery system utilising thermoelectric generators and heat pipes. Appl Therm Eng 126:1185–1190
13. Kim CS, Lee GS, Choi H, Kim YJ, Yang HM, Lim SH, Lee S-G, Cho BJ (2018) Structural design of a flexible thermoelectric power generator for wearable applications. Appl Energy 214:131–138
14. Huu TN, Van TN, Takahito O (2018) Flexible thermoelectric power generator with Y-type structure using electrochemical deposition process. Appl Energy 210:467–476
15. Jung YS, Jeong DH, Kang SB, Kim F, Jeong MH, Lee K-S, Son JS, Baik JM, Kim J-S, Choi KJ (2017) Wearable solar thermoelectric generator driven by unprecedentedly high temperature difference. Nano Energy 40:663–672
16. Deng F, Qiu H, Chen J, Wang L, Wang B (2016) Wearable thermoelectric power generators combined with flexible supercapacitor for low-power human diagnosis devices. IEEE Trans Industr Electron 64(2):1477–1485
17. Lu Z, Zhang H, Mao C, Li CM (2016) Silk fabric-based wearable thermoelectric generator for energy harvesting from the human body. Appl Energy 164:57–63
18. Hyland M, Hunter H, Liu J, Veety E, Vashaee D (2016) Wearable thermoelectric generators for human body heat harvesting. Appl Energy 182:518–524
19. Kim SJ, We JH, Cho BJ (2014) A wearable thermoelectric generator fabricated on a glass fabric. Energy Environ Sci 7(6):1959–1965
20. Zhu W, Deng Y, Wang Y, Shen S, Gulfam R (2016) High-performance photovoltaic-thermoelectric hybrid power generation system with optimized thermal management. Energy 100:91–101
21. Amatya R, Ram R (2010) Solar thermoelectric generator for micropower applications. J Electron Mater 39(9):1735–1740
22. Kajihara T, Makino K, Lee YH, Kaibe H, Hachiuma H (2015) Study of thermoelectric generation unit for radiant waste heat. Mater Today: Proc 2(2):804–813
23. Aranguren P, Astrain D, Rodríguez A, Martínez A (2015) Experimental investigation of the applicability of a thermoelectric generator to recover waste heat from a combustion chamber. Appl Energy 152:121–130
24. Kiziroglou ME, Wright SW, Toh TT, Mitcheson PD, Becker T, Yeatman EM (2013) Design and fabrication of heat storage thermoelectric harvesting devices. IEEE Trans Industr Electron 61(1):302–309
25. Glatz W, Muntwyler S, Hierold C (2006) Optimization and fabrication of thick flexible polymer based micro thermoelectric generator. Sens Actuators, A 132(1):337–345
26. Navone C, Soulier M, Testard J, Simon J, Caroff T (2011) Optimization and fabrication of a thick printed thermoelectric device. J Electron Mater 40(5):789–793
27. Chen A, Madan D, Wright P, Evans J (2011) Dispenser-printed planar thick-film thermoelectric energy generators. J Micromech Microeng 21(10):104006
28. Francioso L, De Pascali C, Farella I, Martucci C, Cretì P, Siciliano P, Perrone A (2011) Flexible thermoelectric generator for ambient assisted living wearable biometric sensors. J Power Sources 196(6):3239–3243
29. Zhu W, Deng Y, Gao M, Wang Y (2015) Hierarchical Bi–Te based flexible thin-film solar thermoelectric generator with light sensing feature. Energy Convers Manag 106:1192–1200
30. Fourmont P, Gerlein LF, Fortier F-X, Cloutier SG, Nechache R (2018) Highly efficient thermoelectric microgenerators using nearly room temperature pulsed laser deposition. ACS Appl Mater Interfaces 10(12):10194–10201
31. Zhou H, Mu X, Zhao W, Tang D, Wei P, Zhu W, Nie X, Zhang Q (2017) Low interface resistance and excellent anti-oxidation of Al/Cu/Ni multilayer thin-film electrodes for Bi_2Te_3-based modules. Nano Energy 40:274–281
32. Jiang J, Chen L, Bai S, Yao Q, Wang Q (2005) Fabrication and thermoelectric performance of textured n-type Bi2 (Te, Se) 3 by spark plasma sintering. Mater Sci Eng, B 117(3):334–338

33. Kim D-H, Kim C, Ha D-W, Kim H (2011) Fabrication and thermoelectric properties of crystal-aligned nano-structured Bi_2Te_3. J Alloy Compd 509(17):5211–5215
34. Zou M, Li J-F, Du B, Liu D, Kita T (2009) Fabrication and thermoelectric properties of fine-grained TiNiSn compounds. J Solid State Chem 182(11):3138–3142
35. Luo W, Yang M, Chen F, Shen Q, Jiang H, Zhang L (2009) Fabrication and thermoelectric properties of $Mg_2Si_{1-x}Sn_x$ ($0 \leq x \leq 1.0$) solid solutions by solid state reaction and spark plasma sintering. Mater Sci Eng: B 157 (1–3):96–100
36. Zhao D, Tian C, Tang S, Liu Y, Jiang L, Chen L (2010) Fabrication of a $CoSb_3$-based thermoelectric module. Mater Sci Semicond Process 13(3):221–224
37. Wang H, Sun X, Yan X, Huo D, Li X, Li J-G, Ding X (2014) Fabrication and thermoelectric properties of highly textured $Ca_9Co_{12}O_{28}$ ceramic. J Alloy Compd 582:294–298
38. Hewitt CA, Kaiser AB, Roth S, Craps M, Czerw R, Carroll DL (2012) Multilayered carbon nanotube/polymer composite based thermoelectric fabrics. Nano Lett 12(3):1307–1310
39. Li S, Toprak MS, Soliman HM, Zhou J, Muhammed M, Platzek D, Müller E (2006) Fabrication of nanostructured thermoelectric bismuth telluride thick films by electrochemical deposition. Chem Mater 18(16):3627–3633
40. Junior OA, Maran A, Henao N (2018) A review of the development and applications of thermoelectric microgenerators for energy harvesting. Renew Sustain Energy Rev 91:376–393
41. Zheng X, Liu C, Yan Y, Wang Q (2014) A review of thermoelectrics research–Recent developments and potentials for sustainable and renewable energy applications. Renew Sustain Energy Rev 32:486–503
42. Elsheikh MH, Shnawah DA, Sabri MFM, Said SBM, Hassan MH, Bashir MBA, Mohamad M (2014) A review on thermoelectric renewable energy: principle parameters that affect their performance. Renew Sustain Energy Rev 30:337–355
43. Snyder GJ, Toberer ES (2008) Complex thermoelectric materials. Nat Mater 7(2):105114Son
44. Wood C (1988) Materials for thermoelectric energy conversion. Rep Prog Phys 51(4):459
45. Saito K, Benenti G, Casati G (2010) A microscopic mechanism for increasing thermoelectric efficiency. Chem Phys 375(2–3):508–513
46. Alam H, Ramakrishna S (2013) A review on the enhancement of figure of merit from bulk to nano-thermoelectric materials. Nano Energy 2(2):190–212
47. Zhao D, Tan G (2014) A review of thermoelectric cooling: materials, modeling and applications. Appl Therm Eng 66(1–2):15–24
48. Dresselhaus MS, Chen G, Tang MY, Yang R, Lee H, Wang D, Ren Z, Fleurial JP, Gogna P (2007) New directions for low-dimensional thermoelectric materials. Adv Mater 19(8):1043–1053
49. Ho CY, Powell RW, Liley PE (1972) Thermal conductivity of the elements. J Phys Chem Ref Data 1(2):279–421. https://doi.org/10.1063/1.3253100
50. Vining CB (1991) A model for the high temperature transport properties of heavily doped n type silicon germanium alloys. J Appl Phys 69(1):331–341. https://doi.org/10.1063/1.347717
51. Savvides N, Goldsmid HJ (1980) Boundary scattering of phonons in fine-grained hot-pressed Ge-Si alloys. I. The dependence of lattice thermal conductivity on grain size and porosity. J Phys C: Solid State Phys 13(25):4657
52. Boukai AI, Bunimovich Y, Tahir-Kheli J, Yu J-K, Goddard WA, Heath JR (2010) Silicon nanowires as efficient thermoelectric materials. Materials for sustainable energy. Co-Published with Macmillan Publishers Ltd, UK. https://doi.org/10.1142/9789814317665_0018
53. Liao CN, Chen C, Tu KN (1999) Thermoelectric characterization of Si thin films in silicon-on-insulator wafers. J Appl Phys 86(6):3204–3208. https://doi.org/10.1063/1.371190
54. Tonkikh AA, Zakharov ND, Eisenschmidt C, Leipner HS, Werner P (2014) Aperiodic SiSn/Si multilayers for thermoelectric applications. J Cryst Growth 392:49–51. https://doi.org/10.1016/j.jcrysgro.2014.01.047
55. Pernot G, Stoffel M, Savic I, Pezzoli F, Chen P, Savelli G, Jacquot A, Schumann J, Denker U, Monch I, Deneke C, Schmidt OG, Rampnoux JM, Wang S, Plissonnier M, Rastelli A, Dilhaire S, Mingo N (2010) Precise control of thermal conductivity at the nanoscale through

individual phonon-scattering barriers. Nat Mater 9(6):491–495. https://doi.org/10.1038/nmat2752
56. Yamashita O, Sadatomi N (2000) Thermoelectric properties of $Si_{1-x}Ge_x (x \leqslant 0.10)$ with alloy and dopant segregations. J Appl Phys 88(1):245–251. https://doi.org/10.1063/1.373648
57. Rowe DM (1978) Thermoelectric power generation. Proc Inst Electr Eng 125(11):1113–1136. https://doi.org/10.1049/piee.1978.0247
58. Tani J-i, Kido H (2007) Thermoelectric properties of Sb-doped Mg_2Si semiconductors. Intermetallics 15(9):1202–1207. https://doi.org/10.1016/j.intermet.2007.02.009
59. Fedorov MI, Zaĭtsev VK, Eremin IS, Gurieva EA, Burkov AT, Konstantinov PP, Vedernikov MV, Samunin AY, Isachenko GN, Shabaldin AA (2006) Transport properties of $Mg_2X_{0.4}Sn_{0.6}$ solid solutions (X = Si, Ge) with p-type conductivity. Phys Solid State 48(8):1486–1490. https://doi.org/10.1134/s1063783406080117
60. Fedorov MI, Pshenay-Severin DA, Zaitsev VK, Sano S, Vedernikov MV (2003) Features of conduction mechanism in n-type $Mg_2Si_{1-x}Sn_x$ solid solutions. In: 22nd International conference on thermoelectrics, La Grand Motte, France, 17–21. pp 142–145. https://doi.org/10.1109/ict.2003.1287469
61. Nemoto T, Akasaka M, Iida T, Sato J, Nishio K, Takei T, Takanashi Y (2006) Characterization of oxide-incorporated n-type Mg_2Si prepared by a spark plasma sintering method. In: 25th International conference on thermoelectrics, 6–10, pp 552–555. https://doi.org/10.1109/ict.2006.331355
62. Gesele G, Linsmeier J, Drach V, Fricke J, Arens-Fischer R (1997) Temperature-dependent thermal conductivity of porous silicon. J Phys D Appl Phys 30(21):2911
63. Yamamoto A, Takazawa H, Ohta T (1999) Random porous silicon. In: Proceedings of the international conference on thermoelectric
64. Song D, Chen G (2004) Thermal conductivity of periodic microporous silicon films. Appl Phys Lett 5:687–689
65. Lee J-H, Grossman J, Reed J, Galli G (2007) Lattice thermal conductivity of nanoporous Si: molecular dynamics study. Appl Phys Lett 91(22):223110
66. Huxtable ST, Abramson AR, Tien C-L, Majumdar A, LaBounty C, Fan X, Zeng G, Bowers JE, Shakouri A, Croke ET (2002) Thermal conductivity of Si/SiGe and SiGe/SiGe superlattices. Appl Phys Lett 80(10):1737–1739. https://doi.org/10.1063/1.1455693
67. Lee SM, Cahill DG, Venkatasubramanian R (1997) Thermal conductivity of Si–Ge superlattices. Appl Phys Lett 70(22):2957–2959. https://doi.org/10.1063/1.118755
68. Borca-Tasciuc T, Liu W, Liu J, Zeng T, Song DW, Moore CD, Chen G, Wang KL, Goorsky MS, Radetic T (2000) Thermal conductivity of symmetrically strained Si/Ge superlattices. Superlattices Microstruct 28(3):199–206
69. Carrier pocket engineering to design superior thermoelectric materials using GaAs/AlAs superlattices (1998)
70. Hochbaum AI, Chen R, Delgado RD, Liang W, Garnett EC, Najarian M, Majumdar A, Yang P (2008) Enhanced thermoelectric performance of rough silicon nanowires. Nature 451(7175):163–167. https://doi.org/10.1038/nature06381
71. Ko DK, Kang Y, Murray CB (2011) Enhanced thermopower via carrier energy filtering in solution-processable Pt–Sb_2Te_3 nanocomposites. Nano Lett 11(7):2841–2844
72. Yang HQ, Miao L, Liu CY, Li C, Honda S, Iwamoto Y, Huang R, Tanemura S (2015) A facile surfactant-assisted reflux method for the synthesis of single-crystalline Sb_2Te_3 nanostructures with enhanced thermoelectric performance. ACS Appl Mater Inter 7(26):14263–14271
73. Das S, Singha P, Deb AK, Das SC, Chatterjee S, Kulbachinskii VA, Kytin VG, Zinoviev DA, Maslov NV, Dhara S (2019) Role of graphite on the thermoelectric performance of Sb_2Te_3/graphite nanocomposite. J Appl Phys 125(19):195105
74. Liu EY, Lin FH, Yang ZR, Liu CJ (2018) A facile energy-saving route of fabricating thermoelectric Sb_2Te_3-Te nanocomposites and nanosized Te. Roy Soc Open Sci 5(10):180698
75. Schaumann J, Loor M, Ünal D, Mudring A, Heimann S, Hagemann U, Schulz S, Maculewicz F, Schierning G (2017) Improving the ZT value of thermoelectrics by nanostructuring: tuning the nanoparticle morphology of Sb_2Te_3 by using ionic liquids. Dalton T 46(3):656–668

76. Mukherjee S, Femi OE, Chetty R, Chattopadhyay K, Suwas S, Mallik RC (2018) Microstructure and thermoelectric properties of Cu_2Te-Sb_2Te_3 pseudo-binary system. Appl Surf Sci 449:805–814
77. Mehta RJ, Zhang Y, Zhu H, Parker DS, Belley M, Singh DJ, Ramprasad R, Borca-Tasciuc T, Ramanath G (2012) Seebeck and figure of merit enhancement in nanostructured antimony telluride by antisite defect suppression through sulfur doping. Nano Lett 12(9):4523–4529
78. Mun H, Lee KH, Yoo SJ, Kim HS, Jeong J, Oh SH, Snyder GJ, Lee YH, Kim YM, Kim SW (2018) Highly fluidic liquid at homointerface generates grain-boundary dislocation arrays for high-performance bulk thermoelectrics. Acta Mater 159:266–275
79. Yang HQ, Chen YJ, Wang XY, Miao L, Li XY, Han XD, Lu X, Wang GY, Zhou XY (2018) Realizing high thermoelectric performance in Te nanocomposite through Sb_2Te_3 incorporation. CrystEngComm 20(47):7729–7738
80. Zheng W, Bi P, Kang H, Wei W, Liu F, Shi J, Peng L, Wang Z, Xiong R (2014) Low thermal conductivity and high thermoelectric figure of merit in p-type Sb_2Te_3/poly (3, 4-ethylenedioxythiophene) thermoelectric composites. Appl Phys Lett 105(2):023901
81. Zhang C, Ng H, Li Z, Khor KA, Xiong Q (2017) Minority carrier blocking to enhance the thermoelectric performance of solution-processed $Bi_xSb_{2-x}Te_3$ nanocomposites via a liquid-phase sintering process. ACS Appl Mater Inter 9(14):12501–12510
82. Lee MH, Kim KR, Rhyee JS, Park SD, Snyder GJ (2015) High thermoelectric figure-of-merit in Sb_2Te_3/Ag_2Te bulk composites as Pb-free p-type thermoelectric materials. J Mater Chem C 3(40):10494–10499
83. Sinduja M, Amirthapandian S, Jegadeesan P, Magudapathy P, Asokan K (2018) Morphological investigations on the growth of defect-rich Bi_2Te_3 nanorods and their thermoelectric properties. CrystEngComm 20(33):4810–4822
84. Sinduja M, Amirthapandian S, Magudapathy P, Srivastava SK, Asokan K (2018) Tuning of the thermoelectric properties of Bi_2Te_3 nanorods using helium ion irradiation. ACS Omega 3(12):18411–18419
85. Du Y, Li J, Xu J, Eklund P (2019) Thermoelectric properties of reduced graphene oxide/Bi_2Te_3 nanocomposites. Energies 12(12):2430
86. Stavila V, Robinson DB, Hekmaty MA, Nishimoto R, Medlin DL, Zhu S, Tritt TM, Sharma PA (2013) Wet-chemical synthesis and consolidation of stoichiometric bismuth telluride nanoparticles for improving the thermoelectric figure-of-merit. ACS Appl Mater Inter 5(14):6678–6686
87. Guo W, Ma J, Yang J, Li D, Qin Q, Wei C, Zheng W (2014) A new strategy for realizing the conversion of "Homo–Hetero–Homo" heteroepitaxial growth in Bi_2Te_3 and the thermoelectric performance. Chem-Eur J 20(19):5657–5664
88. Soni A, Yanyuan Z, Ligen Y, Aik MKK, Dresselhaus MS, Xiong Q (2012) Enhanced thermoelectric properties of solution grown $Bi_2Te_{3-x}Se_x$ nanoplatelet composites. Nano Lett 12(3):1203–1209
89. Zhang Y, Wang H, Kraemer S, Shi Y, Zhang F, Snedaker M, Ding K, Moskovits M, Snyder GJ, Stucky GD (2011) Surfactant-free synthesis of Bi_2Te_3 − Te micro − nano heterostructure with enhanced thermoelectric figure of merit. ACS Nano 5(4):3158–3165
90. Li S, Fan T, Liu X, Liu F, Meng H, Liu Y, Pan F (2017) Graphene quantum dots embedded in Bi_2Te_3 nanosheets to enhance thermoelectric performance. ACS Appl Mater Inter 9(4):3677–3685
91. Son JS, Choi MK, Han MK, Park K, Kim JY, Lim SJ, Oh M, Kuk Y, Park C, Kim SJ (2012) N-type nanostructured thermoelectric materials prepared from chemically synthesized ultrathin Bi_2Te_3 nanoplates. Nano Lett 12(2):640–647
92. Min Y, Roh JW, Yang H, Park M, Kim SI, Hwang S, Lee SM, Lee KH, Jeong U (2013) Surfactant-free scalable synthesis of Bi_2Te_3 and Bi_2Se_3 nanoflakes and enhanced thermoelectric properties of their nanocomposites. Adv Mater 25(10):1425–1429
93. Cheng L, Chen ZG, Yang L, Han G, Xu HY, Snyder GJ, Lu GQ, Zou J (2013) T-shaped Bi_2Te_3–Te heteronanojunctions: epitaxial growth, structural modeling, and thermoelectric properties. J Phys Chem C 117(24):12458–12464

94. Zhang Q, Ai X, Wang L, Chang Y, Luo W, Jiang W, Chen L (2015) Improved thermoelectric performance of silver nanoparticles-dispersed Bi_2Te_3 composites deriving from hierarchical two-phased heterostructure. Adv Funct Mater 25(6):966–976
95. Ivanov O, Yaprintsev M, Lyubushkin R, Soklakova O (2018) Enhancement of thermoelectric efficiency in Bi_2Te_3 via rare earth element doping. Scripta Mater 146:91–94
96. Hong M, Chen ZG, Yang L, Zou J (2016) Enhancing thermoelectric performance of Bi_2Te_3-based nanostructures through rational structure design. Nanoscale 8(16):8681–8686
97. Xu B, Feng T, Agne MT, Zhou L, Ruan X, Snyder GJ, Wu Y (2017) Highly porous thermoelectric nanocomposites with low thermal conductivity and high figure of merit from large-scale solution-synthesized $Bi_2Te_{2.5}Se_{0.5}$ hollow nanostructures. Angew Chem Int Edit 56(13):3546–3551
98. Hao F, Qiu P, Tang Y, Bai S, Xing T, Chu HS, Zhang Q, Lu P, Zhang T, Ren D (2016) High efficiency Bi_2Te_3-based materials and devices for thermoelectric power generation between 100 and 300 °C. Energy Environ Sci 9(10):3120–3127
99. Wu HJ, Yen WT (2018) High thermoelectric performance in Cu-doped Bi_2Te_3 with carrier-type transition. Acta Mater 157:33–41
100. Han MK, Hwang J, Kim SJ (2019) Improved thermoelectric properties of n-type Bi_2Te_3 alloy deriving from two-phased heterostructure by the reduction of CuI with Sn. J Mater Sci Mater El 30(2):1282–1291
101. Li Z, Miao N, Zhou J, Sun Z, Liu Z, Xu H (2018) High thermoelectric performance of few-quintuple Sb_2Te_3 nanofilms. Nano Energy 43:285–290
102. Wanarattikan P, Jitthammapirom P, Sakdanuphab R, Sakulkalavek A (2019) Effect of grain size and film thickness on the thermoelectric properties of flexible Sb_2Te_3 thin films. Adv Mater Sci Eng, 1–7
103. Zhang X, Zeng Z, Shen C, Zhang Z, Wang Z, Lin C, Hu Z (2014) Investigation on the electrical transport properties of highly (00 l)-textured Sb_2Te_3 films deposited by molecular beam epitaxy. J Appl Phys 115(2):024307
104. Kim J, Zhang M, Bosze W, Park SD, Lim JH, Myung NV (2015) Maximizing thermoelectric properties by nanoinclusion of γ-SbTe in Sb_2Te_3 film via solid-state phase transition from amorphous Sb–Te electrodeposits. Nano Energy 13:727–734
105. Zhang Z, Wu Y, Zhang H, Zeng Z, Hu Z (2015) Enhancement of seebeck coefficient in Sb-rich Sb_2Te_3 thin film. J Mater Sci Mater El 26(3):1619–1624
106. Zhang Z, Zhang H, Wu Y, Zeng Z, Hu Z (2015) Optimization of the thermopower of antimony telluride thin film by introducing tellurium nanoparticles. Appl Phys A 118(3):1043–1051
107. Khumtong T, Sukwisute P, Sakulkalavek A, Sakdanuphab R (2017) Microstructure and electrical properties of antimony telluride thin films deposited by RF magnetron sputtering on flexible substrate using different sputtering pressures. J Electron Mater 46(5):3166–3171
108. Dun C, Hewitt CA, Li Q, Xu J, Schall DC, Lee H, Jiang Q, Carroll DL (2017) 2D chalcogenide nanoplate assemblies for thermoelectric applications. Adv Mater 29(21):1700070
109. Shi D, Wang R, Wang G, Li C, Shen X, Nie Q (2017) Enhanced thermoelectric properties in Cu-doped Sb_2Te_3 films. Vacuum 145:347–350
110. Yoo I-J, Song Y, Lim DC, Myung NV, Lee KH, Oh M, Lee D, Do Kim Y, Kim S, Choa Y-H (2013) Thermoelectric characteristics of Sb_2Te_3 thin films formed via surfactant-assisted electrodeposition. J Mater Chem A 1(17):5430–5435
111. Kim J, Lee KH, Kim SD, Lim JH, Myung NV (2018) Simple and effective fabrication of Sb_2Te_3 films embedded with Ag_2Te nanoprecipitates for enhanced thermoelectric performance. J Mater Chem A 6(2):349–356
112. Bendt G, Kaiser K, Heckel A, Rieger F, Oing D, Lorke A, Rodriguez NP, Schierning G, Jooss C, Schulz S (2018) Structural and thermoelectrical characterization of epitaxial Sb_2Te_3 high quality thin films grown by thermal evaporation. Semicond Sci Tech 33(10):105002
113. Shen H, Lee S, Kang J-g, Eom TY, Lee H, Han S (2018) Thickness dependence of the electrical and thermoelectric properties of co-evaporated Sb_2Te_3 films. Appl Surf Sci 429:115–120
114. Goncalves LM, Alpuim P, Rolo AG, Correia JH (2011) Thermal co-evaporation of Sb_2Te_3 thin-films optimized for thermoelectric applications. Thin Solid Films 519(13):4152–4157

115. Thankamma G, Kunjomana AG (2015) Studies on sulfur doping and figure of merit in vapor grown Sb_2Te_3 platelet crystals. J Cryst Growth 415:65–71
116. Zhang Y, Snedaker ML, Birkel CS, Mubeen S, Ji X, Shi Y, Liu D, Liu X, Moskovits M, Stucky GD (2012) Silver-based intermetallic heterostructures in Sb_2Te_3 thick films with enhanced thermoelectric power factors. Nano Lett 12(2):1075–1080
117. Zhang Y, Bahk JH, Lee J, Birkel CS, Snedaker ML, Liu D, Zeng H, Moskovits M, Shakouri A, Stucky GD (2014) Hot carrier filtering in solution processed heterostructures: a paradigm for improving thermoelectric efficiency. Adv Mater 26(17):2755–2761
118. Trung N, Sakamoto K, Toan N, Ono T (2017) Synthesis and evaluation of thick films of electrochemically deposited Bi_2Te_3 and Sb_2Te_3 thermoelectric materials. Materials 10(2):154
119. Lee CW, Kim GH, Choi JW, An KS, Kim JS, Kim H, Lee YK (2017) Improvement of thermoelectric properties of Bi_2Te_3 and Sb_2Te_3 films grown on graphene substrate. Phys Status Solidi-R 11(6):1700029
120. Vieira EMF, Figueira J, Pires AL, Grilo J, Silva MF, Pereira AM, Goncalves LM (2019) Enhanced thermoelectric properties of Sb_2Te_3 and Bi_2Te_3 films for flexible thermal sensors. J Alloy Compd 774:1102–1116
121. Dun C, Hewitt CA, Li Q, Guo Y, Jiang Q, Xu J, Marcus G, Schall DC, Carroll DL (2017) Self-assembled heterostructures: selective growth of metallic nanoparticles on V_2–VI_3 Nanoplates. Adv Mater 29(38):1702968
122. Suh J, Yu KM, Fu D, Liu X, Yang F, Fan J, Smith DJ, Zhang YH, Furdyna JK, Dames C (2015) Simultaneous enhancement of electrical conductivity and thermopower of Bi_2Te_3 by multifunctionality of native defects. Adv Mater 27(24):3681–3686
123. Liu S, Peng N, Bai Y, Ma D, Ma F, Xu K (2016) Self-formation of thickness tunable Bi_2Te_3 nanoplates on thin films with enhanced thermoelectric performance. RSC Adv 6(38):31668–31674
124. Talebi T, Ghomashchi R, Talemi P, Aminorroaya S (2019) Thermoelectric performance of electrophoretically deposited p-type Bi_2Te_3 film. Appl Surf Sci 477:27–31
125. Lin JM, Chen YC, Lin CP (2013) Annealing effect on the thermoelectric properties of Bi_2Te_3 thin films prepared by thermal evaporation method. J Nanomater 2013:1
126. Zeng Z, Yang P, Hu Z (2013) Temperature and size effects on electrical properties and thermoelectric power of bismuth telluride thin films deposited by co-sputtering. Appl Surf Sci 268:472–476
127. Wang Z, Zhang X, Zeng Z, Zhang Z, Hu Z (2014) Two-step molecular beam epitaxy growth of bismuth telluride nanoplate thin film with enhanced thermoelectric properties. ECS Solid State Lett 3(8):99–101
128. Shang HJ, Ding FZ, Deng Y, Zhang H, Dong ZB, Xu WJ, Huang DX, Gu HW, Chen ZG (2018) Highly (00 *l*)-oriented Bi_2Te_3/Te heterostructure thin films with enhanced power factor. Nanoscale 10(43):20189–20195
129. Zhang Z, Wang Y, Deng Y, Xu Y (2011) The effect of (00 *l*) crystal plane orientation on the thermoelectric properties of Bi_2Te_3 thin film. Solid State Commun 151(21):1520–1523
130. Jin Q, Jiang S, Zhao Y, Wang D, Qiu J, Tang DM, Tan J, Sun DM, Hou PX, Chen XQ (2019) Flexible layer-structured Bi_2Te_3 thermoelectric on a carbon nanotube scaffold. Nat Mater 18(1):62–68
131. Kim C, Baek JY, Lopez DH, Kim DH, Kim H (2018) Interfacial energy band and phonon scattering effect in Bi_2Te_3-polypyrrole hybrid thermoelectric material. Appl Phys Lett 113(15):153901
132. Choi H, Jeong K, Chae J, Park H, Baeck J, Kim TH, Song JY, Park J, Jeong KH, Cho MH (2018) Enhancement in thermoelectric properties of Te-embedded Bi_2Te_3 by preferential phonon scattering in heterostructure interface. Nano Energy 47:374–384
133. Yang W, Tsakalakos T, Hilliard J (1977) Enhanced elastic modulus in composition-modulated gold-nickel and copper-palladium foils. J Appl Phys 48(3):876–879
134. Wood T, Burrus C, Miller D, Chemla D, Damen T, Gossard A, Wiegmann W (1984) High-speed optical modulation with GaAs/GaAlAs quantum wells in diode structure. Appl Phys Lett 44(1):16–18

135. Schneider H, Fujiwara K, Grahn H, Kv Klitzing, Ploog K (1990) Electro-optical multistability in GaAs/AlAs superlattices at room temperature. Appl Phys Lett 56(7):605–607
136. Siffalovic P, Majkova E, Chitu L, Jergel M, Luby S, Keckes J, Maier G, Timmann A, Roth S, Tsuru T (2009) Characterization of Mo/Si soft X-ray multilayer mirrors by grazing-incidence small-angle X-ray scattering. Vacuum 84(1):19–25
137. Stearns DG, Rosen RS, Vernon SP (1993) Multilayer mirror technology for soft-x-ray projection lithography. Appl Opt 32(34):6952–6960
138. Bottner H, Nurnus J, Gavrikov A, Kuhner G, Jagle M, Kunzel C, Eberhard D, Plescher G, Schubert A, Schlereth K-H (2004) New thermoelectric components using microsystem technologies. J Microelectromech Syst 13(3):414–420
139. Yao T (1987) Thermal properties of AlAs/GaAs superlattices. Appl Phys Lett 51(22):1798–1800
140. Chen B, Zhang Q-M, Bernholc J (1994) Si diffusion in GaAs and Si-induced interdiffusion in GaAs/AlAs superlattices. Phys Rev B 49(4):2985
141. Feuchter M, Jooss C, Kamlah M (2016) The 3ω-method for thermal conductivity measurements in a bottom heater geometry. Phys Status Solidi (a) 213(3):649–661. https://doi.org/10.1002/pssa.201532466
142. Duan N, Wang X, Li N, Liu H-D, Campbell JC (2006) Thermal analysis of high-power InGaAs–InP photodiodes. IEEE J Quantum Electron 42(12):1255–1258
143. Capinski WS, Maris HJ (1996) Thermal conductivity of GaAs/AlAs superlattices. Physica B 219:699–701
144. Harman T, Spears D, Walsh M (1999) PbTe/Te superlattice structures with enhanced thermoelectric figures of merit. J Electron Mater 28(1):L1–L5
145. Harman T, Taylor P, Spears D, Walsh M (2000) Thermoelectric quantum-dot superlattices with high ZT. J Electron Mater 29(1):L1–L2
146. Harman T, Taylor P, Walsh M, LaForge B (2002) Quantum dot superlattice thermoelectric materials and devices. Science 297(5590):2229–2232
147. Vineis CJ, Shakouri A, Majumdar A, Kanatzidis MG (2010) Nanostructured thermoelectrics: big efficiency gains from small features. Adv Mater 22(36):3970–3980. https://doi.org/10.1002/adma.201000839
148. Mahan GD, Woods LM (1998) Multilayer thermionic refrigeration. Phys Rev Lett 80(18):4016–4019. https://doi.org/10.1103/PhysRevLett.80.4016
149. Rawat V, Sands T (2006) Growth of TiN/GaN metal/semiconductor multilayers by reactive pulsed laser deposition. J Appl Phys 100(6):064901
150. Rawat V, Koh YK, Cahill DG, Sands TD (2009) Thermal conductivity of (Zr, W)N/ScN metal/semiconductor multilayers and superlattices. J Appl Phys 105(2):024909
151. Zebarjadi M, Bian Z, Singh R, Shakouri A, Wortman R, Rawat V, Sands T (2009) Thermoelectric transport in a ZrN/ScN superlattice. J Electron Mater 38(7):960–963
152. Bottner H, Nurnus J, Schubert A, Volkert F (2007) New high density micro structured thermogenerators for stand alone sensor systems. In: 26th International conference on thermoelectrics. IEEE, pp 306–309
153. Nuwayhid RY, Shihadeh A, Ghaddar N (2005) Development and testing of a domestic woodstove thermoelectric generator with natural convection cooling. Energy Convers Manag 46(9–10):1631–1643
154. Leonov V, Vullers R (2009) Wearable thermoelectric generators for body-powered devices. J Electron Mater 38(7):1491–1498
155. Gabriel-Buenaventura A, Azzopardi B (2015) Energy recovery systems for retrofitting in internal combustion engine vehicles: a review of techniques. Renew Sustain Energy Rev 41:955–964
156. Gayner C, Kar KK (2016) Recent advances in thermoelectric materials. Prog Mater Sci 83:330–382
157. Nielsch K, Bachmann J, Kimling J, Bottner H (2011) Thermoelectric nanostructures: from physical model systems towards nanograined composites. Adv Energy Mater 1(5):713–731. https://doi.org/10.1002/aenm.201100207

Chapter 2
Thermoelectric Performance Testing Methods and Devices

The main index for measuring the conversion efficiency of thermoelectric devices is the thermoelectric figure of merit (ZT value) of the thermoelectric material. If the conversion efficiency needs to be increased, the ZT value needs to be increased. According to the expression of ZT, its value is closely related to the Seebeck coefficient, electrical conductivity and thermal conductivity of the material. Because the three physical quantities of Seebeck coefficient, electrical conductivity and thermal conductivity are coupled to each other, it is difficult to adjust synchronously, and the ZT value and thermoelectric conversion efficiency are difficult to be greatly improved. By accurately measuring the three parameters of the material and related carrier characteristics, it is beneficial to adjust the method and conditions of material synthesis and improve the thermoelectric performance of the material.

2.1 Thin-Film Thermal Conductivity Test

Unlike bulk materials, relative to the film, when the heat is transferred inside, the average free path of the carrier and the thickness of the film will be equivalent, so the carrier will scatter at the boundary, resulting in a vertical film the thermophysical parameters of the direction vary. In this case, the traditional methods and devices suitable for testing the thermal properties of bulk materials have been difficult to test for thin-film materials, which poses new challenges to the testing of thermophysical parameters of materials. After years of research, people have developed a variety of effective thin film thermal conductivity measurement methods. Thin film thermal conductivity test methods can be divided into different categories according to different test characteristics. According to the thermal conductivity measurement direction, the test methods for the thermal conductivity of the film can be divided into two categories: one is the measurement of the thermal conductivity in the direction of the parallel film, and the other is the measurement of the thermal conductivity in

Z. Hu and Z. Wu, *Nanostructured Thermoelectric Films*,
https://doi.org/10.1007/978-981-15-6518-2_2

the direction of the vertical film. This section describes the two thermal conductivity testing methods and devices used in the experiment.

2.1.1 3ω Method

2.1.1.1 Test Principles

At the end of the 1980s, Cahill [1] proposed using the 3ω method to test the thermal conductivity of materials. The 3ω method is a transient measurement method that can be used to test the thermal conductivity of the vertical film direction. It is divided into the slope 3ω method and differential 3ω method. Compared to other methods for measuring the film's thermophysical properties, the 3ω method uses the combination of temperature fluctuations of a metal heating wire and a theoretical model of heat conduction of a finite-width heating source to determine the thermal conductivity of the material. It is stable and can effectively reduce the influence of heat radiation on the accuracy of test results, thereby improving the speed and accuracy of the test. After years of continuous development, the 3ω method has become an important method for testing the thermal conductivity of thin films.

Before explaining the principle of the 3ω method thermal conductivity test, the theoretical analysis of temperature fluctuations after a certain frequency AC current is passed through the metal wire. Figure 2.1 is a schematic diagram of the 3ω method thermal conductivity test principle. When an alternating current with a frequency of ω is passed through the metal wire, Joule heat with a frequency of 2ω is generated, and the temperature of the metal wire fluctuates. For pure metal, its resistance has a linear relationship with temperature, so an oscillation resistance with a frequency of 2ω will appear in the metal wire. This oscillation resistance with a frequency of

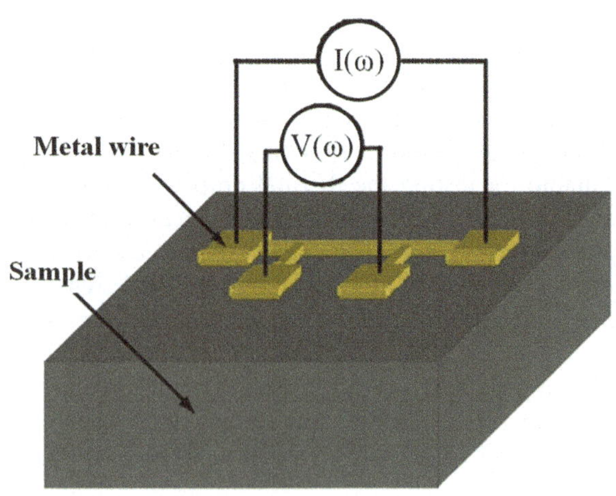

Fig. 2.1 A schematic of the thermal conductivity 3ω test method

2ω interacts with an alternating current with a frequency of ω. Therefore, a voltage with a frequency of 3ω is obtained. By measuring the 3ω voltage, the temperature fluctuation of the metal wire can be obtained. The above process can be expressed in mathematical language.

When an AC current of $I(t) = I_0 cos(\omega t)$ is applied to the metal wire through the two metal electrode pins, the Joule thermal power P (t) generated by it is

$$P(t) = I_0^2 \cos^2(\omega t) \cdot R = \frac{1}{2} I_0^2 R[1 + \cos(2\omega t)]. \quad (2.1)$$

The heating power expression can be divided into two items, one is independent of the heating frequency (DC part) and one is related to the heating frequency (AC part) [2]

$$P(t) = \left(\frac{1}{2} I_0^2 R\right)_{DC} + \left(\frac{1}{2} I_0^2 R \cos(2\omega t)\right)_{2\omega}. \quad (2.2)$$

The heat generated by the current heating will cause the temperature of the metal wire to rise. The temperature change also includes temperature-independent and temperature-dependent terms

$$T(t) = T_{DC} + T_{2\omega} \cos(2\omega t + \varphi). \quad (2.3)$$

where T_{DC} represents the steady-state temperature rise of the metal wire, $T_{2\omega}$ is the amplitude of the temperature fluctuation with a frequency of 2ω, and φ is the phase lag caused by the system heat capacity.

For pure metal, the resistance increases with the temperature. If the resistance temperature characteristic relationship $\delta R/\delta T$ of the metal wire can be measured through experiments, it can be used as both a heating wire and a temperature measuring wire.

Resistance temperature coefficient r_t is defined as the relative change in resistance value, when the temperature changes by 1 °C,

$$\frac{dR}{dT} = R_0 r_t. \quad (2.4)$$

So the expression of the resistance of the metal wire with temperature can be written as

$$R = R_0(1 + r_t T(t)). \quad (2.5)$$

The expression of the resistance of the metal wire can be obtained by Eq. 2.3 and Eq. 2.5 as

$$R(t) = R_0[1 + r_t(T_{DC} + T_{2\omega} \cos(2\omega t + \varphi))]. \quad (2.6)$$

According to Ohm's law $V(t) = I(t) \times R$, the voltage across the metal wire is

$$V(t) = I_0 \cos \omega t \cdot R_0(1+r_t T_{DC} + r_t T_{2\omega} \cos(2\omega t + \varphi)). \tag{2.7}$$

After finishing the above formula, we can find that it contains the 3ω term, which is why the 3ω method thermal conductivity test method is named.

$$V(t) = I_0 R_0(1 + r_t T_{DC}) \cos \omega t + \frac{1}{2} I_0 R_0 r_t T_{2\omega} \cos(\omega t + \varphi) + \frac{1}{2} I_0 R_0 r_t T_{2\omega} \cos(3\omega t + \varphi). \tag{2.8}$$

The above formula shows that the voltage drop on the metal line is composed of three parts. The first two terms indicate the voltage drop caused by Joule heat, when the frequency ω, AC current is applied to the metal line. The last term indicates the 3ω voltage drop. Get the metal wire temperature fluctuation amplitude $T_{2\omega}$

$$T_{2\omega}(\omega) = \frac{2V_{3\omega}}{I_0 R_0 \alpha} = 2 \cdot \frac{V_{3\omega}}{I_0} \frac{dT}{dR} \tag{2.9}$$

Equation 2.9 shows that by measuring the voltage signal at three times the frequency and the temperature coefficient of resistance of the metal wire, the temperature fluctuation of the metal wire as a function of frequency can be determined.

The above process is to analyze the temperature fluctuation of the metal wire itself after an alternating current with a frequency ω is passed through the metal wire. The following is a theoretical analysis of the heat transfer model of the sample to be tested when heated by a finite-width line heating source.

The results of the study by Cahill [1] show that when the current heats the sample through the wire, the temperature change of the wire itself can be related to the thermal conductivity of the material under test. As shown in Fig. 2.2, when an infinitely narrow heating source heats a semi-infinite solid, the equation derived by Carslaw [3] can calculate the sample temperature change caused by the following equation

$$\Delta T = \frac{P}{4\pi l \kappa} \int_{-\infty}^{t} e^{\frac{i\omega t' - r^2}{4D(t-t')}} \frac{dt'}{t - t'}. \tag{2.10}$$

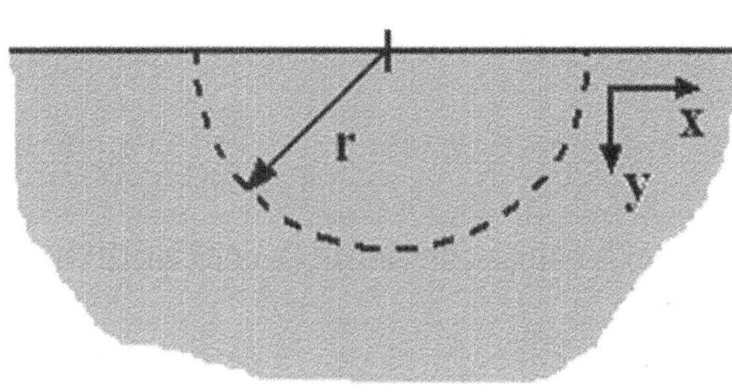

Fig. 2.2 Schematic diagram of heat conduction model when heating by infinite narrow line heating source

The above formula is also applicable to the temperature fluctuation when a periodic AC line heating source with a heating power of *P/l* within a unit length is used to heat the sample, where r is the radial distance $\sqrt{x^2 + y^2}$.

The Laplace transform of the integral term yields a zero-order second-class Bessel function K_0

$$\Delta T = \frac{P}{2\pi l \kappa} K_0(qr). \tag{2.11}$$

The inverse of wave vector q is defined as the wavelength of the heat wave generated by heating, that is,

$$q^{-1} = \sqrt{\frac{D}{i\omega}} \tag{2.12}$$

The heat wave wavelength q^{-1} usually represents the penetration depth of the heat wave, that is, the depth that the heat wave can penetrate the sample in an AC heating cycle. The value of q^{-1} is a very important parameter for measuring different thickness samples using the 3ω method. Equation 2.12 indicates that the penetration depth of thermal waves in the sample decreases with increasing heating frequency, and the penetration depth is related to the thermal diffusion coefficient D of the material. Thermal diffusivity D and thermal conductivity κ can be converted into each other through the material's density ρ and heat capacity C [4]

$$D = \frac{\kappa}{\rho C} \tag{2.13}$$

For the metal wire and the sample to be tested, the temperature change at the contact point is the same. Therefore, the temperature fluctuation of the sample to be measured can be obtained by measuring the temperature at the interface between the metal wire and the sample with a wire width of *2b*, as shown in Fig. 2.3 shown. In the analysis process, it is not necessary to discuss the temperature fluctuation on the sample *y*-axis (vertical sample surface), but only consider the temperature distribution of the sample along the *x*-axis direction (along the sample surface), and ignore the thickness of the metal wire itself. For homogeneous materials that are isotropic in the *x-y* plane, the penetration depth of the thermal wave in the x-axis and y-axis is assumed to be equal. Convert Eq. 2.11 to the Cartesian coordinate system, then the real space coordinate x is replaced by the Fourier space coordinate k

$$\Delta T = \frac{P}{l\pi\kappa} K_0(qx) = \frac{P}{l\pi\kappa} \int_0^{\infty} \frac{\cos(qxt)}{\sqrt{t^2+1}} dt = \frac{P}{l\pi\kappa} \int_0^{\infty} \frac{\cos(kx)}{\sqrt{k^2+q^2}} dk \tag{2.14}$$

Using mathematical relations $e^{ikx} = \cos(kx) + i\sin(kx)$, the Fourier transform can be written as

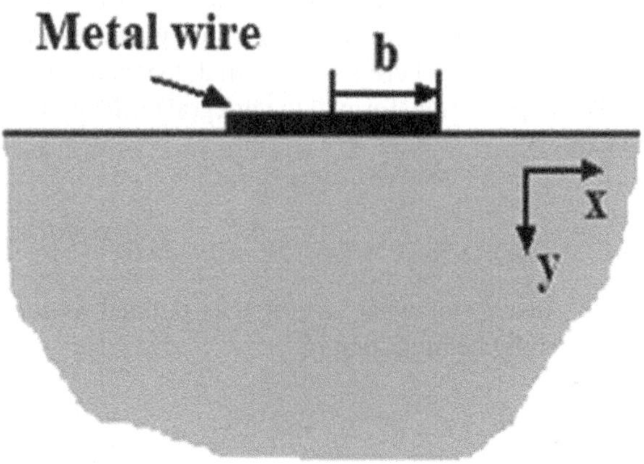

Fig. 2.3 Schematic diagram of heat conduction model when heating with finite width heating source

$$\Delta T(x) = \frac{1}{\sqrt{2\pi}} \int_{-\infty}^{\infty} \Delta T(k) \cdot e^{ikx} dk$$
$$= \frac{1}{\sqrt{2\pi}} \int_{-\infty}^{\infty} \Delta T(k) \cdot \cos(kx) dk + \frac{i}{\sqrt{2\pi}} \int_{-\infty}^{\infty} \Delta T(k) \cdot \sin(kx) dk. \tag{2.15}$$

Since Eq. 2.14 is a symmetric function, the Fourier transform of Eq. 2.15 can be simplified as

$$\Delta T(x) = \sqrt{\frac{2}{\pi}} \int_{0}^{\infty} \Delta T(k) \cos(kx) dk, \tag{2.16}$$

substituting $\Delta T(x)$ for Eq. 2.14 with Eq. 2.16 to obtain

$$\Delta T(k) = \frac{P}{\sqrt{2\pi} l \kappa} \cdot \frac{1}{\sqrt{k^2 + q^2}}. \tag{2.17}$$

Assuming that the heat generated by the metal wire uniformly enters the sample in the range of the line width *2b*, the Fourier transform of a finite-width line heating source with a line width of *2b* can be written

$$\frac{1}{2b} rect\left(\frac{x}{2b}\right) \to \frac{\sin(kb)}{kb}. \tag{2.18}$$

After inverse Fourier transform can be obtained

$$\Delta T(x) = \frac{P}{l \pi \kappa} \int_{0}^{\infty} \frac{\sin(kb) \cos(kx)}{kb \sqrt{k^2 + q^2}} dk. \tag{2.19}$$

Finally, the integration is performed in the range of line width $-b < x<b$ and the average value is obtained in this range

$$\Delta T = \frac{P}{l\pi\kappa}\frac{1}{2b}\int_{-b}^{b}\cos(kx)dx\int_{0}^{\infty}\frac{\sin(kb)}{kb\sqrt{k^2+q^2}}dk = \frac{P}{l\pi\kappa}\int_{0}^{\infty}\frac{\sin^2(kb)}{(kb)^2\sqrt{k^2+q^2}}dk. \tag{2.20}$$

The above formula gives a general expression for the temperature fluctuation of the surface of an object caused by an isotropic sample heated by a finite-width line heating source with a line width of *2b*. This expression ignores the contact heat between the metal wire and the sample to be measured resistance.

(1) **3ω slope method**

When measuring the thermal conductivity of the bulk material, the relationship between the temperature fluctuations caused by the heating of the metal wire and the heating frequency is shown in Fig. 2.4. The curve can be divided into two regions:

(1) In the low frequency region, the penetration depth of the heat wave is much larger than the width of the metal wire *2b*. At this time, the logarithm of the temperature fluctuation and the heating frequency can be approximated by a linear relationship. The constraint condition is approximated under < <1. It can be found from the linear equation that the thermal conductivity of the material is inversely proportional to the slope of the linear curve.
(2) In the high-frequency region, the penetration depth of the heat wave is equal to or smaller than the metal line width *2b*. At this time, the temperature fluctuation curve is an asymptotic line on the horizontal axis.

Since the analytical expression of Eq. 2.20 in this area cannot be known [5], the thermal conductivity coefficient can only be extracted by fitting the temperature

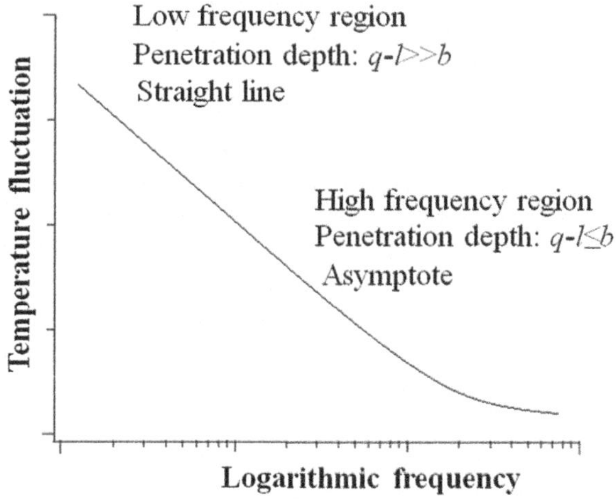

Fig. 2.4 The relationship between the temperature fluctuation of the slope 3ω method and the heating frequency

fluctuation curve obtained from the experiment with Eq. 2.20 with the help of Matlab program.

Cahill [1] derived an approximate expression in the case where the wavelength of the heat wave is much larger than the width of the metal wire in the low frequency region. At this time, the heat wave penetrates deep into the material under test (as shown in the left half of Fig. 2.4). Under constraints, Eq. 2.20 can be approximated as

$$\Delta T = \frac{P}{l\pi\kappa}\left[-\frac{1}{2}\ln(2\omega) - \ln\left(\frac{ib^2}{D}\right) + \ln 2 - \frac{i\pi}{4} - \gamma_R\right] = \frac{P}{l\pi\kappa}\left(-\frac{1}{2}\ln(2\omega) + const\right), \tag{2.21}$$

where Euler's constant $\gamma_R = 0.5772$.

Use the temperature fluctuation expression of Eq. 2.9 instead of Eq. 2.21, and differentiate the frequency on both sides of the equation to get

$$\kappa = \frac{I_0^3 R_0}{4\pi l\left(\frac{dV_{3\omega}}{df}\right)}\frac{dR}{dT}. \tag{2.22}$$

Theoretically, by measuring the third harmonic voltage at two different heating frequencies, the slope of the linear curve can be determined to calculate the thermal conductivity coefficient κ, but the actual measurement is performed by continuously collecting data in a certain low frequency range and using linear fitting then the straight slope of the temperature fluctuation curve is obtained. It is worth noting that the linear approximation is only valid when the penetration depth of the heat wave is much larger than the width of the metal wire, otherwise the relationship between the temperature fluctuation and the logarithm of the frequency is no longer linear.

(2) **Differential 3ω method**

When measuring the thermal conductivity of a thin film sample on a substrate, the relationship between the temperature fluctuation of the metal wire heating and the heating frequency is shown in Fig. 2.5. The curve can also be divided into two regions:

(1) In the high-frequency region, the penetration depth of the heat wave is small. If the penetration depth is less than the thickness t of the film to be measured, the temperature fluctuation curve only contains the thermophysical information of the sample to be measured. At this time, the shape of the curve depends on the metal wire width $2b$. And film thickness t. When the film thickness is equivalent to the width of the metal wire, the temperature fluctuation curve is an asymptotic line on the horizontal axis. When the thickness of the film is much larger than the width of the metal wire, the temperature fluctuation curve contains a linear part.

(2) In the low-frequency region, the wavelength of the heat wave is much larger than the thickness of the film to be measured. If the thickness t of the film is

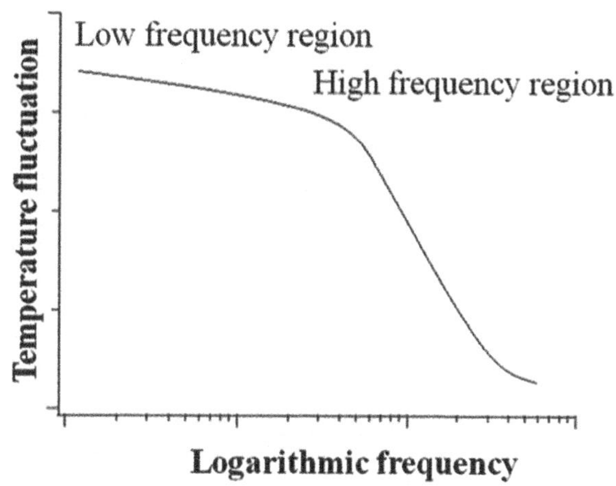

Fig. 2.5 The relationship between the temperature fluctuation of the differential 3ω method and the heating frequency

much smaller than the width of the metal wire *2b*, the heat flow in the film to be measured can be considered to be a one-dimensional vertical film direction, then the temperature fluctuation curve. It can be considered that an offset is added to the temperature fluctuation of the substrate (shown in Fig. 2.6). This offset can be used to calculate the thermal conductivity of the film to be measured.

In the low-frequency measurement region, the temperature fluctuation has a linear relationship with the logarithm of the frequency, and the linear slope is related to the substrate thermal conductivity coefficient $\kappa_{substrate}$ without being affected by the film thermal conductivity coefficient κ_{film}. The total temperature fluctuation curve ΔT_{total} is equal to the sum of the substrate temperature fluctuation $\Delta T_{substrate}$ and the film temperature fluctuation ΔT_{film}

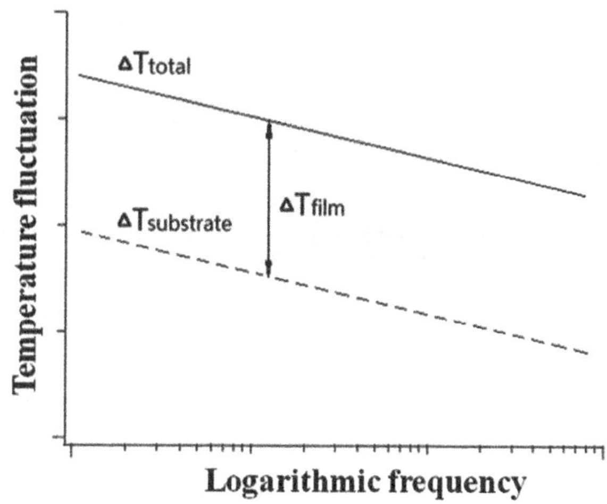

Fig. 2.6 The relationship between the temperature fluctuation in the low frequency region of the differential 3ω method and the heating frequency

$$\Delta T_{total} = \Delta T_{substrate} + \Delta T_{film}. \tag{2.23}$$

The substrate temperature fluctuation $\Delta T_{substrate}$ can be obtained by experimental measurement or theoretical calculation, and the total temperature fluctuation ΔT_{total} can be obtained by measuring the thin film sample on the substrate. After the difference between these two temperature fluctuations, the temperature fluctuation ΔT_{film} of the film to be measured can be obtained. The thickness t of the film is proportional to the film's thermal conductivity κ_{film}

$$\Delta T_{film} = \frac{t}{2bl\kappa_{film}}, \tag{2.24}$$

where 2b and l are the width and length of the metal wire, respectively.

The thermal conductivity of the film to be measured on the base can be calculated from Eq. 2.24. Because the interface contact thermal resistance is ignored in the theoretical model analysis, the measured thermal conductivity of the film includes three parts: the thermal conductivity of the film itself, the film thermal resistance at the interface with the substrate and thermal resistance at the interface between the film and the metal wire. For measuring the thermal conductivity of a thin film on a substrate, two test structures are usually prepared, one of which contains the thin film to be tested and the other without the thin film to be tested, that is, the reference sample. Under the same heating power, the temperature fluctuations of the two are measured separately. After the difference calculation, the temperature fluctuations on the film to be measured are obtained, and then the thermal conductivity of the film to be measured can be obtained by using Eq. 2.24. In the 3ω method, because the heated metal wire is both a heating source and a temperature sensor, if the leakage current in the metal wire enters the sample to be measured, it will cause measurement errors, which limits the differential 3ω method to only be used to measure insulating material's thermal conductivity. For non-insulating materials, an insulating layer must be added between the sample to be tested and the heated metal wire, as shown in Fig. 2.7.

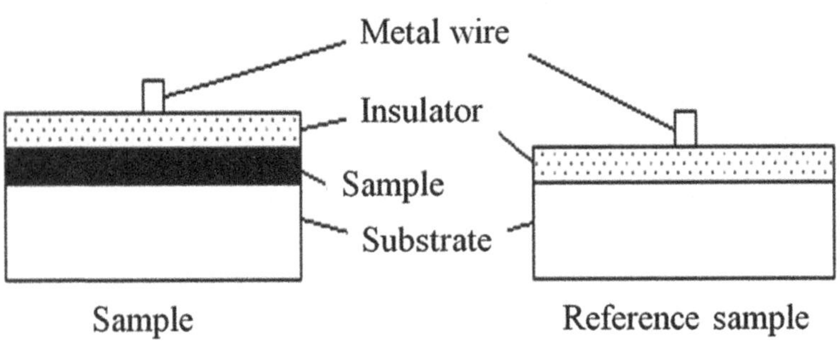

Fig. 2.7 Structure diagram of the differential 3ω method

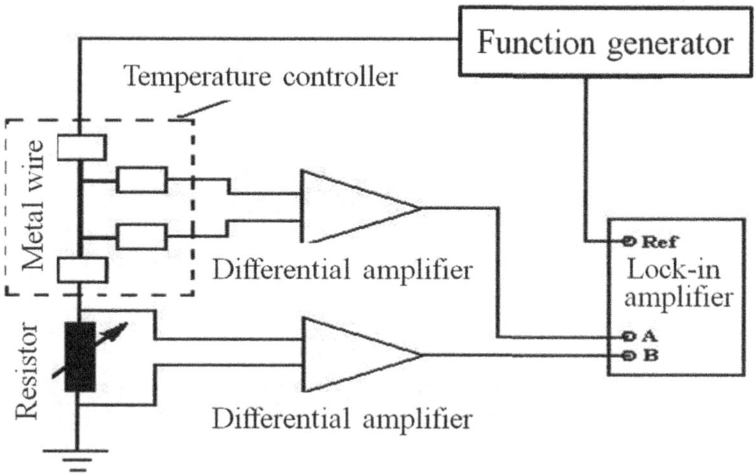

Fig. 2.8 Schematic diagram of the 3ω method thermal conductivity tester

2.1.1.2 Construction and Verification of 3ω Method Thermal Conductivity Tester

(1) Construction of 3ω method thermal conductivity tester

The schematic diagram of the 3ω method thermal conductivity measurement device is shown in Fig. 2.8. Because the fundamental frequency voltage is about three orders of magnitude greater than the 3ω voltage during the measurement, in order to improve the accuracy of the 3ω voltage measurement, a high-precision low temperature coefficient varistor box is usually connected in series in the test circuit. At the same time, in order to stabilize the signal input to the phase-locked amplifier, the voltage signals at both ends of the heating metal wire and the adjustable resistor are entered into two single-gain differential amplifiers, and then into the A and B inputs of the phase-locked amplifier respectively. By adjusting the resistance of the adjustable resistor to make the fundamental frequency voltage across the heating metal wire and the adjustable resistor as equal as possible, the fundamental frequency voltage component can be removed to reduce its impact on the 3ω voltage signal test. A phase-locked amplifier is used to accurately and effectively extract the 3ω voltage across the heated metal wire. In addition, the function generator provides driving signals and reference signals for the entire measuring device.

The 3ω method has a high requirement for the accuracy of the instrument during the measurement of the thermal conductivity. Schematic diagram and experimental setup of 3ω conductivity method are shown in Fig. 2.8 and Fig. 2.9, respectively. The phase-locked amplifier in the picture is a product of the Stanford Research Systems (Model SR830, USA), which can extract and test weak signals in the frequency range of 1–102.4 kHz; the function generator uses an AC and DC current source Keithley (Model 6221, USA) which can provide amplitude AC current signal with a range of 2

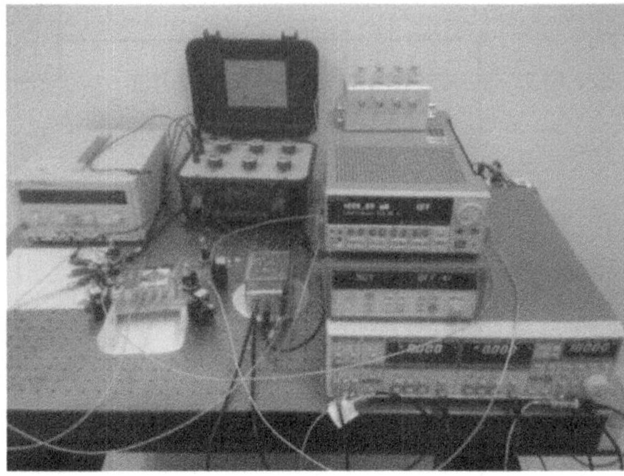

Fig. 2.9 Experimental setup for 3ω method conductivity test

pA-100 mA and a frequency range of 1–100 kHz; the model of the differential amplifier is AD524AD, the model of the adjustable resistor is ZX25P, and the accuracy is 0.01 Ω; the temperature controller is a 50 mm × 50 The thermoelectric module of mm (TEC1-12726, China) the surface can provide heating or cooling temperature range of −10– + 100 °C; the data acquisition instrument (Agilent 34970A, USA) monitors and records the experimental measurement values in real time.

(2) Preparation of SiO_2 thin film standard samples

Since the thermal conductivity of the thin film is affected by external factors such as growth mode and processing technology, it is difficult to reach a consistent conclusion from the test results in different laboratories, consequently it is difficult to calibrate the reliability of the thin film thermal conductivity measurement device. The test of the thermal conductivity of the SiO_2 film is selected below to verify the accuracy of the measuring device. The SiO_2 material was selected because the SiO_2 thin film growth process is well understood, and there are fewer external factors affecting its thermal conductivity, and the thermophysical properties are relatively stable. At the same time, as the key material in the microelectronic process, its research is extensive, so it has become a standard material calibration film for thermal conductivity measurement. In 1994, Cahill [6] tested the thermal conductivity of SiO_2 films in the temperature range of 80–400 K, and found that its thermal conductivity increased with increasing temperature, but all were smaller than the corresponding block thermal conductivity. In 1999, Kim [7] tested the thermal conductivity of SiO_2 films of different thicknesses prepared by thermal oxidation. The test results show that the thermal conductivity is independent of the thickness of the film after considering interface factors. In 2002, Yamane [8] tested the thermal conductivity of SiO_2 films prepared using different methods, and found that its thermal conductivity is related to the porosity of the films. Therefore, this article also selects SiO_2 film as a standard sample to verify the reliability of the measurement device.

The following is the preparation process of the SiO_2 thin film standard sample test structure. First, a Si <100> substrate with a 1 μm thick SiO_2 thin film is selected.

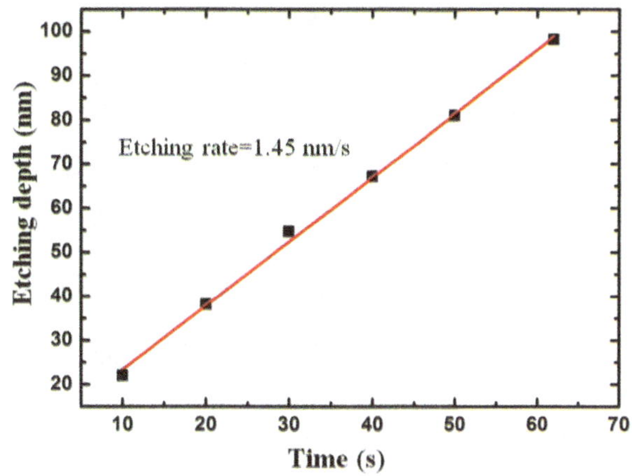

Fig. 2.10 BOE solution etch rate of SiO_2 material

The SiO_2 thin film above it is formed by thermal oxidation. Then use BOE solution ($HF:NH_4 = 1:6$) to chemically etch to obtain SiO_2 films with a thickness of 200 nm, 400 nm, 600 nm, and 800 nm, respectively. Figure 2.10 shows the etching rate of SiO_2 in BOE solution obtained in the previous work.

When using the differential 3ω method to measure the thermal conductivity of a SiO_2 film, in addition to a standard sample of SiO_2 film, another Si <100> substrate with the same but no grown SiO_2 film is needed as a reference sample. Then, a micro-machining process is used to deposit heating metal wires on the surface of the SiO_2 thin film sample and the Si substrate reference sample. The size of the heating metal wire and the choice of material have a significant impact on the test results. Due to the low Debye temperature of Ag, Ag is generally used as the material of the heating wire for measurements below room temperature. When measured above room temperature, Pt and Au are excellent inert metals, are not easily oxidized, and their temperature coefficient of resistance is very stable. Therefore, the material of the heating metal wire is generally Pt or Au. However, the adhesion between these metals and the SiO_2 film is relatively poor, so it is possible to deposit a metal Ti with a thickness of 5–10 nm above the SiO_2 film, and then deposit a heating metal wire. Exactly what kind of metal is used as the material of the heating wire in the experiment, in addition to the test temperature range, the resistance temperature change characteristics of the selected metal and the resistivity characteristics of the metal must be considered. In this experiment, a series of heated metal wire masks with different line widths were designed, as shown in Fig. 2.11.

The micro-machining process is used to deposit heating metal wires on the SiO_2 film, as shown in Fig. 2.12. The specific experimental steps are as follows:

(1) First clean the sample surface with deionized water and blow dry with N_2. Then pre-bake on a hot stage at 90 °C for 3 min, the purpose is to enhance the adhesion between the sample and the photoresist. Next, use a homogenizer (WS-400Bz-6NPP, Laurell, USA) to uniformly apply negative photoresist (AR-N 4450-10, Allresist, Germany) to the sample. Different photoresists require different spin

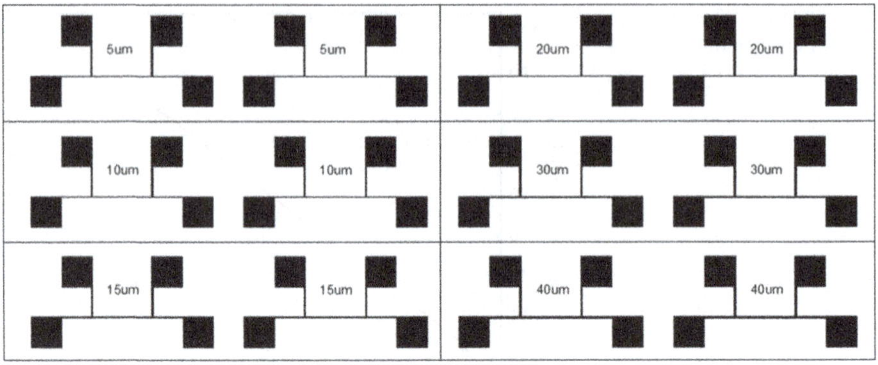

Fig. 2.11 Masks of heated metal wires with different line widths

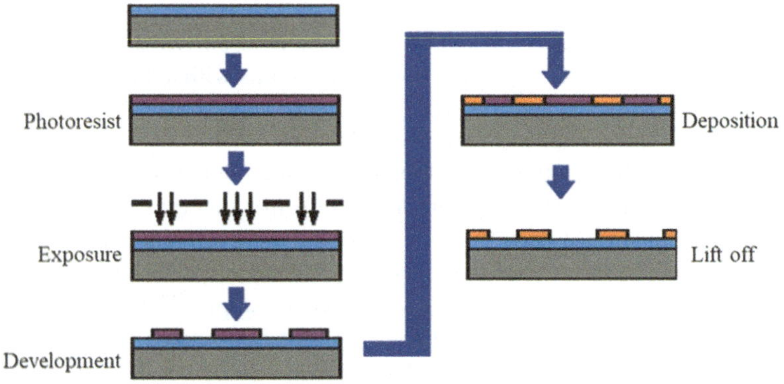

Fig. 2.12 SiO_2 film standard sample preparation process

coatings. The rotating conditions are set to initially rotate at 500 rpm for 5 s at low speed, then jump to 3000 rpm for 30 s at high speed, and finally decelerate to stop. After the photoresist is applied to the surface of the sample, pre-baking is required. The hot stage temperature is set to 90 °C and the pre-baking time is 3 min. The purpose of pre-baking is to remove the solvent in the photoresist, so as to improve the adhesion and uniformity of the photoresist. The sample is cooled to room temperature after the sample is baked before.

(2) After the sample is coated with photoresist, it is aligned with a UV lithography machine (URE-2000/35, Institute of Optoelectronic Technology, Chinese Academy of Sciences, China), and exposed to contact. The exposure allowed the pattern on the photomask to be transferred to the gummed sample. For a negative photoresist, the UV-exposed areas are hardened, making the exposed photoresist difficult to dissolve in the developer solution, while the unexposed areas are removed in the developer solution. After the samples are exposed, they need to be post-baked for 3 min on a hot stage at 90 °C. The purpose of

post-baking is to improve the photoresist adhesion and reduce the standing wave effect during photolithography.

(3) Development is a key step in generating a pattern in the photoresist on the surface of the sample. Use a developer solution (RD6, Futurex, USA) corresponding to the negative photoresist to dissolve the soluble area on the photoresist, and then rinse it with deionized water and blow dry with N_2.

(4) Before coating, use a plasma etching machine (ME-3A, Institute of Microelectronics, Chinese Academy of Sciences, China) to etch the sample to remove the photoresist remaining during the development of the sample surface. The etching gas is O_2. The etch strength and time were set to 100 W and 1 min, respectively. In the experiment, the material of the heating metal wire was selected as Ag. In order to increase the adhesion between the metal and the SiO_2 film, 5 nm Ti was deposited by magnetron sputtering (PVD75, Kurt J. Lesker, USA), and then 200 nm was deposited. Ag.

(5) After the metal is deposited on the sample, the sample is immersed and degummed with acetone and ethanol respectively. This is repeated 3 times. Finally, the sample is rinsed with deionized water and blown dry with N_2.

(3) Analysis of test results of SiO_2 film thermal conductivity

After the above micromachining process steps, the SiO_2 thin film standard samples of the metal heating wire are obtained. The thermal conductivity test of these standard samples is performed at room temperature (20 °C). Figure 2.13 shows the 3ω method thermal conductivity test structure. Picture under a light microscope.

The 3ω method thermal conductivity measurement process is divided into two steps, as follows:

(1) Measure the resistance temperature coefficient dR/dT of the heated metal wire and its initial resistance value R0. Connect the four pins of the heated metal wire to the four corresponding interfaces of the four-wire resistance of the Agilent data acquisition instrument, and attach a K-type thermocouple (Omega)

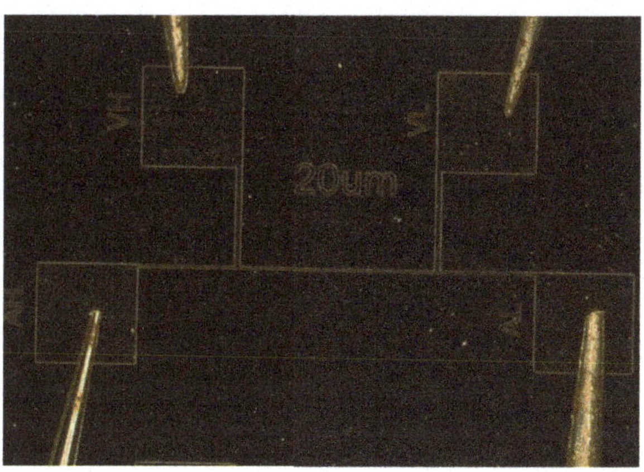

Fig. 2.13 Optical micrograph of 3ω method thermal conductivity test structure

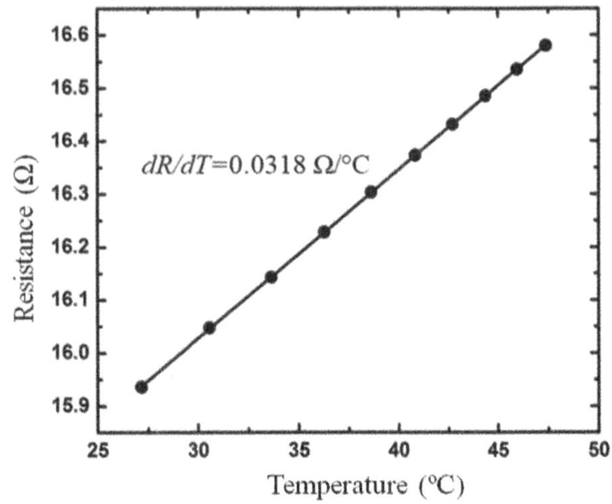

Fig. 2.14 Resistance temperature curve of a 20 μm heating metal wire

to the surface of the sample to be tested for real-time monitoring of the sample surface temperature then turn on the power and use the thermoelectric module to slowly heat the sample. While heating the sample, we use the Agilent 34970A with 34902 high-speed module data acquisition instrument to collect a set of resistance temperature data at 1 Hz speed, and then stop the test after 10 s and repeat twice. The test results show that there is a good linear relationship between the resistance of the metal heating wire and the temperature. Take a 20 μm line of heating metal wire on a 200 nm thick SiO_2 film as an example. Figure 2.14 is the measured resistance temperature curve. After linear fitting of the data points, the resistance temperature coefficient *dR/dT* of the heating metal wire is 0.0318 Ω/°C, and from this, the initial resistance value R0 of the heating metal wire at room temperature (20 °C) can be calculated to be 15.71 Ω.

(2) Measure the 3ω voltage $V_{3\omega}$ of the heated metal wire at different test frequencies. After connecting the test circuit according to the circuit diagram above, first adjust the resistance of the adjustable resistor to be equal to the initial resistance value of the heated metal wire at room temperature. Then Keithley 6221 was used to pass in AC current signal to heat the sample. At this time, the fundamental voltage between the heating metal wire and the adjustable resistor is approximately equal. The fundamental signal can be filtered through the differential input of the lock-in amplifier, and then test the 3ω voltage at a given frequency while change the frequency of the output signal of the AC/DC current source from small to large and test the corresponding 3ω voltage at different frequencies. In the specific operation process, take a 20 μm wire width heating wire as an example, keep the input AC current amplitude $I_{rms} = 50$ mA constant, and then change the frequency of the input current in the range of 100–1000 Hz to measure the heating wire 3ω voltage signal output when different current

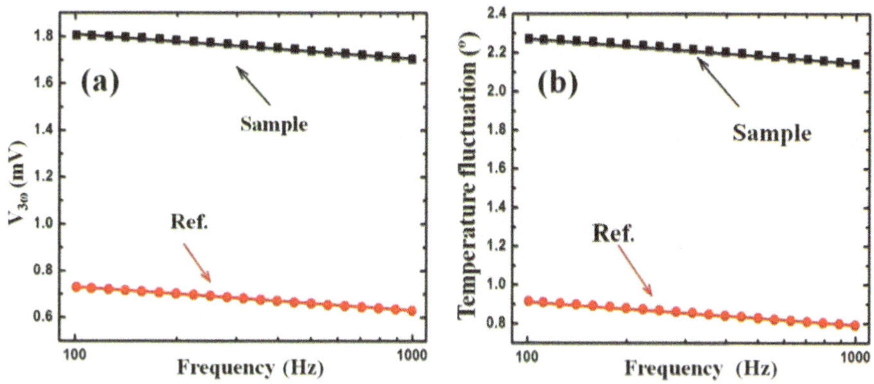

Fig. 2.15 **a** 3ω voltage signal and **b** temperature fluctuation amplitude of SiO_2 thin film samples and reference samples as a function of test frequency

frequency is applied. Taking a heating wire with a 20 μm line thickness on a 200 nm SiO_2 film as an example, Fig. 2.15a shows the relationship between the 3ω voltage signal of the SiO_2 film sample and the reference sample measured with the test frequency. Using Eq. 2.9, the relationship between the temperature fluctuation amplitude of the SiO_2 film sample and the reference sample and the test frequency can be calculated from the 3ω voltage signal, as shown in Fig. 2.15b. After measuring the relationship between the temperature fluctuation amplitude of the SiO_2 film sample and the reference sample as a function of the test frequency, the temperature fluctuation amplitude of the SiO_2 film sample itself can be obtained by differential calculation. Table 2.1 lists the thermal conductivity of SiO_2 thin film standard samples and reference values in the literature measured by heating metal wires with different line widths.

The thermal conductivity of the SiO_2 film measured in Table 2.1 includes the factors of the contact thermal resistance between the sample and the heating metal wire and the sample and the Si substrate. In order to subtract the effect of the contact thermal resistance on the measurement results, use Eq. 2.6 to calculate different thickness SiO_2 The thermal resistance of the film is shown in Fig. 2.16. It can be seen from the figure that the thermal resistance value of SiO_2 thin film samples of different thicknesses has a linear relationship with the thickness. According to Eq. 2.7, the thermal conductivity of the sample can be obtained by using the slope of the fitted straight line. The test results show that at room temperature (20 °C) The intrinsic thermal conductivity of the SiO_2 film is 1.24 W m^{-1} K^{-1}, which is basically consistent with the reference value in the literature.

Table 2.1 Thermal conductivity test results of SiO_2 thin film standard samples

SiO_2 film thickness (nm)	Wire width (μm)	Thermal conductivity ($W\ m^{-1}\ K^{-1}$)	Thermal conductivity reference values ($W\ m^{-1}\ K^{-1}$)
179.5	10	1.38	(1) Kim [7]: Film thickness 488 nm Substrate thickness 380 μm Test temperature 300.7 K Thermal conductivity 1.33 $W\ m^{-1}\ K^{-1}$ (2) Yamane [8]: Film thickness 500 nm Substrate thickness 500 μm Test temperature 298 K Thermal conductivity 1.34 $W\ m^{-1}\ K^{-1}$
	15	1.59	
	20	1.61	
	Average	**1.53**	
364.6	10	1.37	
	15	1.43	
	20	1.22	
	Average	**1.34**	
560.4	10	1.29	
	15	1.24	
	20	1.17	
	Average	**1.23**	
762.2	10	1.36	
	15	1.32	
	20	1.28	
	Average	**1.32**	

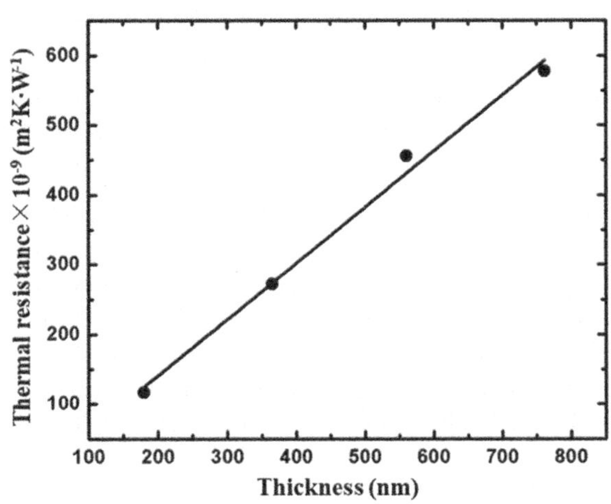

Fig. 2.16 Relationship between thermal resistance and thickness of SiO_2 thin film samples

2.1.2 TDTR Method

The most common measurement method is usually the 3ω method, but the 3ω method requires processes such as electrode processing, introduces additional interfaces, and the process is complicated. When measuring thin films, a high heating frequency is

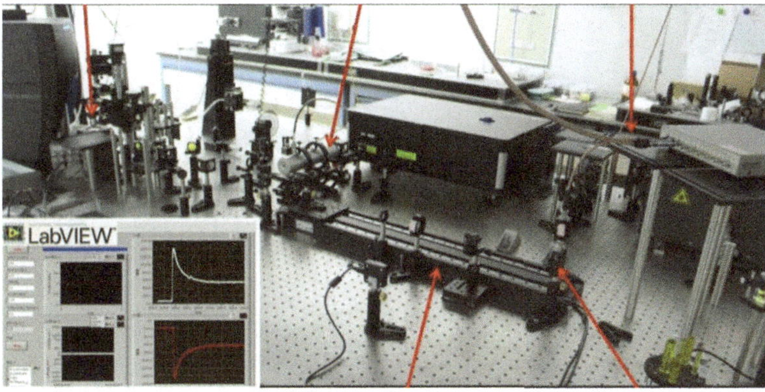

Fig. 2.17 Experimental setup of time-domain picosecond femtosecond pulse method in time domain

required. Therefore, we also use the non-contact time-domain thermoreflectance (TDTR) method to measure the thermal conductivity of the film. This method has little effect on the structure of the film and has high accuracy. The device test device is shown in Fig. 2.17, and the optical path diagram is shown in Fig. 2.18.

Before the test, an Al film of a certain thickness is first deposited on the thin film to be tested as a sensing layer. Since the relationship between the reflectance of the Al film and the temperature is approximately linear, the change in the temperature of the thin film can be derived based on this. After the laser emits a femtosecond laser, it is split under the action of a polarizing beam splitter. The pump light (pump) reaches the surface of Al and heats the Al film. The temperature of the Al film first increases, and then the heat is conducted to the inside of the film to be measured, which makes Its temperature will decrease. The probe light reaches the same position on the Al film surface after a certain time delay, collects the signal intensity of the probe light reflected by the film surface, and obtains the curve of the signal and the delay time. The heat transfer model is selected to fit the obtained curve. The thermal conductivity of the film was obtained afterwards (Fig. 2.19).

2.2 Conductivity and Hall Effect Test

2.2.1 Principle and Method of Conductivity Measurement

The electrical conductivity test of a thin film usually first measures the sheet resistance of the thin film and calculates the electrical conductivity of the thin film based on the thickness of the thin film. There are mainly two methods of van der Pauw and linear four-probe method. The van der Pauw method is applicable to samples of any shape, which can remove the Seebeck effect caused by the temperature difference

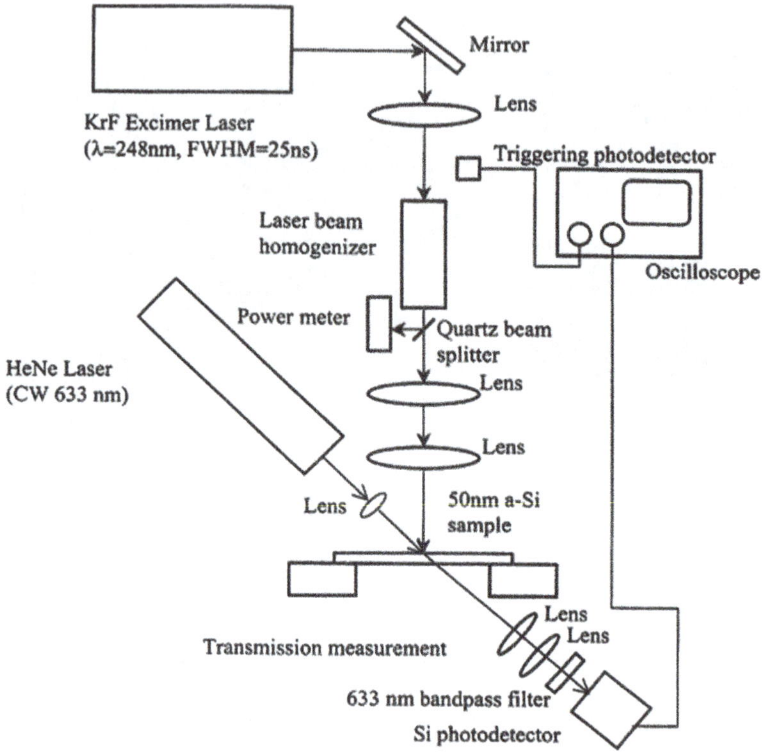

Fig. 2.18 Optical path diagram of the time-domain picosecond femtosecond pulse method

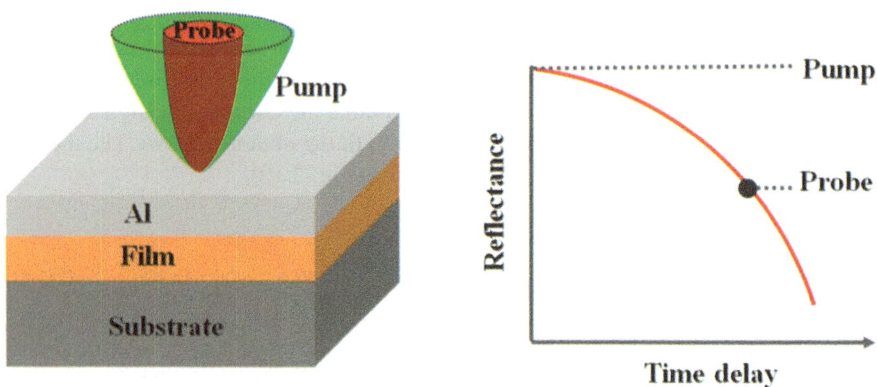

Fig. 2.19 Schematic diagram of TDTR thermal conductivity test

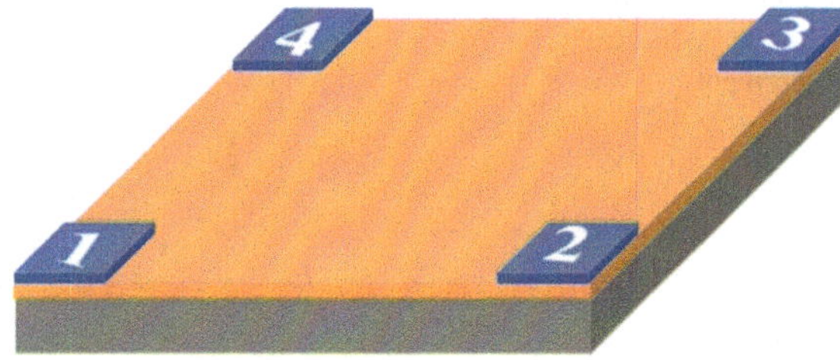

Fig. 2.20 van der Pauw method for measuring film conductivity

in the film, and it is also suitable for the Hall effect test. Therefore, the conductivity test of the film in this paper uses the van der Pauw method for testing, as shown in Fig. 2.20.

First, the current I is applied between the electrodes 1 and 2 to obtain a voltage V_{34} between the electrodes 3 and 4. A reverse current I is passed between the electrodes 1 and 2 to obtain a voltage V_{43} between 4 and 3. Passing the reverse current test can offset the effect of the Seebeck effect of thermoelectric materials on the test. Subsequently, in the same manner, the voltages V_{14}, V_{41}, V_{12}, V_{21}, V_{32}, and V_{23} are obtained in this order. According to the measured voltage and current, the average resistance R_h and R_v of the film in both horizontal and vertical directions can be obtained, as follows

$$R_h = (V_{12} + V_{21} + V_{34} + V_{43}) / 4I, \tag{2.25}$$

$$R_v = (V_{23} + V_{32} + V_{14} + V_{41}) / 4I. \tag{2.26}$$

According to R_h and R_v, the sheet resistance R_s (Ω/sq) of the film can be obtained by the following formula

$$e^{-\pi R_v / R_s} + e^{-\pi R_h / R_s} = 1. \tag{2.27}$$

The thickness t_f of the thin film is determined by a scanning electron microscope or a film thickness meter to obtain the electrical conductivity σ of the thin film

$$\sigma = 1 / \left(R_s \times t_f\right). \tag{2.28}$$

2.2.2 *Principle and Method of Hall Effect Test*

The carrier concentration and mobility in a semiconductor thin film are important factors affecting the electrical/thermal properties of a material. The above parameters of the thin film can be obtained by the Hall effect. The Hall effect test at different

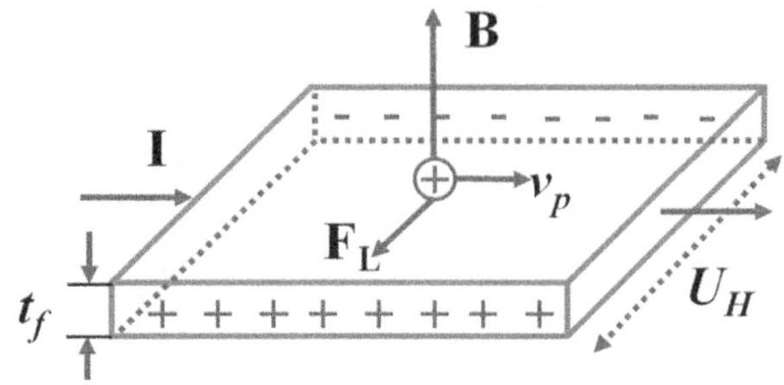

Fig. 2.21 Schematic diagram of Hall effect measurement

temperatures can also be used to study the carrier scattering mechanism in thin film materials.

The Hall effect test is based on the electrical conductivity measurement based on the van der Pauw method. The thin film is placed between parallel plates with a magnetic field perpendicular to the thin film. The magnetic induction intensity is B. A current I is passed through the thin film. There will be a potential difference at the terminal, that is, the Hall voltage U_H. The schematic diagram of the Hall effect test is shown in Fig. 2.21. The model of the Hall effect test device is MMR.

When a current is passed in the film, hole carriers are taken as an example, its drift velocity is v, and it is subjected to Lorentz force F_L under the action of a magnetic field, which causes the charge in the film to deflect and gather at one end of the film to form the Hall voltage U_H, and the carrier concentration n and the carrier mobility μ of the film are obtained. The expressions of each parameter are as follows:

$$U_{\mathrm{H}} = R_{\mathrm{H}} I B / t_f, \tag{2.29}$$

$$n = 1 / R_{\mathrm{H}} q, \text{ and} \tag{2.30}$$

$$\mu = R_{\mathrm{H}} \sigma. \tag{2.31}$$

Among them, R_H is a Hall coefficient and q is a carrier amount. In the Hall voltage measurement, the van der Pauw method is used to perform the analysis in the direction of a magnetic field, and then the magnetic field is reversed to perform the same test, which can eliminate errors caused by influencing factors such as thermoelectric potential.

The above has introduced the conductivity and Hall effect measurement methods at room temperature. Through precise temperature control equipment, the corresponding parameters can be tested at different temperatures. It should be noted that the electrical parameter coefficient tested by this method is a parameter in the plane of the film, that is, the direction parallel to the substrate. The relevant parameters tested by this method in this paper are all in this direction, and the in-plane and out-plane of the film can be estimated based on the anisotropy of the film's electrical parameters.

2.3 Seebeck Coefficient Test

The film is placed on one hot end and the cold end horizontally. By controlling the temperature of the hot and cold ends, a stable temperature difference ΔT is formed at both ends of the film, and the potential difference at two points is collected at the same time. Peltier semiconductor refrigeration chips are used for the hot and cold junctions, and the opposite currents are passed to achieve constant control of the hot and cold junction temperatures. During the test, the temperature difference should be controlled within 5 K. After the temperature difference is stabilized, the current at the hot and cold ends is disconnected, the temperature difference gradually decreases to zero, and the temperature at equilibrium is T. Through real-time acquisition of the temperature drop to zero, the voltage changes with temperature. When the temperature difference is small, the parasitic potential can be regarded as a constant, the potential difference changes linearly with the temperature difference, and the slope is the Seebeck coefficient.

This paper compares the Seebeck coefficients of the static method and the dynamic method after testing by setting up the equipment. Considering that the static method requires higher test control, the dynamic method is used to test the Seebeck coefficient. A Pt100 thermal resistance with 1/3B accuracy is used as a temperature sensor to collect temperature changes in real time. Pt100 thermal resistance wiring uses a four-wire method to remove the influence of line resistance. Both the thermal resistance and the thermoelectric potential difference were collected using an Agilent data acquisition instrument, model 34972A, with an accuracy of $6^{1/2}$. The distance between the two probes is 1 cm. The Seebeck coefficient of each measurement was averaged by measuring 3 times, and the linear fitting coefficient was better than 0.996. When testing the Seebeck coefficient at different temperatures, an oven is used to provide the required environmental test temperature, and the test temperature range is 300–460 K. During the test, high-purity nitrogen with a purity of 99.999% was passed to provide the atmosphere. Constantan standard samples (composition Cu55/Ni45) were used to calibrate the Seebeck coefficient equipment. As shown in Fig. 2.22b,

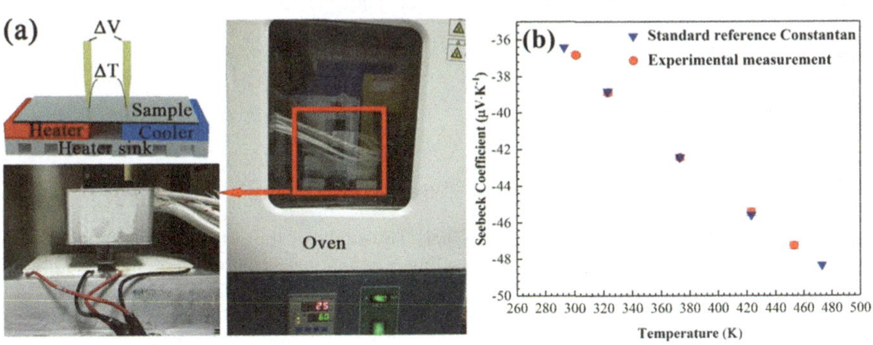

Fig. 2.22 Seebeck coefficient test device and calibration: **a** Experimental setup; **b** Constantan measurement results and calibration values

the test results of constantan standard samples by this device agree well with the calibration values, indicating The reliability of this device. Constantan standard samples and standard values are provided by Linseis. It should be noted that the Seebeck coefficient tested by this method is a parameter in the plane of the film, that is, the direction parallel to the substrate, and the Seebeck coefficient tested in this paper is all in this direction.

2.4 Chapter Summary

This chapter introduces the testing principles and methods of thermoelectric materials, and calibrates the built equipment to provide a guarantee for the subsequent experiments. Human beings have been longing to find an inexhaustible green energy supply like nature does. Thermoelectric devices have demonstrated the possibility to directly converting the ultra-low-grade ambient heat energy on the earth's surface to electrical power. In future, when such advanced energy technologies are wildly available and very low costs after large-scale production employed, we will then obtain truly sustainable, completely environmentally friendly clean energy, and greatly reducing our dependence on petrochemical energy. At present, all energy sources, whether it is oil, natural gas, hydropower, nuclear power, solar energy, wind energy, etc., have a fundamental problem. That is, the energy sources are unevenly and unfairly distributed around the world. These uneven distributions of energy have caused countless disputes and even wars that have brought countless disasters and suffering to the peoples of the world.

Physics tells us that if there is temperature, there exists thermal energy; and if there is a temperature difference, work can be produced. As thermal energy exists everywhere in the world regardless the time, location and weather. Advanced thermoelectric devices system can convert small temperature differences in many scenarios (such as indoor/outdoor temperature differences, water temperature differences between surface and air, etc.) to electric energy that provides an innovative solution for obtaining the affordable and sustainable green energy.

References

1. Cahill DG (1990) Thermal conductivity measurement from 30 to 750 K: the 3 omega method. Rev Sci Instrum 61(2):802–808
2. Yao DJ, Chien HC, Liu YC, Tseng MH (2004) The study of thin films thermal conductivity measurement by 3 omega method. In: Proceeding international symposium on nanotechnology and energy. Hsinchu, Taiwan
3. Carslaw HS, Jaeger JC (1959) Conduction of heat in solids. Oxford University Press, Oxford
4. Callister WD (2003) Materials science and engineering: an introduction. Wiley, Inc
5. Cahill DG, Fischer HE, Klitsner T, Swartz ET, Pohl RO (1989) Thermal conductivity of thin films: measurements and understanding. J Vac Sci Technol A: Vac, Surf, Films 7(3):1259–1266. https://doi.org/10.1116/1.576265

6. Cahill DG, Allen TH (1994) Thermal conductivity of sputtered and evaporated SiO_2 and TiO_2 optical coatings. Appl Phys Lett 65(3):309–311
7. Kim JH, Feldman A, Novotny D (1999) Application of the three omega thermal conductivity measurement method to a film on a substrate of finite thickness. J Appl Phys 86(7):3959–3963
8. Yamane T, Nagai N, Katayama S, Todoki M (2002) Measurement of thermal conductivity of silicon dioxide thin films using a 3 omega method. J Appl Phys 91(12):9772–9776. https://doi.org/10.1063/1.1481958

Chapter 3
Si-Based Multilayer Films

Si is stable in performance, non-toxic and non-polluting. It is the most widely used semiconductor material and a kind of thermoelectric material. However, Si has a high thermal conductivity. How to reduce its thermal conductivity and improve thermoelectric performance is an important issue facing Si-based thermoelectric materials. In order to reduce the thermal conductivity of Si-based materials, Si can be combined with other materials to prepare two-dimensional films or form one-dimensional nanowires. In addition, it is one of the important ways to reduce the thermal conductivity of Si-based materials by constructing periodic nano multilayer films to produce nano-scale effects and interface effects.

3.1 $Si/Si_{0.75}Ge_{0.25}$ Multilayer Film

3.1.1 Film Preparation

Amorphous $Si/Si_{0.75}Ge_{0.25}$ nm multilayer films were prepared and studied by magnetron sputtering. The model of the magnetron sputtering equipment was PVD75 (Kurt J. Lesker, USA). A $Si/Si_{0.75}Ge_{0.25}$ multilayer thin film sample was prepared by alternately depositing Si and $Si_{0.75}Ge_{0.25}$ targets with a purity of 99.999% at room temperature. The substrate was Si<100> and the resistivity was 1000 Ω cm. The base distance of the internal target of magnetron sputtering is ~20 cm. Figure 3.1 shows the relationship between the sputtering rate and sputtering power of Si and $Si_{0.75}Ge_{0.25}$ materials. The sputtering rate and sputtering power of Si and $Si_{0.75}Ge_{0.25}$ are basically linear. Before coating, first the Si<100> substrate was ultrasonically cleaned with acetone and ethanol solution for 10 min, and then washed with ultra-pure water and blown with dry N_2. After that, the substrate was transferred to a magnetron sputtering vacuum chamber and fixed. On the substrate table. When the vacuum reaches the background vacuum, the coating is started. During the deposition process, the sputtering power of Si and $Si_{0.75}Ge_{0.25}$ are controlled at 100 W, and the

Z. Hu and Z. Wu, *Nanostructured Thermoelectric Films*,
https://doi.org/10.1007/978-981-15-6518-2_3

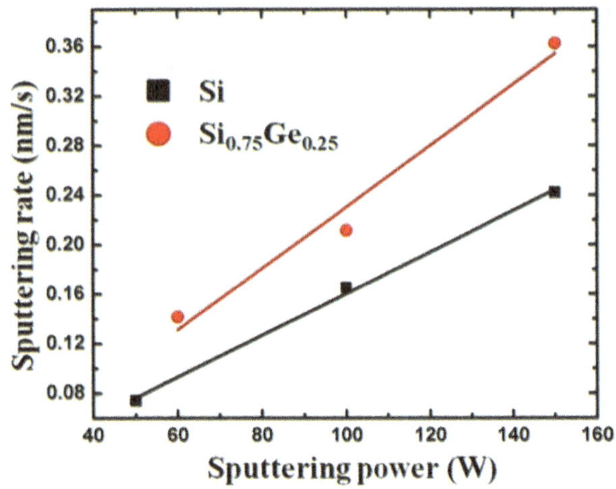

Fig. 3.1 Relationship between sputtering rate and sputtering power of Si and $Si_{0.75}Ge_{0.25}$ materials

Table 3.1 Preparation parameters of $Si/Si_{0.75}Ge_{0.25}$ multilayer film

Sampler number	Base vacuum (Torr)	Sputtering air pressure and sputtering power	Si/SiGe thickness ratio (nm/nm)	Number of cycles	Number of interfaces	Total thickness (nm)
#1	1.9×10^{-7}	3 mTorr Si 100 W $Si_{0.75}Ge_{0.25}$ 100 W	~1.5/~1	200	399	540.3
#2	1.4×10^{-7}		~3/~2	100	199	515.0
#3	3×10^{-7}		~6/~4	50	99	514.8
#4	3.2×10^{-7}		~12/~8	25	49	508.0
#5	2×10^{-7}		~30/~20	10	19	507.2

working pressure of Ar is 3 mTorr. The sputtering rates are ~0.17 Å/s and ~0.22 Å/s, respectively, which are consistent with the linear fitting values as shown in Fig. 3.1. During the sputtering process, to ensure uniform film thickness, the substrate was kept at a rotation rate of 20 rms.

Table 3.1 lists the preparation parameters of the $Si/Si_{0.75}Ge_{0.25}$ multilayer film. A total of 5 $Si/Si_{0.75}Ge_{0.25}$ multilayer films with thicknesses ranging from 2.5 nm to 50 nm were prepared, of which Si and $Si_{0.75}Ge_{0.25}$ respectively. The thickness ratio of is controlled at 3:2, and the total thickness of the samples is around 500 nm.

3.1.2 Structural Characterization

X-ray diffractometer (D\max-2200, Rigaku, Japan) and scanning electron microscope (ULTRA55-36-69, Zeiss, Germany) were used to characterize the structure of the prepared samples. Figure 3.2 is a small-angle XRD diffraction pattern of a

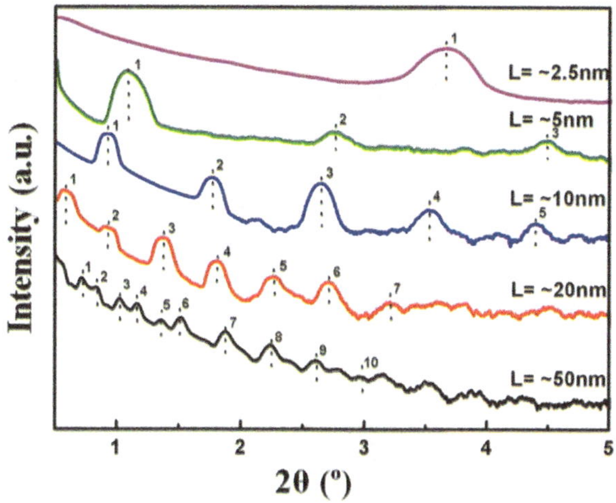

Fig. 3.2 Small-angle XRD diffraction pattern of Si/$Si_{0.75}Ge_{0.25}$ multilayer film

Si/$Si_{0.75}Ge_{0.25}$ multilayer film. The X-ray diffractometer uses Cu Kα rays with a wavelength $\lambda = 1.54056$ Å and a tube pressure of 40 kV. Sampling is performed by continuous scanning. The scanning speed is 0.5°/min, the step width is 0.002°, and 2θ is 0.5°–5°. It can be seen from the figure that as the period thickness increases, the number of diffraction peaks also increases. When the period thickness of the sample is 50 nm, the number of diffraction peaks reaches 10, which indicates that the prepared by magnetron sputtering The Si/$Si_{0.75}Ge_{0.25}$ multilayer film has a flat interface (Fig. 3.3).

The period thickness d is given by

$$d = \frac{\lambda}{2}\sqrt{\frac{(n+1)^2 - n^2}{\sin^2\theta_{n+1} - \sin^2\theta_n}}, \tag{3.1}$$

where λ is the wavelength of X-rays, n is the number of XRD diffraction peaks, θ_n is the angle corresponding to the XRD diffraction peaks. It is worth noting that the Si/$Si_{0.75}Ge_{0.25}$ multilayer film with a period of 2.5 nm has only a small-angle XRD diffraction peak, so it is connected to the origin through a straight line through this point. The Bragg diffraction formula was used to calculate the periodic thicknesses of samples #1, #2, #3, #4, and #5, which were 2.4, 5.8, 10.0, 19.6, and 48.8 nm, which were basically consistent with the experimental design values.

Figure 3.4 is a SEM image of a cross section of a Si/$Si_{0.75}Ge_{0.25}$ multilayer film (sample #5) with a thickness of 50 nm at different magnification periods. Figure 3.4a clearly shows that there are 10-cycles of a Si/$Si_{0.75}Ge_{0.25}$ multilayer film with a period thickness of 50 nm, and the interface of the multilayer film is very flat. From Fig. 3.4b, it can be seen that the thickness of the Si layer and the $Si_{0.75}Ge_{0.25}$ layer in the Si/$Si_{0.75}Ge_{0.25}$ multilayer film (sample #5) with a periodic thickness of 50 nm is 29.2 nm and 18.6 nm respectively, which is consistent with the design value. The Si

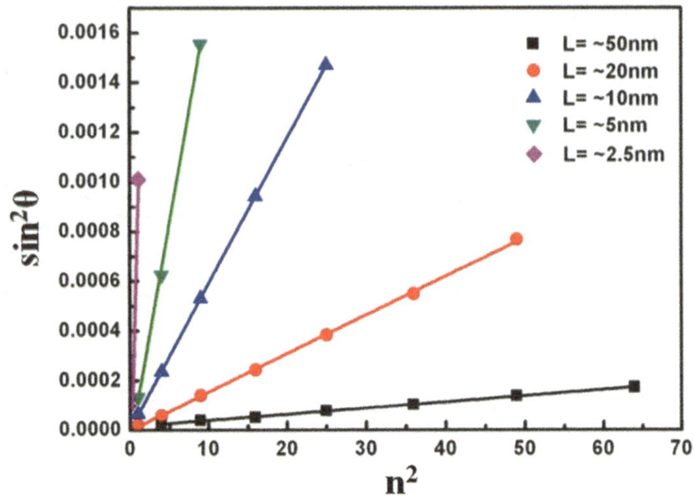

Fig. 3.3 Shows the fitted line between the angle of the small-angle XRD diffraction peak and the corresponding number of diffraction peaks. Using the Bragg diffraction formula, the periodic thickness of the $Si/Si_{0.75}Ge_{0.25}$ multilayer film can be calculated

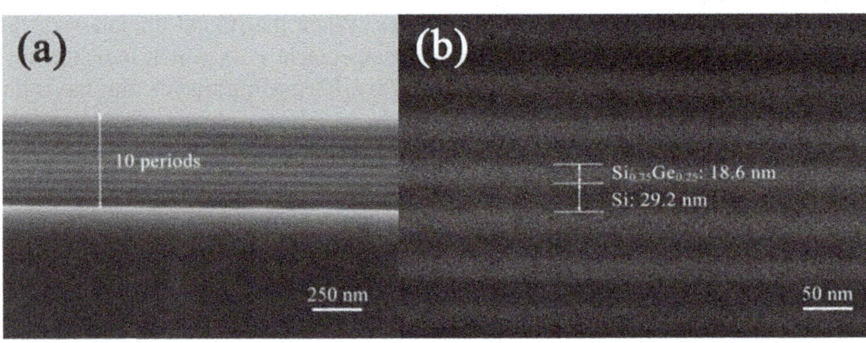

Fig. 3.4 SEM image of $Si/Si_{0.75}Ge_{0.25}$ multilayer film (Sample #5)

layer and $Si_{0.75}Ge_{0.25}$ layer in the figure show that the light and darkness are caused by the different atomic weight.

3.1.3 Thin Film Thermal Conductivity

When using the differential 3ω method to measure the thermal conductivity of amorphous $Si/Si_{0.75}Ge_{0.25}$ nm multilayer films, in addition to the thin film sample to be measured, another piece of the same but not grown $Si/Si_{0.75}Ge_{0.25}$ is required. A thin film Si substrate was used as a reference sample. Since the Si/Si$_{0.75}$Ge$_{0.25}$ multilayer film is a semiconductor, in order to prevent leakage current, a 200 nm thick SiO_2 film is deposited by a magnetron sputtering method before the heating metal wire is deposited, and then a test structure is prepared by a micromachining process. The material of the metal heating wire is Ag and the line width is 20 μm.

Taking a Si/Si$_{0.75}$Ge$_{0.25}$ multilayer film with a periodic thickness of 50 nm (Sample #5) as an example, we first use a 20 μm line width metal heating wire to test the resistance temperature coefficient of the metal wire as shown in Fig. 3.5a. The linear coefficient of resistance is 0.0371 Ω/°C and the initial resistance value of the metal wire at room temperature (20 °C) is 17.47 Ω. Then the 3ω voltage of the heating metal wire has been measured at different test frequencies. During the test, the current amplitude is 60 mA and the test frequency range is 100–1000 Hz. Figure 3.5b is the measured temperature fluctuation curve of the Si/Si$_{0.75}$Ge$_{0.25}$ multilayer film sample and the reference sample from which the thermal conductivity of the Si/Si$_{0.75}$Ge$_{0.25}$ multilayer film can be calculated. Table 3.2 shows the results of the thermal conductivity tests of Si/Si$_{0.75}$Ge$_{0.25}$ multilayer films.

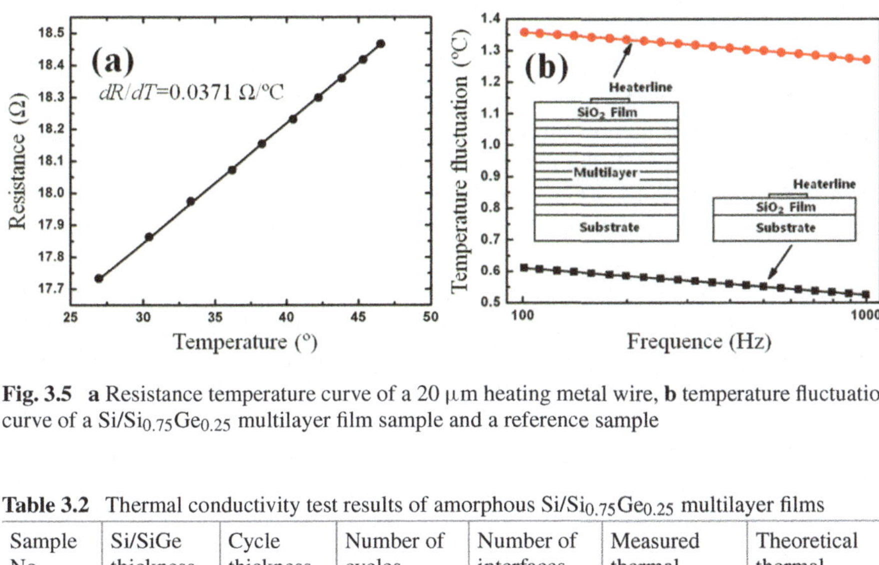

Fig. 3.5 **a** Resistance temperature curve of a 20 μm heating metal wire, **b** temperature fluctuation curve of a Si/Si$_{0.75}$Ge$_{0.25}$ multilayer film sample and a reference sample

Table 3.2 Thermal conductivity test results of amorphous Si/Si$_{0.75}$Ge$_{0.25}$ multilayer films

Sample No.	Si/SiGe thickness ration (nm/nm)	Cycle thickness (nm)	Number of cycles	Number of interfaces	Measured thermal conductivity ($W\ m^{-1}\ K^{-1}$)	Theoretical thermal conductivity ($W\ m^{-1}\ K^{-1}$)
#1	~1.5/~1	2.4	200	399	0.98	1.06
#2	~3/~2	5.8	100	199	1.09	
#3	~6/~4	10.0	50	99	1.09	
#4	~12/~8	19.6	25	49	0.85	
#5	~30/~20	48.8	10	19	1.12	

According to the classical heat conduction formula, the expression of the thermal conductivity coefficient κ_{eff} of a multilayer film composed of two materials alternately can be written as

$$\kappa_{eff} = \frac{d_1 + d_2}{d_1/\kappa_1 + d_2/\kappa_2}, \tag{3.2}$$

where the subscripts 1 and 2 represent two kinds of materials, κ is a thermal conductivity coefficient constituting a multilayer thin film material, and d is a thickness constituting the multilayer thin film material. It is worth noting that the classical thermal conduction model ignores the interface thermal resistance. The thermal conductivity of the amorphous Si film tested in the preliminary exploration experiment was 1.44 W m^{-1} K^{-1}, and Moon reported [1] the test result of the thermal conductivity of amorphous Si (1.5 W m^{-1} K^{-1}) is basically the same, but much smaller than the thermal conductivity of crystalline Si [2] (150 W m^{-1} K^{-1}). The measured thermal conductivity of the amorphous $Si_{0.75}Ge_{0.25}$ film is 0.76 W m^{-1} K^{-1}, which is smaller than the thermal conductivity of the crystalline $Si_{0.75}Ge_{0.25}$ film [3] (6.7 W m^{-1} K^{-1}) of which may be due to the limitation of the lattice vibration during the propagation of heat in the amorphous material, resulting in a decrease in its thermal conductivity. Table 3.3 lists the measurements of the thermal conductivity of the amorphous Si film and amorphous $Si_{0.75}Ge_{0.25}$ film is compared with that of the corresponding crystalline Si film and crystalline $Si_{0.75}Ge_{0.25}$ film. Therefore, according to Eq. 3.2, the theoretical value of the thermal conductivity of the prepared amorphous Si/$Si_{0.75}Ge_{0.25}$ multilayer film should be 1.06 W m^{-1} K^{-1}. This value is basically consistent with the experimentally measurement of the amorphous Si/$Si_{0.75}Ge_{0.25}$ multilayer film.

Figure 3.6 shows the measured thermal conductivity of the amorphous Si/$Si_{0.75}Ge_{0.25}$ nm multilayer film compared with the thermal conductivity of the Si/Ge superlattice and SiGe alloy film. Each film sample was measured 5 times during the experiment, where the error bars represent the standard deviation. From the comparison in the figure, it can be found that the thermal conductivity of the amorphous Si/$Si_{0.75}Ge_{0.25}$ nm multilayer film is significantly smaller than the thermal conductivity of the Si/Ge superlattice or SiGe alloy film reported in the literatures. In the Si/Ge superlattice structure, as the number of interfaces increases, its thermal conductivity will gradually increase, but it is smaller than the thermal conductivity of SiGe alloy thin films, and the amorphous Si/$Si_{0.75}Ge_{0.25}$ with 5 different thicknesses is prepared. The thermal conductivity of nanometer multilayer film is basically the

Table 3.3 Thermal conductivity of amorphous Si and amorphous $Si_{0.75}Ge_{0.25}$ materials

Materials	Thermal conductivity (W m^{-1} K^{-1})
Amorphous Si	1.44
Crystal Si	150
Amorphous $Si_{0.75}Ge_{0.25}$	0.76
Crystal $Si_{0.75}Ge_{0.25}$	6.7

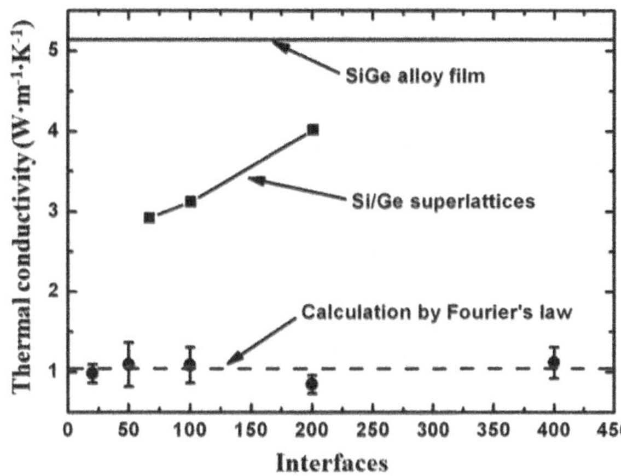

Fig. 3.6 Comparison of thermal conductivity test results of amorphous Si/$Si_{0.75}Ge_{0.25}$ nm multilayer films

same, and there is no direct relationship with the number of interfaces, and it is consistent with the theoretical value obtained using the classic thermal conduction formula (Table 3.2).

Compared with SiGe alloy films, the thermal conductivity of Si/Ge superlattices increases with the decrease of the cycle thickness due to the increase of dislocation density and residual stress [4]. In the BiTe superlattice, studies have shown that there is a certain period thickness that minimizes its longitudinal thermal conductivity [5]. In the Si/SiGe superlattice and GaAs/AlAs superlattice, the longitudinal thermal conductivity decreases as the thickness of the cycle decreases. In superlattice or alloy materials, the constraint of the mean-free-path of phonons is mainly due to interface scattering or alloy scattering. In amorphous Si/$Si_{0.75}Ge_{0.25}$ nm multilayer films, the decrease in the mean-free-path of phonons is mainly due to structural disorder rather than scattering mechanisms. According to the Debye model [6], the average free path Λ of a phonon can be calculated from the thermal conductivity κ, the specific heat capacity C, and the phonon group velocity v_p

$$\kappa = C v_p \Lambda / 3 \tag{3.3}$$

Using the thermal conductivity of amorphous Si and amorphous $Si_{0.75}Ge_{0.25}$ obtained in previous measurements, the average free path of phonons at room temperature is calculated to be 0.407 and 0.238 nm as shown in Table 3.4. The value is smaller than the layer thickness of the designed amorphous Si/$Si_{0.75}Ge_{0.25}$ multilayer film. Therefore, it is speculated that the average phonon free path is smaller than the layer thickness of the multilayer film due to the disorder of the amorphous Si/$Si_{0.75}Ge_{0.25}$ multilayer film structure, so that the measured longitudinal thermal conductivity is consistent with the theoretical value. There is no significant decrease in thermal conductivity occurred.

Table 3.4 Mean-free-path of amorphous Si and amorphous $Si_{0.75}Ge_{0.25}$ phonon

Materials	Thermal conductivity κ ($W\ m^{-1}\ k^{-1}$)	Heat capacity C ($\times 10^6\ J\ m^{-3}\ K^{-1}$)	Phonon group speed v ($m\ s^{-1}$)	Phonon average free path λ (nm)
Amorphous Si	1.44	1.66	6600	0.407
Amorphous $Si_{0.75}Ge_{0.25}$	0.76	1.67	5700	0.238

3.2 $Si/Si_{0.75}Ge_{0.25}$ + Au Multilayer Film

In this section, Si $Si_{0.75}Ge_{0.25}$ + Au periodic multilayer films are prepared by magnetron sputtering. On the basis of the previous section, insert different Au layers to investigate their effect on the thermal conductivity of multilayer films.

3.2.1 Film Preparation

The magnetron sputtering system (PVD75, Kurt J. Lesker, USA) was used to prepare the required samples. At room temperature, Si and $Si_{0.75}Ge_{0.25}$ targets with a purity of 5 N were alternately deposited on each other to obtain a multilayer thin film sample $Si/Si_{0.75}Ge_{0.25}$. The Si<100> substrate was used with a resistivity of 1000 Ω cm.

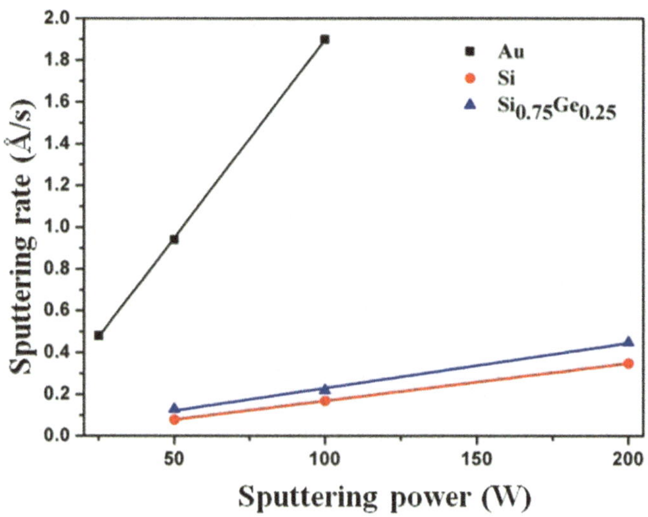

Fig. 3.7 Relationship between sputtering rate and sputtering power for Si, $Si_{0.75}Ge_{0.25}$ and Au targets

The distance between the target and the substrate is kept at ~20 cm. Then, a 4 N A target was used to replace the $Si_{0.75}Ge_{0.25}$ layer in the $Si/Si_{0.75}Ge_{0.25}$ multilayer film sample, and the number of substitutions was 0, 2, 5, 10 respectively. Through a large number of previous experiments, a linear relationship between the sputtering rate of Si, $Si_{0.75}Ge_{0.25}$, and Au with the sputtering power have been determined as shown in Fig. 3.7. Before sputtering coating, in order to remove the surface oxide layer, the substrate Si<100> was immersed in a B.O.E solution ($HF:NH_4 = 1:6$) for five minutes, and then ultrasonically cleaned with acetone and ethanol solutions for 10 min then was rinsed with ultrapure water and blown dry with nitrogen. Next, the substrate was placed in a magnetron sputtering chamber and fixed on a sampler holder. The background vacuum of the coating was set to 5×10^{-7} Torr. When the coating started, the coating pressure is maintained at 3×10^{-3} Torr. The sputtering power of Si and $Si_{0.75}Ge_{0.25}$ was exacted at 100 W and Au was at 25 W, the corresponding coating rates are ~0.17, ~0.22, and ~0.48 Å/s, which are also consistent with the linear fitting values in Fig. 3.7. In order to maintain uniform film thickness throughout the sputtering process, the substrate was rotated at a fixed rotation rate of 20 rpm.

Samples of the prepared Au-doped multilayer films are shown in Fig. 3.8. Table 3.5 lists the preparation parameters of the multilayer film. A total of 4 samples were prepared, which were $Si/Si_{0.75}Ge_{0.25}$ and the Au layer instead of 2 layers, 5 layers, and 10 layers of $Si_{0.75}Ge_{0.25}$, where the thickness of the Si layer is about 12 nm, the

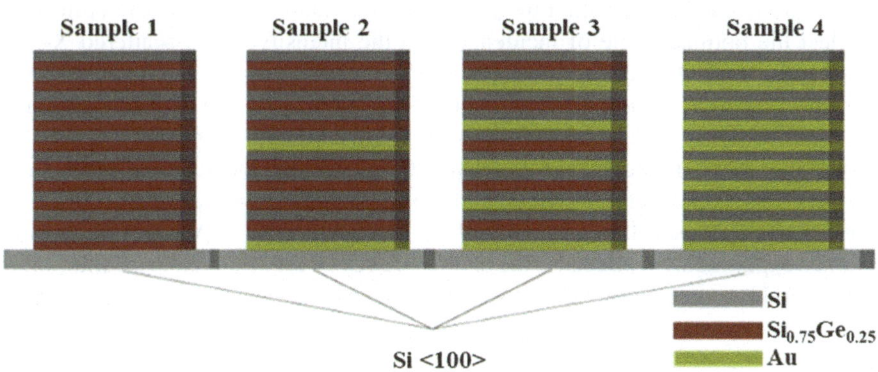

Fig. 3.8 Schematic diagram of $Si_{0.75}Ge_{0.25}$ thin film sample after replacing Au with different layers

Table 3.5 Experimental parameters of $Si/Si_{0.75}Ge_{0.25}$ and $Si/Si_{0.75}Ge_{0.25}$ + Au multilayer films

Samples	Base vacuum	Sputtering Power (W)	Thickness ratio (nm)			Coating time (min)		
			Si	SiGe	Au	Si	SiGe	Au
$Si/Si_{0.75}Ge_{0.25}$; $Si/Si_{0.75}Ge_{0.25}$ + Au	5×10^{-7} Torr 3 mTorr	Si:100 $Si_{0.75}Ge_{0.25}$:100 Au:25	12	10	10	12	2	2.75

thickness of the SiGe layer and the Au layer is about 10 nm, the total thickness is controlled at about 220 nm, and the multilayer film period is 10.

3.2.2 *Structural Characterization*

3.2.2.1 Characterization of Grazing Incidence Small Angle Scattering (GISAXS)

Using a synchrotron radiation source, the sample is subjected to grazing incidence small-angle X-ray scattering to characterize its periodic multilayer film structure (GISAXS). As shown in Figs. 3.9, through a small incident angle α_I, the incident X-rays The light beam is incident on the sample surface, and the reflected light beam also has a corresponding scattering pattern (Fig. 3.9).

By grazing incidence small-angle X-ray scattering, not only can the layered periodic structure be characterized, but also structural information of the surface film and the nanoparticles inside the film can be measured, such as internal particle size, shape, spacing, and distribution. The experiment was performed at the BL14B1 beamline station of the Shanghai Synchrotron Radiation Light Source (SSRF). In the experiment, using a high-density X-ray source to enter the sample shows that the position of the sample can be controlled by a sample stage that can be precisely controlled to select the optimal angle of incidence, and the intensity of the scattered X-rays collected by a CCD detector with 2048 × 2048 pixels. Because the incident X-ray energy is relatively large, in order to avoid damage to the CCD detector, an X-ray beam interceptor is used to weaken the strong reflected light. In the experiment, the wavelength of the incident X-ray light was 0.124 nm, the incident angle was 0.5°, and the distance between the sample and the CCD detector was 2042 mm.

Figure 3.10 is the GISAXS image of the $Si/Si_{0.75}Ge_{0.25}$, Si/Au multilayer films obtained experimentally. Because the incident X-rays were scattered in the multilayer film, several Bragg scattering peaks were seen in the reciprocal space. It can be clearly

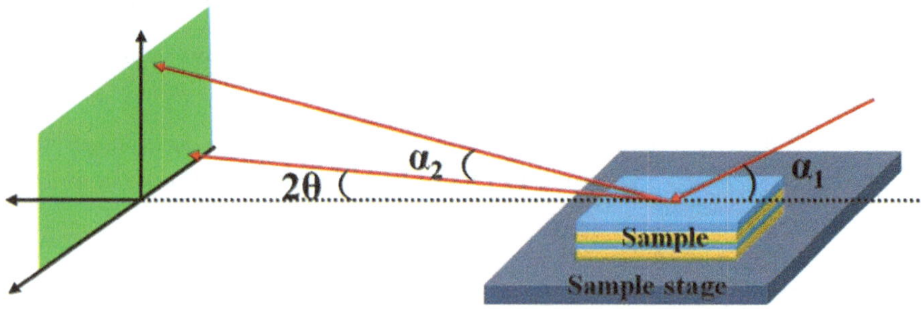

Fig. 3.9 GISAXS principle diagram, α_2 relative to the film surface's exit angle, 2θ relative to the incident surface's scattering angle

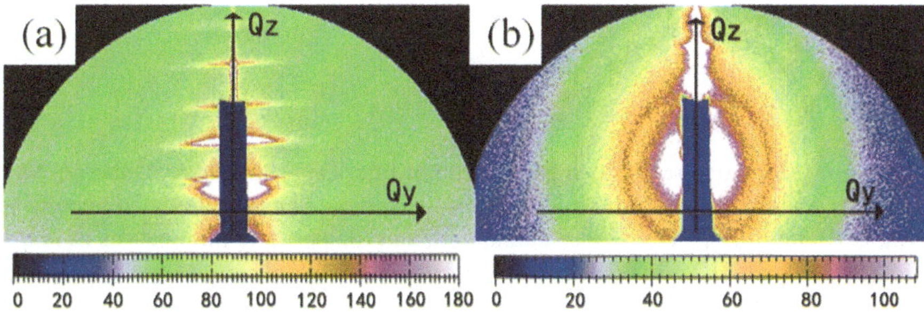

Fig. 3.10 GISAXS images of **a** Si/Si0.75Ge0.25 and **b** Si/Au multilayer films

seen from the figure that the distribution of the Bragg scattering peaks are on both sides of the Y axis. The Z-axis direction was distributed in parallel. Because the scattering intensity was symmetrically distributed along the Y axis, only the Z axis was analyzed and discussed, which indicated that the multilayer film prepared by magnetron sputtering was a periodic layered structure. The rectangular integral of the entire Z axis was used to study the multilayer structure and periodic thickness of the sample. As shown in Fig. 3.11, it is the result of $Si/Si_{0.75}Ge_{0.25}$, and the Si/Au sample is one-dimensional integrated along the Z axis. The small picture in the upper right corner is the fitted straight line of Q_Z vs n.

According to Eq. 3.4, the periodic thickness of the sample can be determined by the peak position of the scattering peak

$$Q_Z = \frac{2\pi n}{d}, \tag{3.4}$$

where Q_Z is the component of the wave vector along the Z direction, n is the order of the Bragg scattering peaks, and d is the periodic thickness of the sample. Equation 3.4

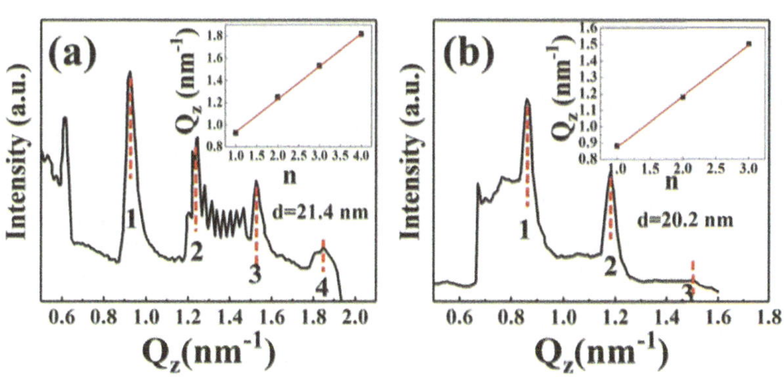

Fig. 3.11 **a** $Si/Si_{0.75}Ge_{0.25}$ and **b** Si/Au multilayer thin film one-dimensional rectangular integral plot and Q_Z vs n fitted straight line

can be transformed into

$$d = \frac{2\pi}{k}, \tag{3.5}$$

where is the slope of the Q_Z vs n fitted straight line. From the slope of the Q_Z vs n fitted straight line, the periodic thickness of the sample can be calculated according to Eq. 2.5. The $Si/Si_{0.75}Ge_{0.25}$ was calculated, and the periodic thicknesses of Au/Si were 21.4 nm and 20.2 nm, respectively. Since the number of sample cycles prepared is 10, the total thickness of the sample is calculated to be 214 and 202 nm, which is in good agreement with the cycle thickness of the designed multilayer film.

3.2.2.2 X-ray Diffraction (XRD) Characterization

An X-ray diffractometer (D\max-2200, Rigaku Corporation) was also used to characterize the periodic layered structure of the sample. The X-rays are Cu Kα rays with a wavelength of 1.54 Å, the scanning angle is 0.5°–5°, the acceleration voltage is 40 kV, the working current is 40 mA, and the scanning rate is 0.002 deg/min. Figure 3.12 is the small-angle X-ray diffraction pattern (LAXRD) of the $Si/Si_{0.75}Ge_{0.25}$ (left) and Si/Au (right) multilayer films, where the small image inserted in the upper right corner is $\sin^2\theta$ and the corresponding number of diffraction peaks n^2 fitted straight line. The periodic thickness of the prepared Si/SiGe + Au multilayer film can be calculated using the Bragg diffraction formula.

$$d = \frac{\lambda}{2}\sqrt{\frac{(n+1)^2 - n^2}{\sin^2\theta_{n+1} - \sin^2\theta_n}}, \tag{3.6}$$

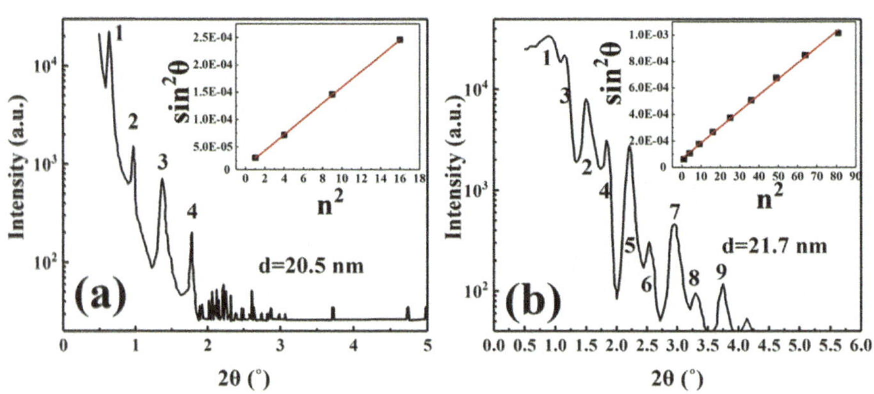

Fig. 3.12 LAXRD images of **a** $Si/Si_{0.75}Ge_{0.25}$ and **b** Si/Au multilayer films and $\sin^2\theta$ vs n^2 fitted straight lines

where is the periodic thickness, λ is the wavelength of the X-ray, n is the number of X-ray diffraction peaks, and θ_n is the angle corresponding to the diffraction peak. Because $\sin^2\theta$ and the square n^2 of the corresponding number of diffraction peaks can be fitted into a straight line, and the slope of this straight line is given as

$$k = \frac{\Delta \sin^2 \theta}{\Delta n^2} = \frac{\sin^2 \theta_{n+1} - \sin^2 \theta_n}{(n+1)^2 - n^2}, \tag{3.7}$$

where k is the slope of the fitted straight line. Equation 2.6 can be transformed into

$$d = \frac{\lambda}{2} \frac{1}{\sqrt{k}}, \tag{3.8}$$

the periodic thickness of the sample can be obtained from the above formula. The Bragg diffraction formula was used to calculate the Si/$Si_{0.75}Ge_{0.25}$ (left) and Si/Au (right) samples with a periodic thickness of 20.5 and 21.7 nm, which basically matched the design values.

3.2.2.3 Scanning Electron Microscope (SEM) Characterization

The vertical structure of the sample was examined by a field emission scanning electron microscope (ULTRA55-36-69, Zeiss, Germany). Figure 3.13 is a SEM image of the Si/Au sample. The sample has a total of 20 layers, which were obtained by alternating deposition of Au and Si. From this SEM figure, it can be seen that the thickness of the prepared structure and design is similar. The thicknesses are 9.8 nm and 12.1 nm, respectively, and the period thickness is 21.9 nm. Because of the different atomic numbers, the Si layer looks darker than the Au layer.

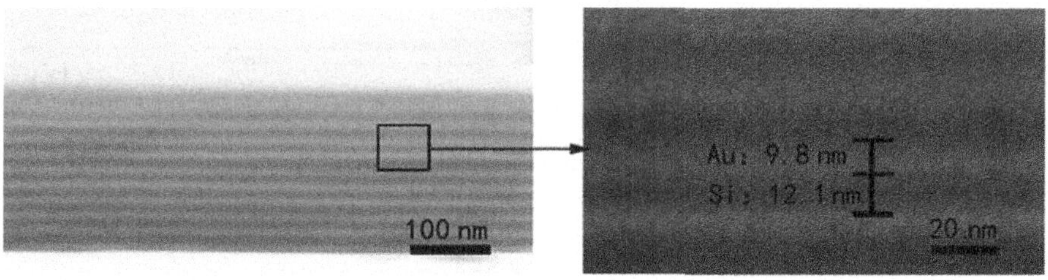

Fig. 3.13 SEM images of Si/Au multilayer films

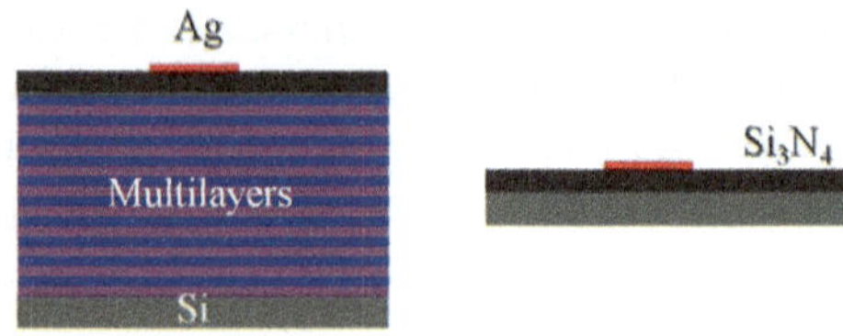

Fig. 3.14 Schematic diagram of test structure of multilayer film to be tested (left) and reference sample (right)

3.2.3 Thin Film Thermal Conductivity

Before testing the thermal conductivity of the film, the corresponding test structure must first be prepared using micromachining techniques, as shown in Fig. 3.14. When the longitudinal thermal conductivity of the sample is tested by the differential 3ω method, a sample that does not include the deposited film to be tested is used as a reference sample. The selected substrates are N<100>Si with a resistivity of 1000 Ω cm. Then, a metal heating wire is prepared on top of both using micromachining technology. If the sample is a conductive sample, an insulating layer has to be deposited before the heating wire's deposition, where the metal heating wire used is Ag with a thickness of ~200 nm, and the insulating layer is Si_3N_4 with a thickness of ~250 nm. At the same time, a layer of Ti with a thickness of about 5 nm is deposited between the metal wire layer and the insulating layer as a connection layer mainly to improve the bonding between the two layers and prevent the film from falling off. Through UV exposure lithography, metal wire test units with different line widths and line lengths can be obtained. Among them, a metal heating wire with a line width of 20 μm and a length of 2 mm is used as the test unit. In order to ensure the accuracy of the experiment, this type of metal wire is used for testing throughout the experiment.

After the preparation of the test sample structure is completed, a formal longitudinal thermal conductivity test is performed. First, the metal wire resistance temperature coefficient of the metal heating wire with a width of 20 μm and a length of 2 mm

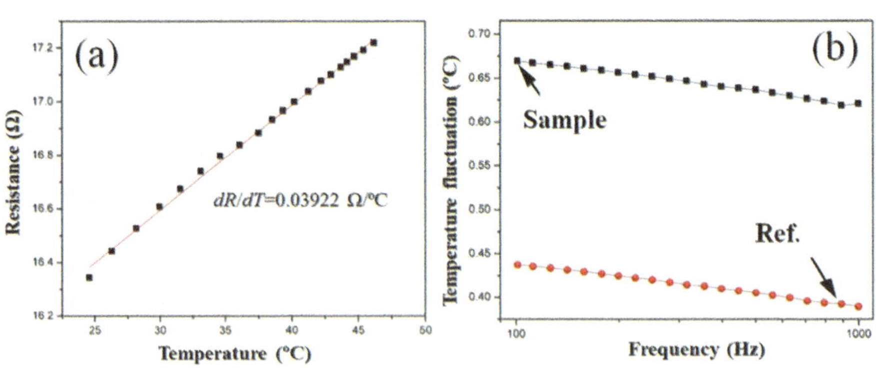

Fig. 3.15 **a** Resistance temperature coefficient of a metal heating wire with a width of 20 μm and a length of 2 mm, and **b** The relationship between the amplitude and frequency of the temperature fluctuation of the test sample and the reference sample

is tested, as shown in Fig. 3.15a. The linear fitting method can obtain a resistance temperature coefficient of 0.03922 Ω/°C, and can further calculate the resistance value of the metal heating wire at room temperature (20 °C) is 16.20398 Ω.

After testing the resistance temperature coefficient of the metal heating wire, test the $V_{3\omega}$ voltage of the metal heating wire. The effective value of the input current during the test is 50.00 mA, and the test frequency is 100–1000 Hz. Figure 3.15b is the measured temperature fluctuation amplitude-frequency curve of the Si/Si$_{0.75}$Ge$_{0.25}$ sample and the reference sample. In order to verify the stability of the test, four multilayer film samples were prepared, each of which was tested three times. At room temperature, for multilayer film samples replacing 0, 2, 5, and 10 Au layers, The derivative values are 0.94, 0.97, 1.02, and 1.31 W m^{-1} K^{-1}.

Two theoretical models were used to analyze the thermal conductivity of the multilayer films. The first is the classic heat conduction model. When using this model to calculate, we consider the sample as a bulk material regardless of the thermal resistance between its interfaces, then the thermal conductivity calculated using the classic thermal conduction model can be expressed by the following formula:

$$\kappa_I = \frac{d_1 + d_2}{\frac{d_1}{\kappa_1} + \frac{d_2}{\kappa_2}}, \tag{3.9}$$

where d_I and d_2 are the thicknesses of the X (another material) and Si layers, respectively, κ_I and κ_2 are the thermal conductivity of X and Si, respectively.

The second is a dual temperature model. The electrical and thermal properties of the metal-nonmetal interface in multilayer films are of great significance for many modern electronic devices and the devices that directly convert thermal energy into electrical energy. The main energy carriers in metals are electrons, while in semiconductors they are phonons, which means that the non-equilibrium effect caused by the interface will definitely lead to the interaction and energy transfer between electrons and phonons. In fact, at the metal-nonmetal interface, especially the multilayer structure, the coupling of electrons and phonons plays a very important role. And there are mainly two possible ways of energy transmission at the metal-nonmetal contact interface:

1. The thermal resistance R_{ep} is formed by the coupling of electrons and phonons in the metal layer, and then the phonons in the metal are coupled with the phonons in the semiconductor to form the thermal resistance R_{pp} [7–9];
2. At the interface between metal and semiconductor, through non-harmonic interaction, the electrons in the metal directly couple with the phonons in the semiconductor to form the thermal resistance R_I.

Here, the entire heat conduction process is mainly analyzed based on the first energy transmission method, so there is a coupling effect of electrons and phonons in the metal layer. The temperature of the electrons is T_e and the temperature of the phonons is T_p as shown in Fig. 3.16. For the metal-semiconductor interface, the vertical direction is the x-axis. The energy transmission equation of each metal layer is as follows:

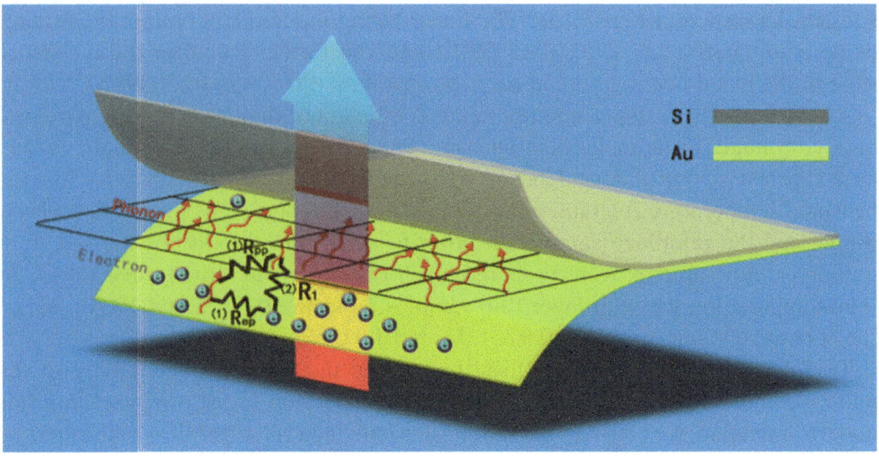

Fig. 3.16 Metal-semiconductor interface

$$\kappa_e d^2 T_e / dx^2 - g(T_e - T_p) = 0, \tag{3.10a}$$

$$\kappa_p d^2 T_p / dx^2 - g(T_e - T_p) = 0, \tag{3.10b}$$

where κ_e is the thermal conductivity of electrons in the metal layer, κ_p is the thermal conductivity of phonons in the metal layer, and g is the electrophonon coupling constant after taking into account the interaction of electrophonons. The main energy carriers in the non-metallic layer are still phonons, so according to the Fourier law of heat conduction, the corresponding energy transfer equation can be obtained:

$$d^2 T / dx^2 = 0, \tag{3.11}$$

where the lattice temperature is T, it can be seen from this formula that when the electron gas and phonon gas in the metal layer are in equilibrium ($T_e = T_p$). Then Eq. 3.11 becomes the same as Eq. 3.10a, 3.10b, which illustrates the dual temperature model and the difference between the Fourier's laws. Subtracting Eq. 3.10a and 3.10b solves $T_e - T_p$:

$$T_e - T_p = a \sinh(x/\delta) + b \cosh(x/\delta). \tag{3.12}$$

where T is the intrinsic electron-phonon coupling length in the metal layer, which can be specifically expressed as follows:

$$\delta = \sqrt{\kappa_\beta / g}, \tag{3.13a}$$

$$\kappa_\beta = \kappa_e \kappa_p / \kappa_e + \kappa_p. \tag{3.13b}$$

Bring Eq. 3.12 into Eqs. 3.10a and 3.10b, respectively, to get the temperature of electrons and phonons in the metal layer:

$$T_e = c + hx + \frac{\kappa_\beta}{\kappa_e} a \sinh(x/\delta) + b\cosh(x/\delta), \tag{3.14a}$$

$$T_p = c + hx + \frac{\kappa_\beta}{\kappa_p} a \sinh(x/\delta) + b\cosh(x/\delta), \tag{3.14b}$$

Among them, a, b, c, and h depend on the constants of the boundary conditions. The two-layer system is a simple model relative to the multilayer structure, so when the two layers are calculated first, the equations are solved by the boundary conditions:

$$T_e(x) = T_0 - \frac{(T_0 - T_1)}{\eta(\kappa_e + \kappa_p)}[x - \delta\frac{\sinh(x/\delta)}{\cosh(d_1/\delta)}], \tag{3.15a}$$

$$T_p(x) = T_0 - \frac{(T_0 - T_1)}{\eta(\kappa_e + \kappa_p)}[x + \frac{\kappa_e}{\kappa_p}\delta\frac{\sinh(x/\delta)}{\cosh(d_1/\delta)}], \tag{3.15b}$$

$$T(x) = T_1 + \frac{(T_0 - T_1)}{\eta\kappa_2}[d_2 - x], \tag{3.15c}$$

$$\eta = \frac{d_1}{\kappa_e + \kappa_p} + \rho_1 + \frac{d_2}{\kappa_2} + \frac{\kappa_\beta\delta}{\kappa_p\kappa_p}\tanh(d_1/\delta). \tag{3.15d}$$

Among them, T_0 and T_1 are the initial and end temperatures, respectively, and the phonon interface thermal resistance of the ρ_1 two-layer system. Under steady-state conditions, the effective thermal conductance of a two-layer thin-film system can be obtained by Fourier's law:

$$\kappa = \frac{d_1 + d_2}{\frac{d_1}{\kappa_e+\kappa_p} + \frac{d_2}{\kappa_2} + \rho_1 + \frac{\kappa_\beta\delta}{\kappa_p\kappa_p}\tanh(d_1/\delta)}, \tag{3.16a}$$

$$R_1 = \frac{d_1}{\kappa_e + \kappa_p}, \tag{3.16b}$$

$$R_2 = \frac{d_2}{\kappa_2}, \tag{3.16c}$$

$$R_{ep} = \frac{\kappa_\beta\delta}{\kappa_p\kappa_p}\tanh(d_1/\delta), \tag{3.16d}$$

where R_1 and R_2 are the thermal resistance of the metal part and the thermal resistance of the semiconductor part, respectively, and R_{ep} is the electrical-phonon coupling thermal resistance caused by the interaction of electrons and phonons in the metal layer.

Similar to the treatment of a two-layer thin film system, when the total number of layers is N and the last layer is metal (N is an even number), the total thermal conductivity value can be shown as follows:

$$\kappa = \frac{d_N}{(\frac{N+1}{2})\frac{d_1}{\kappa_e+\kappa_p} + (\frac{N-1}{2})\frac{d_2}{\kappa_2} + \sum_{n=1}^{N-1} R_{n,n+1}}. \quad (3.17)$$

where d_N and $R_{n,n+1}$ are obtained by the following formulas:

$$d_N = \frac{(N+1)d_1 + (N-1)d_2}{2}, \quad (3.18a)$$

$$R_{n,n+1} = \rho_{n,n+1} + \frac{\kappa_\beta}{\kappa_p}\begin{cases} \frac{\delta}{\kappa_p}\tanh(d_1/\delta), n = 1, N-1 \\ \frac{2\delta}{\kappa_p}\tanh(d_1/\delta), n \neq 1, N-1 \end{cases}. \quad (3.18b)$$

If the last layer is a non-metallic layer, the effective thermal conductivity of the multilayer film is still as shown in Eq. 3.14a, 3.14b. As the effective thermal conductivity of the film eventually approaches an asymptotic line as the number of layers increases, the following is obtained The effective thermal conductivity of the multilayer film:

$$\kappa_{II} = \frac{d_1 + d_2}{\frac{d_1}{\kappa_e+\kappa_p} + \frac{d_2}{\kappa_2} + 2\rho_2 + 4\frac{\kappa_e\delta}{\kappa_p(\kappa_p+\kappa_e)}\tanh(d_1/2\delta)}. \quad (3.19)$$

Among them, the phonon interface thermal resistance in the ρ_2 multilayer film, the last term in the denominator is the electro-phonon coupling thermal resistance due to the electro-phonon interaction. In previous studies, it was known that the thermal conductivity of the Si/$Si_{0.75}Ge_{0.25}$ multilayer film is smaller than its own superlattice and alloy film values, and the reason is that each layer is amorphous. Moreover, the thermal conductivity values of the Si film and the $Si_{0.75}Ge_{0.25}$ film are 1.44 W m^{-1} K^{-1} and 0.76 W m^{-1} K^{-1}, respectively.

Tables 3.6 and 3.7 show the calculation results using the classic heat conduction model and the dual temperature model, and their experimental values. Here take $g_{Au} = 2.4 \times 10^{16}$ W m^{-1} K^{-1}, $\kappa_e = 315.83$ W m^{-1} K^{-1}, $\kappa_p = 2.17$ W m^{-1} K^{-1}. For samples 2 and 3, the 100 nm and 34 nm Si/$Si_{0.75}Ge_{0.25}$ multilayer films are considered as a whole, and the thermal conductivity is the experimentally measured thermal conductivity of sample 1, which is 0.94 W m^{-1} K^{-1} (Table 3.6).

Figures 3.17 illustrates the influence of the metal-nonmetal interface on the thermal conductivity of the entire multilayer film. When the number of metal-nonmetal interfaces increases, the thermal conductivity calculated at the classical

Table 3.6 Calculation parameters and experimental values based on a classic thermal conduction model

Samples	d_1 (nm)	d_2 (nm)	κ_1 (W m^{-1} K^{-1})	κ_2 (W m^{-1} K^{-1})	κ_I (W m^{-1} K^{-1})	κ (W m^{-1} K^{-1})
1	10	12.00	0.76	1.44	1.02	0.94
2	10	100.00	318.00	0.94	1.03	0.97
3	10	34.00	318.00	0.94	1.22	1.02
4	10	12.00	318.00	1.44	2.63	1.31

Table 3.7 Calculation of parameters based on a dual temperature model

Samples	d_1 (nm)	d_2 (nm)	κ_p (W m^{-1} K^{-1})	κ_e (W m^{-1} K^{-1})	κ_2 (W m^{-1} K^{-1})	κ_{II} (W m^{-1} K^{-1})
2	10	100	2.17	315.83	0.94	0.96
3	10	34	2.17	315.83	0.94	0.99
4	10	12	2.17	315.83	1.44	1.31

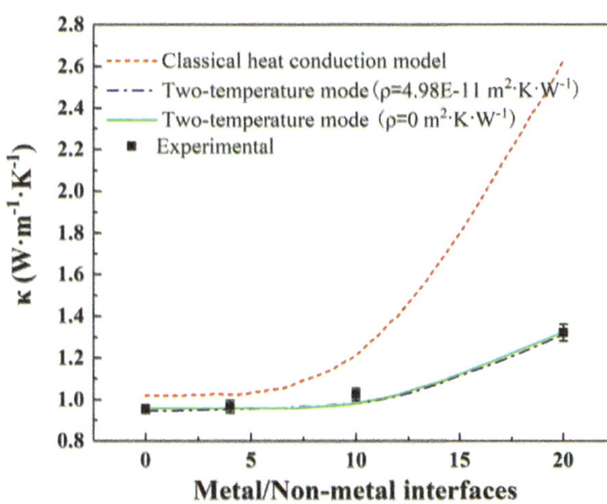

Fig. 3.17 Relationship between calculated values of thermal conductivity theory, experimental values and the number of metallic and non-metallic interfaces

thermal conduction theory is significantly improved, mainly due to the introduction of a high thermal conductivity Au layer. It can be seen that the simple classical heat conduction model is not suitable for the layered results with metallic/non-metal interfaces. At the same time, it was found that although a high thermal conductivity Au layer was incorporated into the multilayer film, for samples 2 and 3, the dual temperature model was applied, and they basically had little effect on the effective thermal conductivity of the film, remaining at 1.00 W m^{-1} K^{-1}. This value is 33% (3.00 W m^{-1} K^{-1}) of the Si–Ge superlattice value, and the Si/Ge alloy is worth 20% (5.00 W m^{-1} K^{-1}). Moreover, for sample 4, when the $Si_{0.75}Ge_{0.25}$ layer was completely replaced by the Au layer, the thermal conductivity calculated using the dual temperature model (1.31 W m^{-1} K^{-1}) was only half of the calculated value using the classic thermal conduction model (2.63 W m^{-1} K^{-1}). This is precisely because when the original $Si_{0.75}Ge_{0.25}$ layer is replaced with an Au layer, a metal/nonmetal interface will inevitably be formed, and the electro-phonon coupling thermal resistance formed at the interface will cause the thermal conductivity value to be less than the value calculated by classical thermal conduction. The experimental values also agree well with the two-temperature model, indicating that electro-phonon coupling plays an important role in the heat conduction process of multilayer films. The existence of electro-phonon coupling resistance at the metal-nonmetal interface is also experimentally verified.

3.3 X($Si_{0.75}Ge_{0.25}$/Au/Cr/Ti)/Si Multilayer Film

In this section, magnetron sputtering is used to prepare X ($Si_{0.75}Ge_{0.25}$/Au/Cr/Ti)/Si periodic multilayer film. Insert different metal layers into Si, and examine its effect on the thermal conductivity of the multilayer film.

3.3.1 Film Preparation

In this section, magnetron sputtering is used to prepare X (X = $Si_{0.75}Ge_{0.25}$, Au, Cr, Ti)/Si multilayer film. The substrate used is a silicon wafer with N<100>, whose resistivity is 1000 Ω cm. Before coating, the substrate was sonicated in acetone and absolute ethanol for 10 min to remove the attachment on the surface of the substrate, and then the substrate was placed in BOE solution (HF:NH4 = 1:6) for 1 min to remove The oxide and nitride layers on the surface are then rinsed with ultrapure water and blown dry with nitrogen. The diameter of the target is 50 mm, where the Au, Cr, and Ti targets are metal targets with a thickness of 6 mm, while $Si_{0.75}Ge_{0.25}$ and Si are semiconductor targets with a thickness of 3 mm. All targets were securely bound onto a 3 mm Cu base to improve its heat dissipation. The sample substrate was a silicon N<100> wafer with the resistivity of 1000 Ω cm, and the distance between the target and the substrate (target base distance) is ~20 cm. Before coating, first sonicate the substrate in acetone and absolute ethanol for 10 min to remove the adherence on the surface of the substrate, and then immerse the substrate in a BOE solution ($HF:NH_4$ = 1:6) for 1 min to remove the oxide layer and nitride layer on the surface, then washed with ultrapure water and blown dry with nitrogen. The background vacuum during coating is 5×10^{-7} Torr, the sputtering pressure used is 3 mTorr (1 Torr = 133 Pa), and the substrate rotation rate is 20 rpm. Figure 3.18 shows the relationship between the sputtering rate and power of several materials.

Table 3.8 lists the preparation parameters of X/Si multilayer films. A total of 4 multilayer films were prepared, of which SiGe/Si is a semiconductor-semiconductor interface multilayer film, and Au/Si, Cr/Si, Ti/Si. It is a metal-semiconductor interface

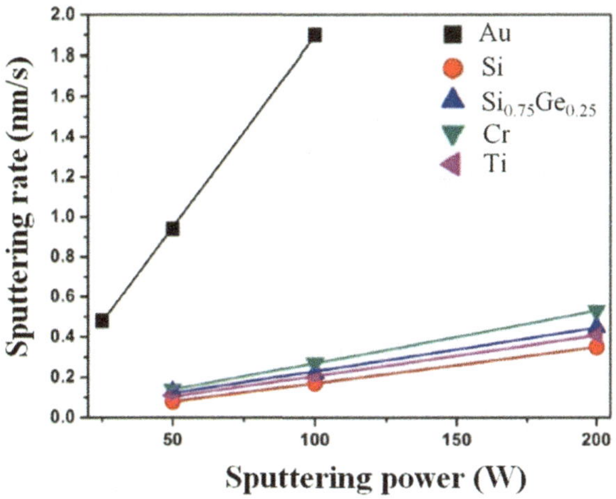

Fig. 3.18 Relationship between sputtering rate and sputtering power for Au, Si, $Si_{0.75}Ge_{0.25}$, Cr and Ti materials

Table 3.8 Preparation parameters and thickness of X/Si multilayer films

Sample	Sputtering air pressure and sputtering power	Deposition rate (Å/s)	Periodic thickness measurement (nm)	Total thickness measurement (nm)
#1 $Si_{0.75}Ge_{0.25}$/Si	3 mTorr Si:100 W $Si_{0.75}Ge_{0.25}$:100 W Au:25 W Cr:100 W Ti: 200 W	0.22/0.17	21.580	211.5
#2 Au/Si		0.48/0.17	20.501	209.9
#3 Cr/Si		0.27/0.17	23.686	237.5
#4 Ti/Si		0.41/0.17	19.886	203.5

multilayer film. The thickness ratio of X/Si is controlled at 2:3, the number of cycles is 10, and the total thickness of the sample is controlled at 200 nm (Fig. 3.19).

3.3.2 *Structural Characterization*

Synchrotron microfocus X-ray source was used to measure the film thickness by grazing incidence small angle scattering (GISAXS) on the sample, as shown in the figure below.

The slope of the straight line is fitted, and the periodic thickness of the sample can be obtained from the slope of the straight line fitted by Q_Z vs n. The periodic thicknesses of $Si_{0.75}Ge_{0.25}$/Si, Au/Si, Cr/Si and Ti/Si were calculated to be 21.580 nm, 20.501 nm, 23.686 nm, and 19.886 nm, respectively. Since the number of sample cycles prepared was 10, the total thickness of the samples was calculated to be 215.80,

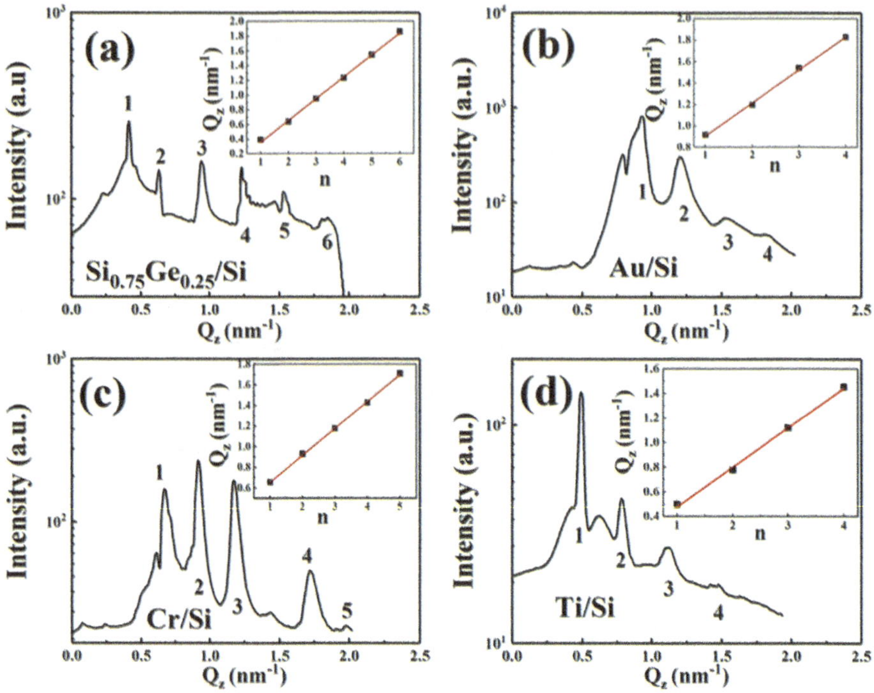

Fig. 3.19 One-dimensional rectangular integral and Q_Z vs n of **a** $Si_{0.75}Ge_{0.25}/Si$, **b** Au/Si, **c** r/Ti, and **d** Ti/Si multilayer thin film samples along the Z axis fits a straight line

205.01, 236.86, and 198.86 nm. They are in agreement with the results obtained by the step scanner (211.5, 209.9, 237.5 and 203.5 nm).

The vertical structure of the sample was examined by a field emission scanning electron microscope (ULTRA55-36-69, Zeiss, Germany). Figure 3.20a is a schematic diagram of the sample structure. The sample has a total of 20 layers, which were obtained by alternate deposition of X and Si. Figure 3.20b is a SEM image of the Cr/Si sample. The thickness of the structure and design is similar, and is consistent with the thickness measured by the step meter. The thickness of the Cr layer and the Si layer are 9.4 and 14.2 nm, and the periodic thickness is 23.6 nm. Because of the different atomic numbers, the Si layer looks darker than the Cr layer.

3.3.3 Thin Film Thermal Conductivity

In this section, the magnetron sputtering method was used to prepare X (X = $Si_{0.75}Ge_{0.25}$, Au, Cr, and Ti)/Si multilayer films with similar periodic thicknesses

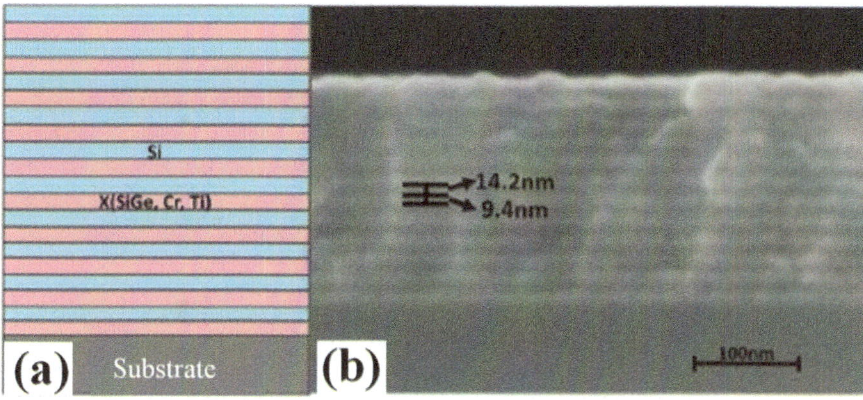

Fig. 3.20 Schematic diagram of **a** periodic X/Si thin film sample, **b** cross-section SEM image of Cr/Si multilayer film, magnification is 100,000 times

Table 3.9 Thermal conductivity and 3ω test values calculated using the classic heat conduction model and the dual temperature model

Sample	D_I (NM)	D_2 (NM)	κ_1 (W m^{-1} K^{-1})		κ_2	κ_I	κ_{II}	κ_M
			κ_e	κ_p	(W m^{-1} K^{-1})			
#1 $Si_{0.75}GE_{0.25}$/Si	8.5	13.1	0.76		1.44	1.065	–	1.01
#2 Au/Si	8.2	12.3	317			2.393	1.358	1.31
			314.78	2.22				
#3 Cr/Si	9.4	14.2	93.7			2.369	0.819	0.89
			93.044	0.656				
#4 Ti/Si	7.9	12.0	21.9			2.289	0.376	0.44
			21.747	0.153				

and the same number of cycles, and the differential 3ω method was used to heat these films. Conductance coefficient is tested, and the classic heat conduction model and dual temperature model are used for calculation. The calculation parameters and results are shown in Table 3.9 and Fig. 3.21. The coefficient of conductance was tested, and the measurement results were 1.01, 1.31, 0.89 and 0.44 W m^{-1} K^{-1}.

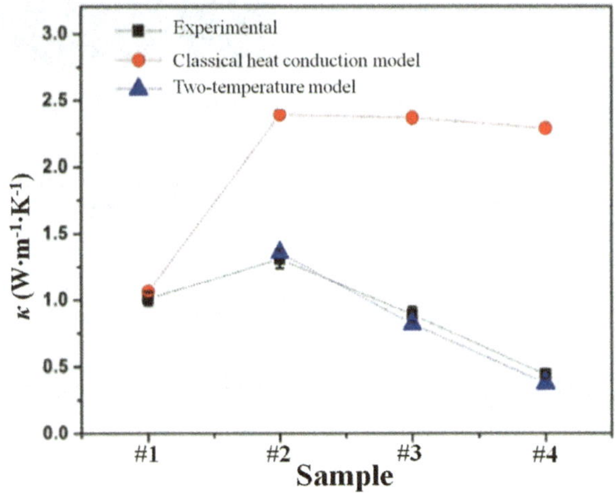

Fig. 3.21 Experimental results and calculated values

3.4 Si/Au Multilayer Film

In this section, Au/Si periodic multilayer films are prepared by magnetron sputtering. Insert Au layers of different thicknesses into Si, and examine their effect on the thermal conductivity of multilayer films.

3.4.1 Film Preparation

In this section, Si/Au multilayer films are prepared by magnetron sputtering. The substrate used is a silicon wafer with N<100>, whose resistivity is 1000 Ω cm. Before coating, first sonicate the substrate in acetone and absolute ethanol for 10 min to remove the attachment on the surface of the substrate, and then place the substrate in BOE solution (HF:NH4 F = 1:6) for 5 min to remove The oxide and nitride layers on the surface are then rinsed with ultrapure water and blown dry with nitrogen. Samples were prepared by alternately depositing 4 N Au targets and 5 N Si targets at room temperature. The targets used are all 50 mm in diameter, where the Au target is a metal target with a thickness of 6 mm, and Si is a semiconductor target with a thickness of 3 mm, and is bound to a Cu base of 3 mm to improve its heat dissipation. The substrate used is a silicon wafer with N<100>, the resistivity is 1000 Ω cm, and the distance between the target and the substrate (target base distance) is ~20 cm. Before coating, the substrate was first sonicated in acetone and absolute ethanol for 10 min to remove the adherence on the substrate surface, and then place the substrate in a BOE solution (HF:NH_4 = 1:6) and soak for 5 min to remove the oxide layer and nitride layer on the surface are then washed with ultrapure water and blown dry with nitrogen to fix the substrate on the substrate holder. The background vacuum during coating was 5×10^{-7} Torr, the sputtering pressure was 3 mTorr (1 Torr =

Table 3.10 Si/Au multilayer film preparation parameters

Sample	Sputtering vacuum	Sputtering power	Thickness (nm)	
			Si	Au
Si/Au	5×10^{-7} Torr 3 mTorr	Si:100 W Au:25 W	12	1, 3, 5, 10, 20, 40

133 Pa), and the substrate rotation rate was 20 rpm. Table 3.10 lists the preparation parameters of the Au/Si multilayer film. A total of 6 multilayer films were prepared with the number of cycles of 10. The thickness of the Au layer was 1, 3, 5, 10, 20, 40 nm.

3.4.2 Structural Characterization

3.4.2.1 GISAXS Characterization of Si/Au Multilayer Films

The sample was characterized by grazing incidence small-angle X-ray scattering using a synchrotron radiation source to characterize its periodic multilayer film structure (GISAXS). The test was performed at the BL14B1 beamline station in Shanghai Synchrotron Radiation Source (SSRF). In the experiment, a high-density X-ray source is used to enter the sample surface. The position of the sample can be controlled by a sample stage that can be precisely controlled to select the optimal angle of incidence. At the same time, the intensity of the scattered X-rays was collected by a CCD detector with 2048 × 2048 pixels. Because of the incident X-ray energy was relatively large, in order to avoid damage to the CCD detector, an X-ray beam interceptor is used to weaken the strong reflected light. The wavelength of X-ray incident light is 0.124 nm, and the size of the incident angle varies with the thickness of the thin film sample. The distance between the sample and the CCD detector is 2042 mm.

It can be clearly seen from the Fig. 3.22 that there are distributions of Bragg scattering peaks on both sides of the Z axis, and they are distributed in parallel along the Z axis direction. Because the scattering intensity is symmetrically distributed along the Z axis, only the components in the Z axis direction are studied for analysis and discussion. Generally, the rectangular integral of the entire Z axis was used to study the multilayer structure and periodic thickness of the sample, as shown in Fig. 3.23 (Fig. 3.24).

The thickness of the multilayer film sample calculated using the Bragg scattering formula is 12.38, 13.46, 18.01, 20.24, 32.09, 51.61 nm, which is very consistent with the periodic thickness of the designed multilayer film.

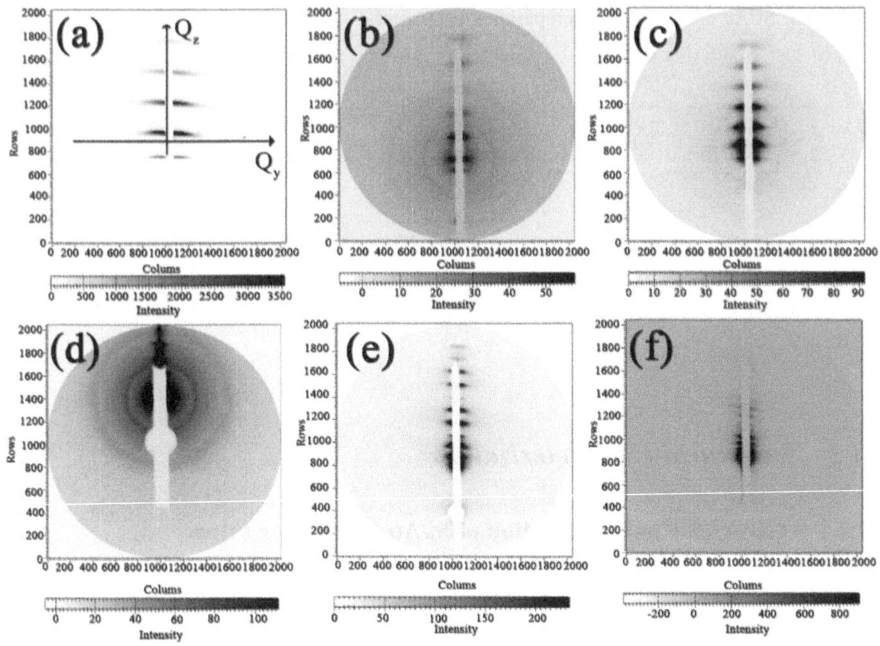

Fig. 3.22 GISAXS image of the prepared multilayer film sample. **a** Si/Au (1 nm), **b** Si/Au (3 nm), **c** Si/Au (5 nm), **d** Si/Au (10 nm), **e** Si/Au (20 nm) and **f** Si/Au (40 nm)

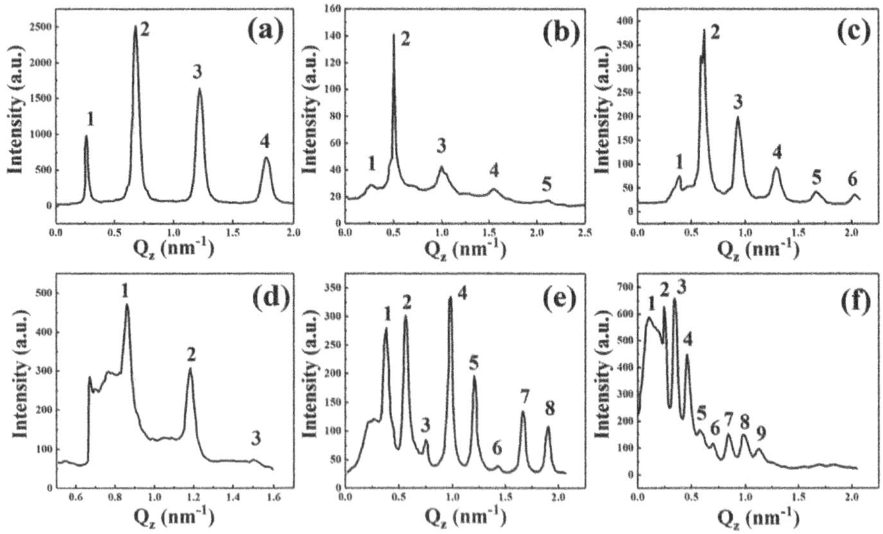

Fig. 3.23 One-dimensional rectangular integral of the **a** Si/Au (1 nm), **b** Si/Au (3 nm), **c** Si/Au (5 nm), **d** Si/Au (10 nm), **e** Si/Au (20 nm) and **c** Si/Au (40 nm) multilayer thin film samples along the Z axis

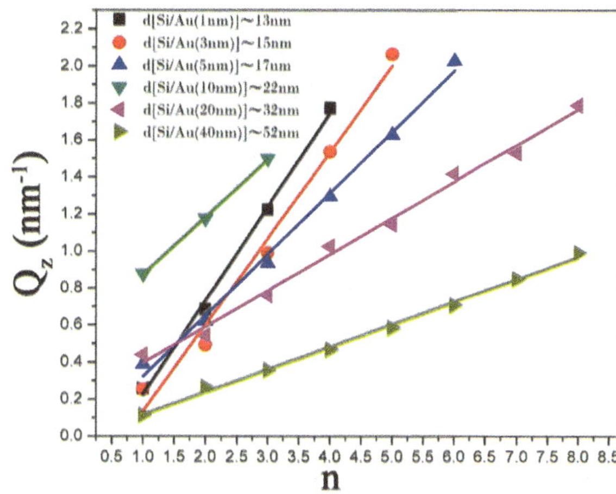

Fig. 3.24 Q_Z vs n fitted straight line for Si/Au multilayer films

3.4.2.2 SEM Characterization of Si/Au Multilayer Films

The vertical structure of the sample was examined by a field emission scanning electron microscope (ULTRA55-36-69, Zeiss, Germany). Figure 3.25 is a schematic diagram of the sample. The sample has a total of 20 layers, which were obtained by

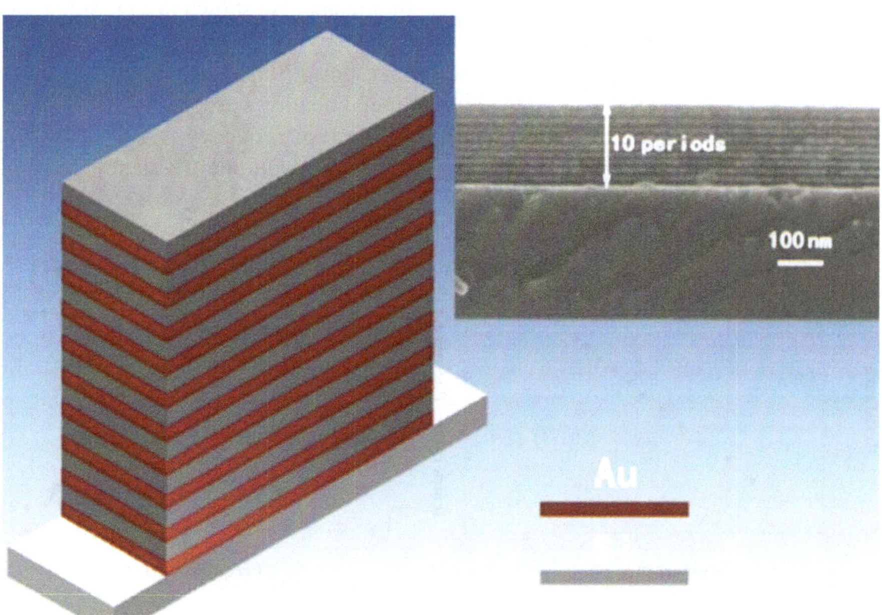

Fig. 3.25 Schematic diagram of a periodic Si/Au multilayer film and a longitudinal SEM image of the Si/Au (5 nm) multilayer film

alternate deposition of Au and Si. The inset in the upper right corner is a SEM image of the Si/Au (5 nm) sample. The thickness of the structure and design are similar. The thickness of the Au layer and the Si layer are about 5 nm and 12 nm, respectively, and the periodic thickness is 17 nm. Because of the different atomic numbers, the Si layer looks darker than the Au layer.

3.4.3 *Thin Film Thermal Conductivity*

3.4.3.1 Thermal Conductivity of Si/Au Multilayer Films

The differential 3ω method was used to test the thermal conductivity of the prepared samples. Because the prepared sample is conductive, it is also necessary to prepare a ~200 nm thick Si_3N_4 layer as an insulating layer between the sample and the metal heating wire. At the same time, it is necessary to prepare another silicon wafer that is the same but without the thin film to be tested as a reference sample. The preparation of the test sample also uses a micro-processing process, and through the steps of homogenization, photolithography, development, coating, priming and peeling, and the Ag resistance wires are prepared as metal heating wires.

The test used a metal heating wire with a width of 20 μm and a length of 2 mm. Taking the sample Si/Au (5 nm) multilayer film as an example, the resistance temperature coefficient of the metal heating wire is first tested. As shown in Fig. 3.26a, the resistance temperature coefficient is 0.03751 Ω/°C by linear fitting. The resistance value of the metal heating wire at room temperature (20 °C) can be calculated to 16.66727 Ω. Then test the $V_{3\omega}$ voltage of the metal heating wire, the effective value of the current during the test is 50.3 mA, and the test frequency is 100–1000 Hz. Figure 3.26b is the measured temperature fluctuation amplitude-frequency curve of

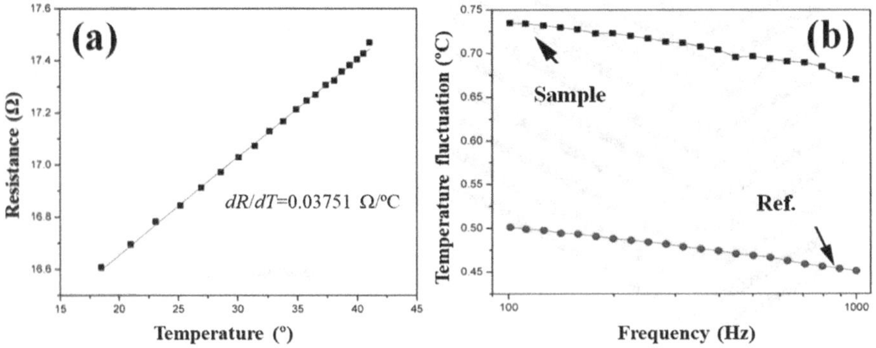

Fig. 3.26 **a** Resistance temperature coefficient of a metal heating wire with a width of 20 μm and a length of 2 mm, and **b** The relationship between the amplitude and frequency of the temperature fluctuation of the test sample and the reference sample

the Si/Au (5 nm) sample and the reference sample, and the thermal conductivity of the sample is calculated. In order to verify the stability of the test, six multilayer film samples were prepared, and each sample was tested 3 times. The measured thermal conductivity values at room temperature as the thickness increased were 0.67, 0.60, 0.62, 1.31 1.55, and 2.28 W m^{-1} K^{-1}, it is found that when the thickness of the Au layer is 10 nm or less, the thermal conductivity of the multilayer film decreases sharply.

3.4.3.2 Thermal Conductivity Analysis of Si/Au Multilayer Films

Two theoretical models are used to analyze the thermal conductivity of the prepared materials, which are the classic heat conduction model and the dual temperature model. Table 3.11 lists the parameters calculated using the classic heat conduction model and the dual temperature model, and their experimental values. Here the electric phonon coupling constant $g_{Au} = 2.4 \times 10^{16}$ W m^{-3} K^{-1} is taken. When the thickness of the Au layer is greater than or equal to 10 nm, the electron and phonon thermal conductance in the metal are $\kappa_e = 315.83$ W m^{-1} K^{-1} and $\kappa_p = 2.17$ W m^{-1} K^{-1}, respectively.

Figure 3.27 illustrates more vividly the effect of ultra-thin Au layer on the thermal conductivity of the entire multilayer film in metal-nonmetal multilayer films. Due to the high thermal conductivity of the Au layer, with the increase of the thickness of the Au layer, the thermal conductivity calculated by the classic heat conduction model has significantly improved (black curve). However, it is found that when the Au layer is greater than or equal to 10 nm, the thermal conductivity calculated by the dual temperature model is obviously not as much improved as the classical thermal conductivity value. Moreover, the thermal conductivity (2.11 W m^{-1} K^{-1}) calculated

Table 3.11 Parameters and experimental values calculated based on the two models

Sample	d_1	d_2	κ_p	κ_e	κ_2	κ_{II}	κ_I	κ
	(nm)		(W m^{-1} K^{-1})					
Si/Au (1 nm)	1	12	0.169	317.83	1.44	0.65	1.56	0.67
Si/Au (3 nm)	3		0.439	317.56		0.70	1.80	0.60
Si/Au (5 nm)	5		0.644	317.356		0.75	2.04	0.62
Si/Au (10 nm)	10		2.17	315.83		1.32	2.63	1.31
Si/Au (20 nm)	20		2.17			1.47	3.81	1.55
Si/Au (40 nm)	40		2.17			2.11	6.15	2.28

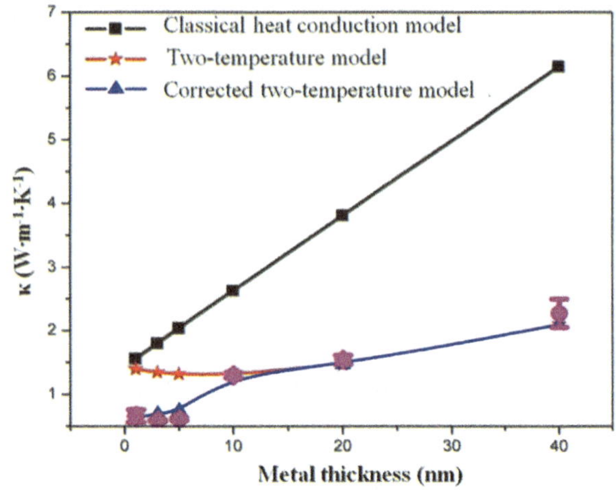

Fig. 3.27 Relationship between calculated values of thermal conductivity theory, experimental values and metal film thickness

for the Si/Au (40 nm) multilayer film using the dual temperature model is 34% of the classic thermal conductivity model (6.15 W m^{-1} K^{-1}). This is mainly because the thermal resistance formed at the metal-nonmetal interface reduces the thermal conductivity and plays an important role in the heat conduction process. However, it was found that the dual temperature model proposed by Ordonez-Miranda et al. [9] One cannot simply use the Au parameter value of the block to analyze the case where the Au layer film thickness is less than 10 nm.

From the experimental values, it can be seen that the critical thickness of the thermal properties of the Au layer bulk and thin film is 10 nm, and the thermal conductivity of the thin film is greatly reduced compared to the bulk value. For this reason, below 10 nm, the relationship between phonon thermal conductance and thickness function in metal is introduced, such as Eq. 3.20, and the corresponding corrected dual temperature model is obtained

$$\frac{1}{\kappa_p} = \frac{5.4598}{l_1} + 0.46023. \quad (3.20)$$

The theoretical value calculated by the corrected dual temperature model is in good agreement with the thermal conductivity measured by the 3ω method. Below 10 nm in the Au layer, a lower thermal conductivity value of ~0.6 W m^{-1} K^{-1} is obtained, which is 60% of amorphous $Si/Si_{0.75}Ge_{0.25}$ multilayer film, 42% of amorphous Si film. This means that under the ultra-thin Au layer, the main heat transfer in the Au layer is not the considered electrons, but a small number of phonons. It is this small number of phonons that causes greater film thermal resistance. A lower thermal conductivity was obtained. The thermal resistance to the metal layer can be expressed as

$$R_{metal} = \frac{l_1}{\kappa_p + \kappa_e} + 2\left(\frac{\kappa_e}{\kappa_p}\right)\left(\frac{\delta}{\kappa_p + \kappa_e}\right)\left(\frac{e^{l_1/\delta} - 1}{e^{l_1/\delta} + 1}\right) = \begin{cases} 2\delta/\kappa_p, & (l_1 > \delta) \\ l_1/\kappa_p, & (l_1 \ll \delta) \end{cases}. \quad (3.21)$$

A very critical parameter in this formula is the electrophonon coupling length $\delta = \left[\kappa_e \kappa_p / G(\kappa_e + \kappa_p)\right]^{1/2}$. It can be seen that when the thickness of the metal layer is smaller than the coupling length of the electrophonon, the approximate total thermal resistance of the metal layer at this time is $R_{metal} = l_1/\kappa_p$, which also demonstrate that in the ultra-thin Au layer the phonon thermal conductance plays a leading role in heat conduction. In the experiment, the thickness of the Au layer is almost comparable to the length of the electrophonon coupling, and the thermal resistance value is a constant value. Therefore, it can be explained in Fig. 3.27, when the Au layer thickness is less than 10 nm, that the thermal conductivity of the multilayer film is basically unchanged. Many previous studies have emphasized the role of this small portion of phonons in the thermal conduction of metal thin films. The detailed calculation parameter values are listed in Table 3.11.

3.5 Chapter Summary

1. Amorphous $Si/Si_{0.75}Ge_{0.25}$ nm multilayer films with periodic thicknesses of 2.5, 5, 10, 20, and 50 nm were prepared by magnetron sputtering, and these films were tested by differential 3ω method. The longitudinal thermal conductivity coefficients are 0.98, 1.09, 1.09, 0.85, 1.12 W m^{-1} K^{-1}. The results show that the thermal conductivity of amorphous $Si/Si_{0.75}Ge_{0.25}$5 nm multilayer films is significantly smaller than that of Si/Ge superlattice or SiGe alloy films.

The average free path of phonons of amorphous Si and amorphous $Si/Si_{0.75}Ge_{0.25}$ calculated by Debye model is 0.407 nm and 0.238 nm, respectively, which are smaller than the period of the prepared $Si/Si_{0.75}Ge_{0.25}$ multilayer film. thickness. The theoretical value of the thermal conductivity of the $Si/Si_{0.75}Ge_{0.25}$ multilayer film obtained using the Fourier thermal conduction theory is basically consistent with the experimental measurement value. This indicates that in the amorphous $Si/Si_{0.75}Ge_{0.25}$ multilayer film system, due to the disordered structure causes the average free path of phonons to be smaller than the thickness of the multilayer film. Therefore, the layered structure of amorphous materials cannot effectively reduce its longitudinal thermal conductivity.

2. Use magnetron sputtering to replace 0, 2, 5, and 10 $Si_{0.75}Ge_{0.25}$ multilayer films with Au layers in $Si/Si_{0.75}Ge_{0.25}$ multilayer films, and prepare ($Si_{0.75}Ge_{0.25}$/Au/Cr/Ti) Si multilayer film, and the thermal conductivity of the film was measured using the differential 3ω method. When replacing layers 2 and 5, the thermal conductivity did not change significantly despite the incorporation of the high thermal conductivity Au layer. The thermal conductivity coefficients of the prepared samples were analyzed using the classic thermal conductivity model and the dual temperature model, and compared with the test values. It was found that for the amorphous multilayer film ($Si_{0.75}Ge_{0.25}$/Si), the classic thermal conductivity model was used to calculate. The values are basically consistent with the experimental values; for metal-nonmetal multilayer films, the thermal

conductivity calculated by the dual temperature model is basically consistent with the experimental values. Due to the effect of electron-phonon coupling, the overall effective thermal resistance of the metal-semiconductor multilayer film system is increased, resulting in a decrease in the thermal conductivity of the multilayer film.

3. Si/Au multilayer films with different Au layer thicknesses were prepared by magnetron sputtering, with Au thicknesses of 1 nm, 3 nm, 5 nm, 10 nm, 20 nm, and 40 nm, respectively, and the thermal properties of the films were measured using the differential 3ω method. When the thickness of the Au layer is greater than or equal to 10 nm, the thermal conductivity measured at room temperature is in good agreement with the dual temperature model, which indicates that the electro-phonon coupling plays an important role in the heat conduction process at the metal-nonmetal interface. When the thickness of the Au layer is less than 10 nm, the functional relationship between the phonon thermal conductivity in the metal and the thickness of the film is introduced to obtain a correspondingly corrected two-temperature model. The experimentally measured thermal conductivity value is ~0.6 W m^{-1} K^{-1} coincides with the corrected model again. When the thickness of the metal layer is smaller than the coupling length of the electrophonon, the approximate total thermal resistance of the metal layer at this time is $R_{metal} = l_1/\kappa_p$, indicating that in the ultra-thin Au layer, the phonon thermal conductance plays a leading role in thermal conduction. In the experiments, the thickness of the Au layer is almost comparable to the coupling length of electrophonons. A small number of phonons in the metal play a leading role in the entire metal film thermal conduction process, resulting in a relatively high film thermal resistance. This means that in the metal-nonmetal multilayer film, a thin film material with a lower thermal conductivity can be obtained simply by introducing an ultra-thin metal layer.

References

1. Moon S, Hatano M, Lee M, Grigoropoulos CP (2002) Thermal conductivity of amorphous silicon thin films. Int J Heat Mass Transf 45(12):2439–2447. https://doi.org/10.1016/s0017-9310(01)00347-7
2. Ruf T, Henn RW, Asen-Palmer M, Gmelin E, Cardona M, Pohl HJ, Devyatych GG, Sennikov PG (2000) Thermal conductivity of isotopically enriched silicon. Solid State Commun 115(5):243–247. https://doi.org/10.1016/s0038-1098(00)00172-1
3. Levinshtein ME, Rumyantsev SL, Shur MS (2001) Properties of advanced semiconductor materials GaN, AlN, InN, BN, SiC, SiGe. Wiley
4. Chen G, Borca-Tasciuc T, Liu WL, Liu JL, Zeng TF, Song DW, Moore CD, Wang KL, Goorsky MS, Radetic T, Gronsky R, Koga T, Dresselhaus MS (2000) Thermal conductivity of symmetrically strained Si/Ge superlattices. Superlattices Microstruct 28(3):199–206
5. Venkatasubramanian R (2000) Lattice thermal conductivity reduction and phonon localization-like behavior in superlattice structures. Phys Rev B 61(4):3091–3097
6. Chen G (1998) Thermal conductivity and ballistic-phonon transport in the cross-plane direction of superlattices. Phys Rev B 57(23):14958–14973

7. Majumdar A, Reddy P (2004) Role of electron-phonon coupling in thermal conductance of metal-nonmetal interfaces. Appl Phys Lett 84(23):4768–4770. https://doi.org/10.1063/1.1758301
8. Hopkins PE, Kassebaum JL, Norris PM (2009) Effects of electron scattering at metal-nonmetal interfaces on electron-phonon equilibration in gold films. J Appl Phys 105(2). https://doi.org/10.1063/1.3068476
9. Ordonez-Miranda J, Alvarado-Gil JJ, Yang R (2011) The effect of the electron-phonon coupling on the effective thermal conductivity of metal-nonmetal multilayers. J Appl Phys 109(9). https://doi.org/10.1063/1.3585824

Chapter 4
Sb_2Te_3-Based Multilayer Films

Sb_2Te_3 Multilayer Films Sb_2Te_3 is one of the best thermoelectric materials near room temperature. The research on Sb_2Te_3 thermoelectric materials was introduced earlier, mainly for nanocomposite and low-dimensional. However, research on its periodic multilayer structure is rare, mainly Sb_2Te_3/Bi_2Te_3 superlattice structure. In the previous chapter, we introduced the study of Si and metal multilayer films, and discussed the interface and thermal conductivity between semiconductors and metals. In this chapter, the metal is introduced into Sb_2Te_3 for research, and its thermal conductivity and thermoelectric properties are investigated.

4.1 Au/Sb_2Te_3 Multilayer Film

In this section, Au and Sb_2Te_3 periodic multilayer films are prepared by magnetron sputtering. By inserting gold layers of different thicknesses in Sb_2Te_3, the effect of Au layer on the thermal conductivity of multilayer thin films was investigated.

4.1.1 Film Preparation

The thin film thermoelectic materials are prepared by means of alternate coating by magnetron sputtering. Multi-layer thin film samples Au/Sb_2Te_3 were prepared by alternately depositing pure Au targets with 4 N purity and Sb_2Te_3 alloy targets alternately. Similarly, a silicon wafer of type N<100> was used as the material substrate, and the resistivity was 5000 Ω cm, and the target pitch was controlled to maintain 20 cm. The substrate was sonicated in acetone and absolute ethanol for 10 min, then immersed in BOE for 1 min to remove the surface oxide layer, then washed with ultrapure water and blown dry with nitrogen, and finally baked in an oven for 15 min. In addition, because Sb_2Te_3 is also a semiconductor target, it was placed on the No. 1

Z. Hu and Z. Wu, *Nanostructured Thermoelectric Films*,
https://doi.org/10.1007/978-981-15-6518-2_4

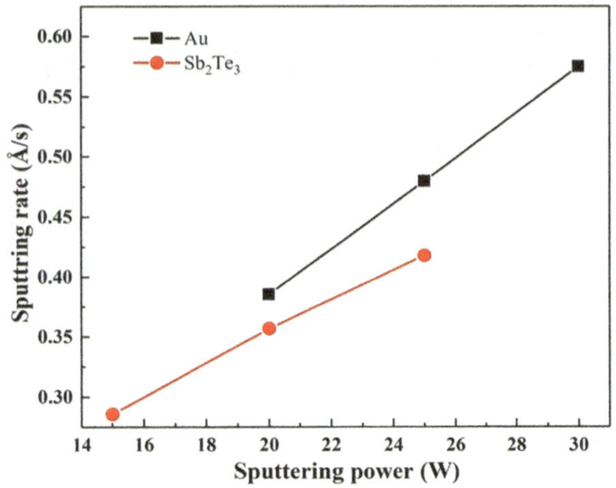

Fig. 4.1 Relationship between sputtering power and sputtering speed of Au and Sb_2Te_3 targets

Table 4.1 Experimental parameters of Au/Sb2Te3 multilayer film preparation

Sample	Sputtering vacuum	Sputtering power	Thickness (nm)	
			Sb_2Te_3	Au
Au/Sb_2Te_3	4×10^{-7} Torr 3 m Torr	Sb_2Te_3: 20 W Au: 25 W	13	0, 1, 3, 5, 10, 20

target holder (radio frequency target), and the gold target was also placed on the No. 2 target holder (DC target). The background vacuum was set to 4×10^{-7} Torr, coating pressure was 3 m Torr, Au sputtering power was 25 W and the Sb_2Te_3 power was 20 W. At this time, it can be known from the magnetron sputtering crystal reading that the corresponding coating rates are ~0.51 Å/s and ~0.36 Å/s, respectively. This reading is basically consistent with the fitted value in Fig. 4.1. The detailed parameters are listed in Table 4.1. Finally, a total of 6 sets of Au/Sb_2Te_3 multilayer samples were prepared, all of which were 10 cycles, and the thickness of Sb_2Te_3 was 13 nm in each cycle. The thickness of Au was changed to 0, 1, 3, 5, 10 and 20 nm, while with 0 nm as the thermal conductivity reference value for pure Sb_2Te_3.

4.1.2 Structural Characterization

The structure was characterized by thermal field emission scanning electron microscope (Zeiss, Germany, ULTRA55-36-69 maximum magnification is 100,000 times) and Fig. 4.2b–d are SEM images of multilayer film samples at 80,000 times. The thickness of the gold layer from left to right is 5 nm, 10 nm, and 20 nm, respectively. It can be seen from the figure that the gold layer and the antimony telluride layer are bright and dark. The reason is that the two materials are different in conductivity and

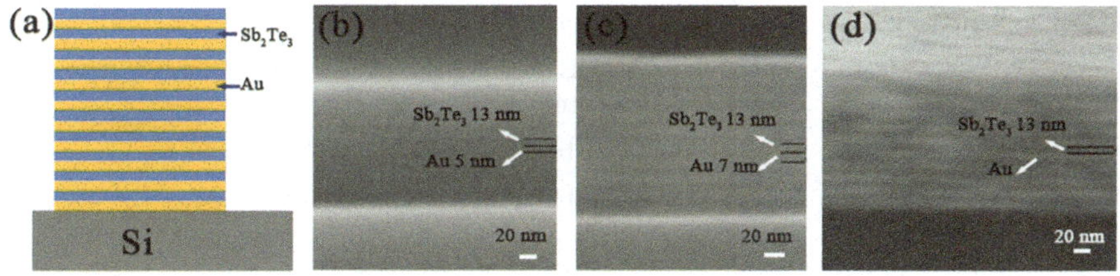

Fig. 4.2 **a** Schematic diagram of Au/Sb_2Te_3 thin film samples; **b–d** are the thickness of the gold layer, respectively cross-sectional view of a scanning electron microscope at 80,000 times for 5, 10, and 20 nm

atomic weight. According to the scale, it can be estimated that its thickness basically matches the designed thickness. And as the thickness of the gold layer increases, the boundary between the gold layer and the antimony telluride layer becomes more apparent, and its resolution becomes higher.

In addition to scanning electron microscopy for structural characterization, a more detailed structural characterization of the periodic thickness of the samples was also performed using an X-ray diffractometer (D\max-2200 Rigaku Corporation). The X-rays used in the experiment are Cu Kα rays, with a wavelength of 1.54 Å, an electron acceleration voltage of 40 kV, and a working current of 40 mA. The small angle of the scanning angle of 0.5°–5° is used to measure the thickness of the multilayer film. The scan rate of the angle is 0.002 deg/min. Figure 4.3 shows the results of the small-angle X-ray diffraction (LAXRD) test of multilayer samples. Table 4.2 shows the measurement results. The figure shows the diffraction peaks of the samples at 1, 3, 5, 10, and 20 nm. Figure 4.4 is a detailed calculation chart of Au thickness of 1 nm. The main chart is based on the 1 nm diffraction peak chart, and the small chart inserted in the upper right corner is a linear fit of $\sin^2\theta$ and the corresponding number of diffraction peaks n^2. When the thickness of the gold layer is 1 m, the thickness of

Fig. 4.3 Small angle X-ray diffraction (LAXRD) test results of multilayer samples

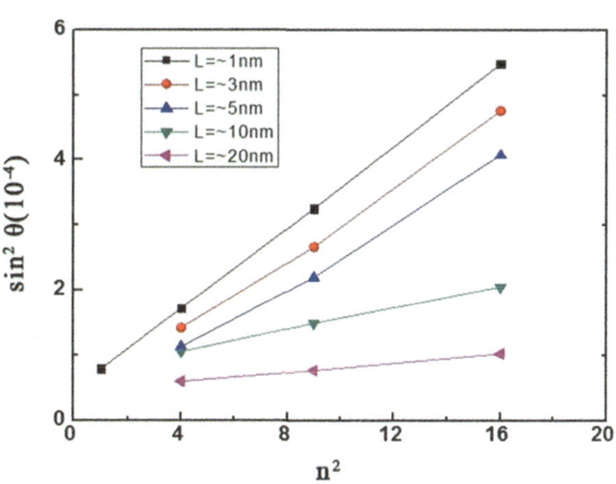

Table 4.2 Periodic thickness of multi-layer samples measured by LAXRD

Sample Au/Sb_2Te_3 (nm)	Measured period thickness (nm)	Expected period thickness (nm)	Difference ratio (%)
Au 1	13.44	14	4.00
Au 3	15.52	16	3.00
Au 5	17.48	18	2.88
Au 10	22.44	23	2.43
Au 20	34.12	33	3.39

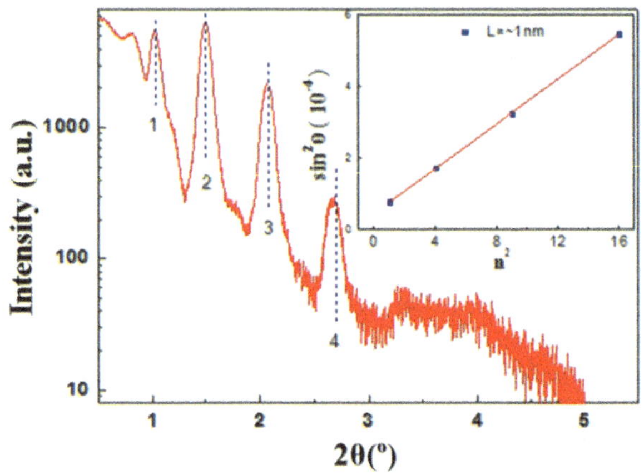

Fig. 4.4 Small-angle XRD images of Sb_2Te_3/Au (1 nm) multilayer thin film samples and sin2θ vs n2 fitted straight lines

a single cycle can be calculated from the Bragg diffraction formula. The thickness of the single layer Sb_2Te_3/Au (1 nm) is 13.44 nm, which is close to the designed 14 nm (Au 1 nm + Sb_2Te_3 13 nm). Table 4.2 is the single cycle thickness of the film under different gold layer thicknesses.

4.1.3 Thin Film Thermal Conductivity

4.1.3.1 Thermal Conductivity Test of Multilayer Films

After the above treatment, the 3ω method of the sample can be used to test the thermal conductivity. First, the resistance thermal temperature coefficient was tested. A wire with a line width of 20 μm and a length of 2 mm was selected as the test standard. Taking the measurement result of the thermal resistance of the reference sample as an example, Figs. 4.5 are the linear fitting values obtained by the thermal conductivity tester. According to the fitted figure, the resistance value at room temperature (20 °C) is 22.425 Ω, and the thermal temperature coefficient of resistance is 0.056 Ω/°C.

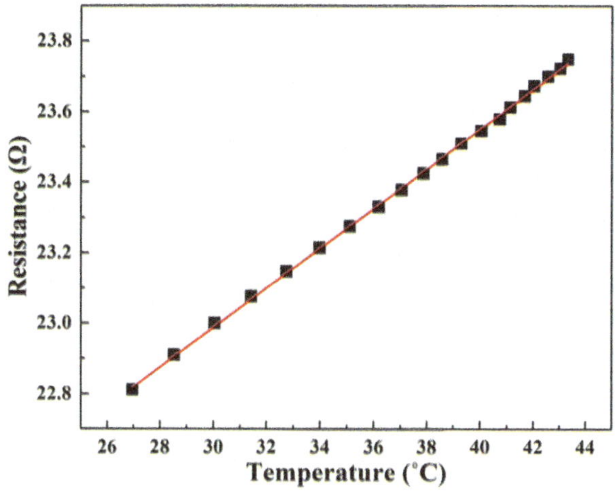

Fig. 4.5 Thermal resistance temperature coefficient of metal wire of reference sample R1

Figure 4.6 is the amplitude-frequency curve of temperature fluctuation of the Au/Sb_2Te_3 sample and the reference sample with a gold layer thickness of 5 nm measured. In order to verify the stability of the test, six multilayer film samples were prepared, each sample was tested 3 times, and the multilayer samples were tested. The test results are obtained in Table 4.3. It is found that the result decreases first and then increase, and the thermal conductivity is the smallest when the thickness of the gold layer is 5 nm.

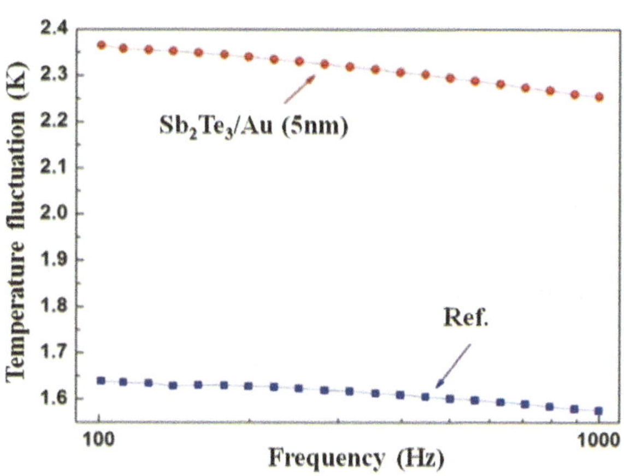

Fig. 4.6 Relationship between the amplitude and frequency of temperature fluctuations of the test sample and the reference sample

Table 4.3 Thermal conductivity SAXRD test results of multilayer samples

Sample Au thickness (nm)	0	1	3	5	10	20
κ (W m^{-1} K^{-1})	1	0.85	0.5	0.45	0.55	0.72

4.1.3.2 Thermal Conductivity Analysis of Multilayer Films

Two theoretical models are used to analyze the thermal conductivity of the prepared materials, which are the classic heat conduction model and the dual temperature model. Because the gold layer in this series is less than 20 nm, improved parameters are used in the two-temperature model for calculations. Table 4.4 shows the detailed parameters used in the dual temperature model (Fig. 4.7).

Finally, Table 4.5 is the value obtained through these two models. κ_I is the classic heat conduction model, and κ_{II} is the improved dual temperature model value. Figures 4.5 show a comparison of the classic heat conduction model (red dot curve), the dual temperature model (blue triangle curve), and the experimental values (black squares). It can be seen from the figure that the classical thermal conductivity model is also not suitable for the Au/Sb_2Te_3 material system. The reason is mainly due to the interface thermal resistance. Compared with the classical heat conduction model, the theoretical value of the dual temperature model is closer to the experimental value, especially when the thickness of the gold layer is greater than 5 nm. However, it can also be seen from this figure that the thermal conductivity value increases sharply when the thickness of the gold layer is less than 5 nm, which is very different from the theoretical value of the dual temperature model. Therefore, a simple two-temperature

Table 4.4 Detailed parameters of the dual temperature model

d_1 (nm)	d_2 (nm)	κ_e ($Wm^{-1}K^{-1}$)	κ_p ($Wm^{-1}K^{-1}$)	κ_2 ($Wm^{-1}K^{-1}$)	ρ_2 (m^2K/W)	g (W/m^3 K)
1	13	319.83	0.17	1	1.064×10^{-8}	2×10^{16}
3		319.57	0.43			
5		319.36	0.64			
10		319.01	0.99			
20		317.76	2.24			

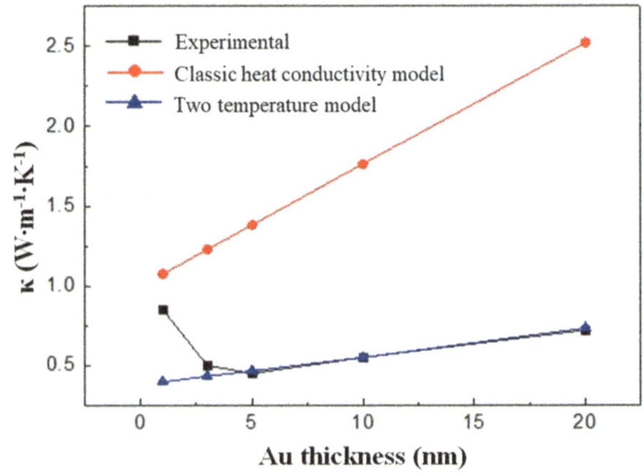

Fig. 4.7 Comparison of classic heat conduction model, dual temperature model and experimental values

Table 4.5 Classic heat conduction model, dual temperature model and experimental values

Sample	Sb_2Te_3/Au (1 nm)	Sb_2Te_3/Au (3 nm)	Sb_2Te_3/Au (5 nm)	Sb_2Te_3Au (10 nm)	Sb_2Te_3/Au (20 nm)
Measured value	0.85	0.5	0.45	0.55	0.72
κ_I (W m^{-1} K^{-1})	1.08	1.23	1.38	1.76	2.52
κ_{II} (W m^{-1} K^{-1})	0.40	0.43	0.47	0.55	0.73

model analysis may not fit this experiment exactly. Therefore, a new model needs to be introduced to explain what happens when Au is less than 5 nm.

First guess: gold will agglomerate under very thin conditions, making it a discontinuous film, which will result in the situation shown in Fig. 4.8. The white layer is the Sb_2Te_3 layer and the gold layer is the Au layer. When the Au layer is discontinuous, the originally continuous gold layer (light yellow) will become an island-like discontinuous gold layer. Therefore, based on the dual temperature model, another model is introduced, which is the sieve model. According to the previous two-temperature model, in the case of a normal multilayer film, the phonons in the semiconductor (white) are transferred from the upper layer to the metal layer (light yellow) with energy during the heat transfer process, and the electrons and harmonies of the metal layer. The ions are coupled to pass through the electrons and phonons in the metal to the underlying semiconductor. In the sieve model, due to the discontinuity of the gold layer, its discontinuity will cause some of the phonons to contact the metal and some of the phonons to directly contact the underlying semiconductor when the phonon heats down, skipping the metal semiconductor coupling occurs at this interface. This is like a sieve, which directly filters out the metal-semiconductor contact part. The total thermal conductance κ1I can be found by the following formula

$$\kappa_{II} = p \times \kappa_{TTM} + (1 - p) \times \kappa_{Sb_2Te_3}. \quad (4.1)$$

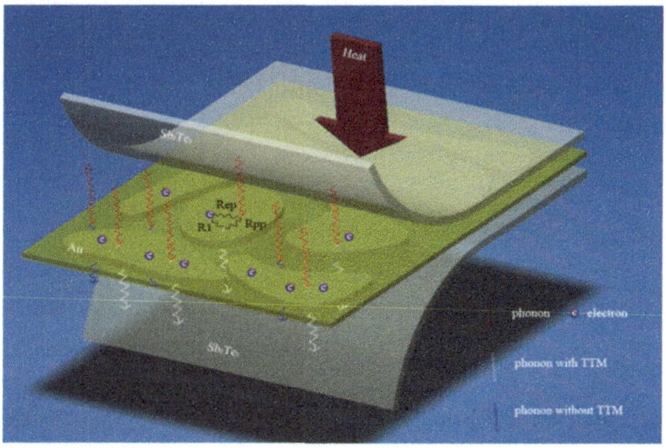

Fig. 4.8 Schematic diagram of sieve model

KTTM is the thermal conductivity in the dual temperature model κ_{Sb2Te3} is the thermal conductivity of antimony telluride material P is the percentage covered by the gold layer ($0 \leq P \leq 1$). According to the conjecture, when the thickness of the gold layer is less than 5 nm, the gold layer agglomerates and discontinuities occur. As a result, the interface thermal resistance only occurs in some places, and the areas not covered by the gold layer are still transferred to the lower layer by ordinary phonon heat transfer methods, resulting in an increase in thermal conductivity.

4.2 M(Au/Ag/Cu/Pt/Cr/Mo/W/Ta)/Sb_2Te_3 Multilayer Film

4.2.1 Film Preparation

In this section a multi-layer thin film structure was prepared and studied by magnetron sputtering. Before preparing the film to be tested, the substrate needs to be processed. Firstly, the substrate is separately placed. Before preparing the film to be tested, the substrate needs to be processed. It needs to be ultrasonically cleaned in an ethanol solution for 10 min, and then the substrate is placed in the BOE immersed in the solution ($HF:NH_4 = 1:6$) for 1 min to remove the oxide layer on the surface of the substrate, then rinse with ultrapure water and blow dry with nitrogen, and then place the substrate in the sample stage for coating.

4.2.2 Structural Characterization

4.2.2.1 SEM Characterization

In order to get a preliminary understanding of the microstructure characteristics prepared by the magnetron sputtering process, in this section first performed SEM characterization on different semiconductor and metal thermoelectric multilayer films. The obtained microstructure cross-sectional morphology diagrams are shown in Fig. 4.9 of which the multilayer semiconductor metal thin film exhibits a clear periodic structure, and the interlayer interface between the layers can be clearly observed.

4.2.2.2 TEM Characterization

In order to clearly obtain the cross-sectional characteristics of the sample, the cross-sectional morphology of the sample needs to be TEM characterized. In order to obtain a high-resolution TEM electron microscope image, the cross-section of the sample needs to be cut and thinning. The Focused Ion beam (FIB) system, that is a

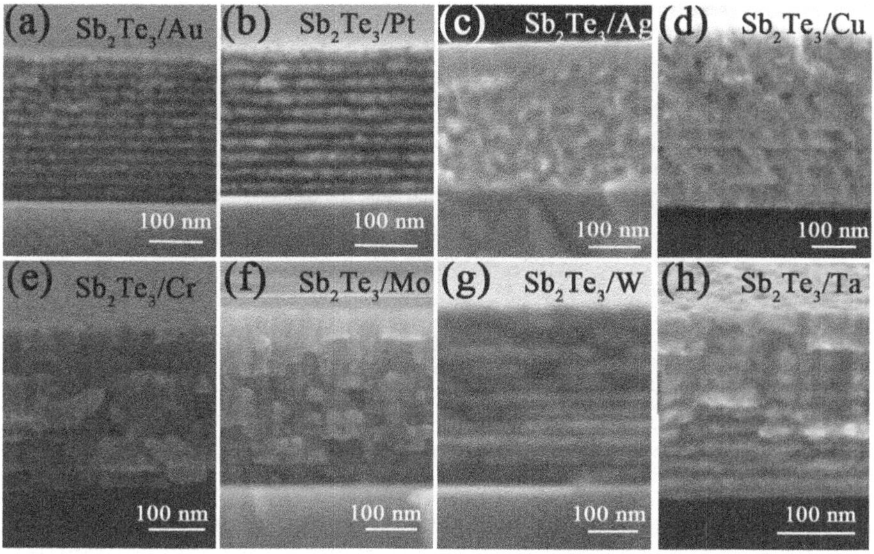

Fig. 4.9 SEM scan of the cross section of Sb2Te3/M (Au, Pt, Ag, Cu, Cr, Mo, W, Ta) (15 nm/15 nm)

micro-cutting instrument using electric lens to focus the ion beam into a very small size, was used to prepare samples. At present, the ion beam of a commercial system uses a liquid metal ion source (Liquid Metal Ion Source, MIS) such as gallium (Ga) because gallium has a low melting point, low vapor pressure, and good oxidation resistance. A typical ion beam microscope includes a liquid metal ion source, electric lens, scanning electrode, secondary particle detector, axially moving test piece base, vacuum system, anti-vibration and magnetic field device, electronic control panel, and computer. An electric field (Suppressor) and liquid phase metal ion source can make liquid gallium form a fine tip under a negative electric field (Extractor) and export the gallium ion beam. Under the general working voltage, the tip current density is about 1 Angstrom 10^{-8} Amp/cm^2. Focusing with an electric lens, the size of the ion beam can be determined after a series of Automatic Variable Aperture (AVA), and then the secondary focus is applied to the surface of the test piece. Physical collisions are used to achieve the purpose of cutting. The thinning pattern of the focused ion beam is obtained as shown in Fig. 4.10. TEM images of Sb_2Te_3/Au and Sb_2Te_3/W multilayer films at different resolutions are shown in Figs. 4.11 and 4.12.

4.2.2.3 Characterization of Spherical Aberration Corrected Field Emission Scanning Transmission Electron Microscope

In order to obtain and obtain the microstructure morphology at the atomic scale, a spherical aberration correction field emission scanning transmission electron microscope was employed to characterize the microstructure of the sample. The cold

Fig. 4.10 FIB thinning before TEM high resolution (Sb_2Te_3/W sample)

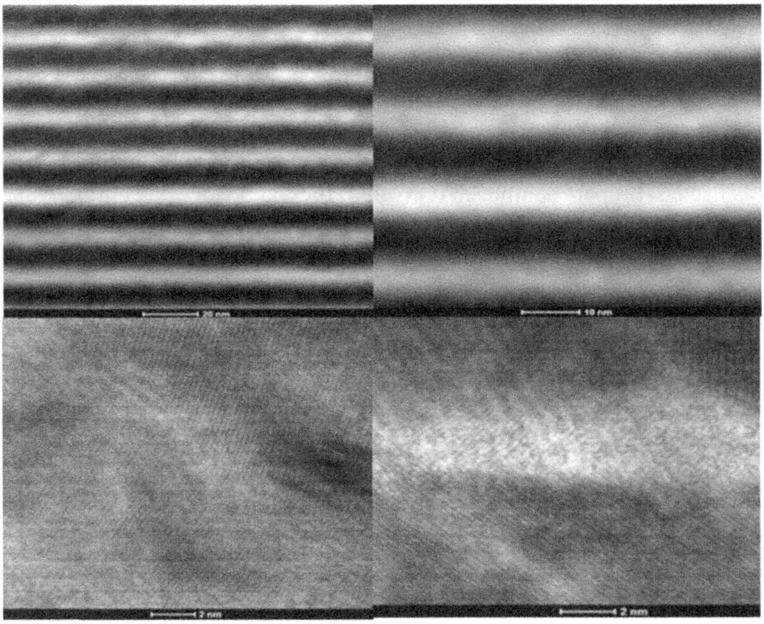

Fig. 4.11 TEM images of Sb_2Te_3/Au multilayer films at different resolutions

field emission electron gun and fully automated spherical aberration correction technology were used to ensure the ultra-high resolution while greatly simplifies the operation achieving the perfect combination of high-resolution observation and user friendly. The spherical aberration corrected field emission scanning transmission electron microscope has a resolution of Angstrom level (Fig. 4.13), which is

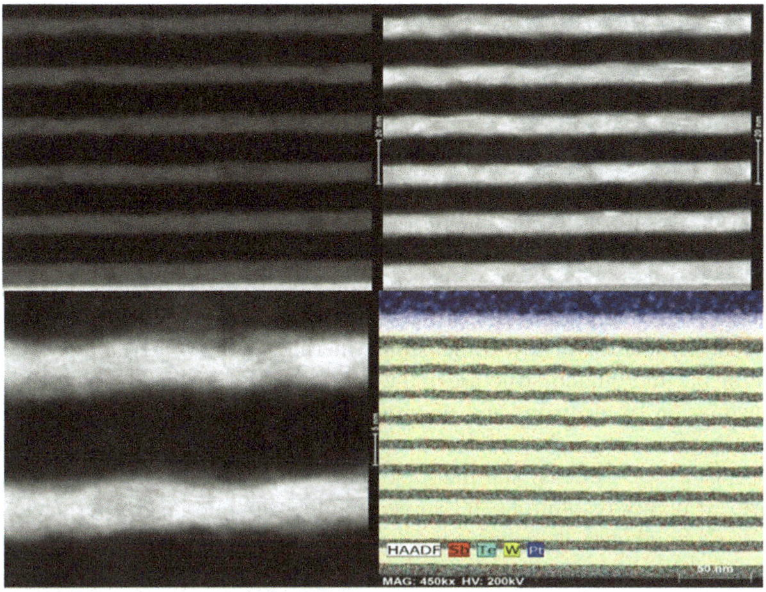

Fig. 4.12 TEM cross-sectional morphology of Sb_2Te_3/W multilayer films at different resolutions

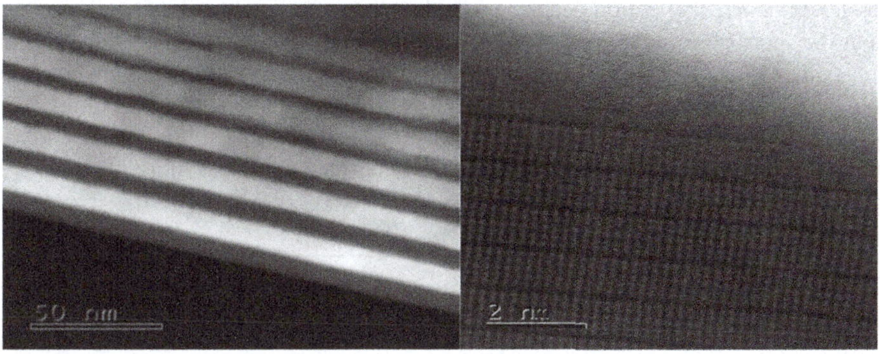

Fig. 4.13 Micromorphology of Sb_2Te_3/W at atomic scale

suitable for observation of metals, ceramics, semiconductors, nanomaterials, etc. In this experiment, the spherical aberration corrected field emission scanning transmission electron microscope (Titan G2 80-200 Chemi STEM, FEI) was used to observe the cross-sectional morphology of Sb_2Te_3/W multilayer films. As shown below.

4.2.3 Thin Film Thermal Conductivity

In this chapter, the cross-section thermal conductivity of different thermoelectric films was measured using the time-domain laser thermal reflection method (TDTR), and the phonon electron transport at the interface of the thermoelectric multilayer film was analyzed in detail. As shown in Fig. 4.14.

Studies have shown that the electron-phonon coupling and interface microstructures (such as interface roughness and lattice mismatch) at the interface of thermoelectric multilayer films affect the phonon scattering mechanism. This phonon anisotropic scattering mechanism further affects the characteristics of thermal transport and the efficiency of low-dimensional thermoelectric materials. By precisely controlling the micro/nano-structure of the interface, it is possible to precisely control the transmission of electrons and phonons, thereby further improving the energy conversion efficiency of thermoelectric materials. From the existing research, it is shown that the two-dimensional thermoelectric multilayer film has a clear layered structure. At higher resolutions, it can be seen that the interlayer interface exhibits atomic-scale roughness. Roughness will cause phonons to scatter between interfaces, which will result in large interface thermal resistance, which will affect the effective thermal conductivity of the film. According to the wave theory, the scattering behavior of phonons at the interface is determined by the phonon's wavelength and interface roughness. When the phonon's wavelength is much larger than the interface roughness, mirror reflection occurs; when the phonon's wavelength is much smaller than the interface roughness diffuse reflection occurs. Studies of Kapitza thermal resistance (interface thermal resistance) by reserachers have found that interface roughness is a key factor affecting Kapitza thermal resistance. From the simulation results of phonon and interface interactions, due to the redistribution of boundary phonon energy by diffuse reflection, at rough interface more and more phonons are

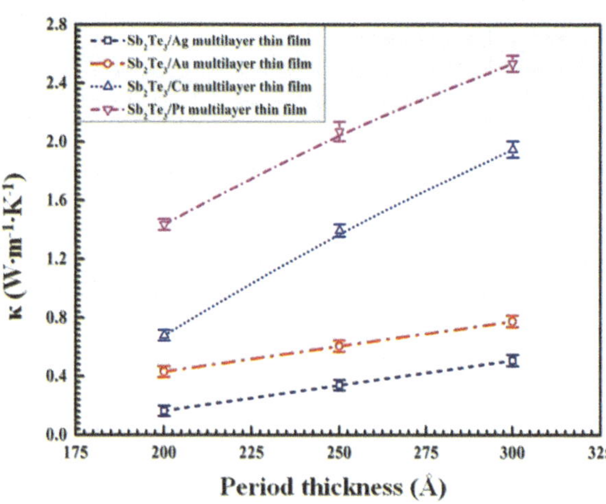

Fig. 4.14 Cross-section thermal conductivity of Sb_2Te_3/(Au, Ag, Cu, Pt) thermoelectric multilayer films

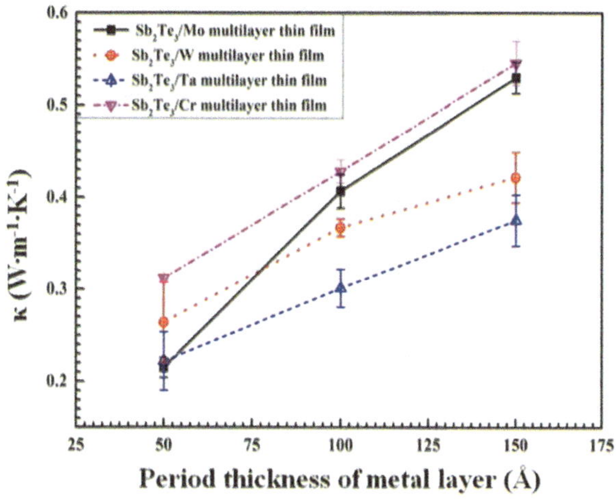

Fig. 4.15 Cross-section thermal conductivity of Sb_2Te_3/(Cr, Mo, W, Ta) thermoelectric multilayer films

scattered back, which increases the interface thermal resistance and reduces the effective thermal conductivity of two-dimensional thermoelectric materials. Therefore, by rationally optimizing the micro/nano structure of the two-dimensional thermoelectric material, its thermal conductivity can be effectively reduced (Fig. 4.15).

In addition, the thermal conductivity of Si-based, Sb_2Te_3, and metal-made multilayer films are compared in this article, as shown in Fig. 4.16.

In order to study the effect of temperature on the effective thermal conductivity of multilayer films, the thermal conductivity of multilayer films was measured at different temperatures using the time-domain picosecond femtosecond pulse method. At room temperature 300 K, the curve of the obtained thermal conductivity as a function of temperature is shown in Fig. 4.17.

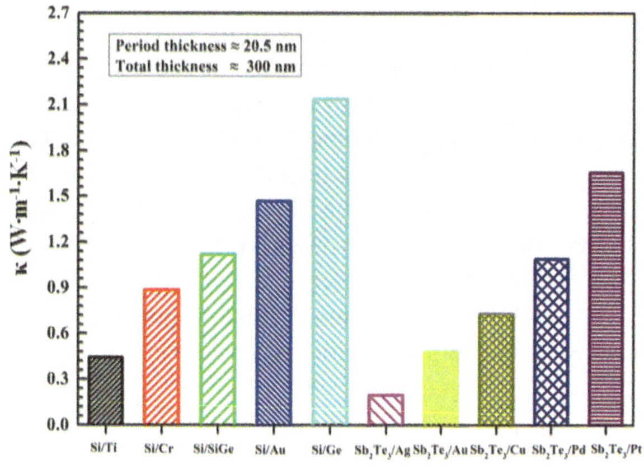

Fig. 4.16 Comparative analysis of Sb_2Te_3/metal multilayer films and Si/metal multilayer films

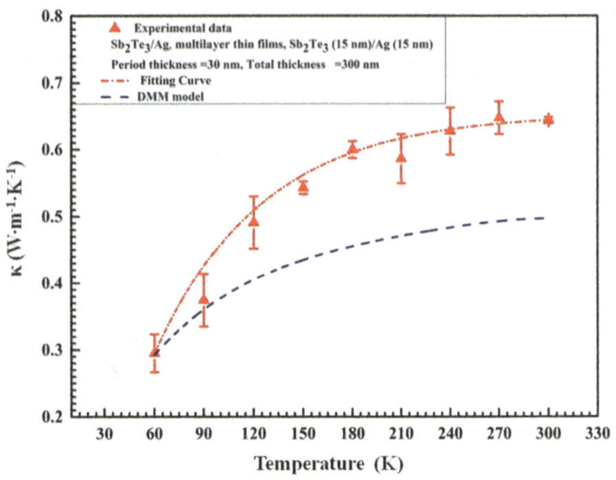

Fig. 4.17 Variation curve of Sb_2Te_3/Ag multilayer films with temperature

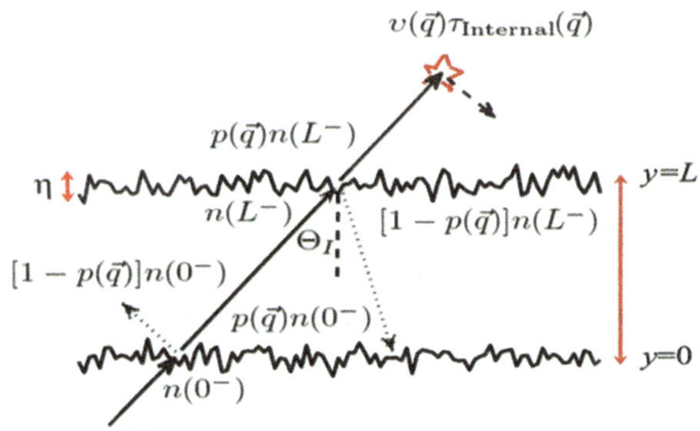

Fig. 4.18 Phonon scattering mechanism in multilayer films

It can be seen from Fig. 4.18 that the Sb_2Te_3/Ag multilayer thermoelectric film exhibits a lower thermal conductivity value, which is 0.61 W m^{-1} K^{-1}. In addition, it can be seen from the figure that the thermal conductivity of a multilayer thermoelectric film increases with increasing temperature. The ultra-low thermal conductivity obtained is mainly due to the transmission characteristics between interfaces, especially the phonon scattering mechanism. In order to further explain the interfaces' phonon transmission characteristics, especially with the change of temperature, the phonon mismatch model (DMM) is used to simulate the phonon transmission characteristics between interfaces, and compared with the experimental results.

According to the phonon diffusion model, the heat flow between interfaces and the boundary thermal conductivity can be described as

$$q_{net,DMM} = \frac{k_b^4}{8\pi^2\hbar^3}\left\{T_1^4\sum_j c_{1,j}\int_0^{\frac{\hbar\omega}{k_bT_1}}\frac{\alpha_{1\to2}\cdot z^3}{e^z-1}dz - T_2^4\sum_j c_{2,j}\int_0^{\frac{\hbar\omega}{k_bT_2}}\frac{\alpha_{2\to1}\cdot z^3}{e^z-1}dz\right\}, \tag{4.2}$$

$$h_{1\to 2} = \frac{\partial q_{net,DMM}}{\partial T} = \frac{1}{4}\sum_{j=1}^{3}\int_{\omega} c_{1,j}(\omega)\alpha_{1\to 2}(\omega)\hbar\omega Dos_1(\omega)\frac{\partial f_0}{\partial T}d\omega, \quad (4.3)$$

where $c_{1,j}$ and $c_{2,j}$ are the group velocities of the phonon in j mode, h is Planck's constant, $Dos(\omega)$ is the density of phonon states, f_0 is the Bose–Einstein distribution function, and $\alpha_{1\to 2}$ is the phonon transmission coefficient from interface 1 to interface 2. Its mathematic expression can be expressed as

$$\alpha_{1\to 2}(\omega) = \frac{\sum_j c_{2,j}(\omega)Dos_2(\omega)\delta_{\omega,\omega'}}{\sum_j c_{1,j}(\omega)Dos_1(\omega)\delta_{\omega,\omega'} + \sum_j c_{2,j}(\omega)Dos_2(\omega)\delta_{\omega,\omega'}}. \quad (4.4)$$

In order to simplify the calculation, this paper uses the Dubai approximation model of phonon dispersion. After the interface thermal resistance is obtained, the total effective thermal conductivity can be obtained by the formula $k_e = (d_1 + d_2)\big/(2R_{1\to 2})$. The comparison between the obtained results and the experimental results is shown in the figure above. It can be seen from the figure that the simulation results are consistent with the experimental results. It should be noted that the model is mainly based on the approximation of single crystals. In the actual thin film preparation, the crystals are mainly polycrystalline. Therefore, the experimental data and theoretical predictions have always been biased. Nevertheless, the phonon diffusion mismatch model can also provide clear physical images and indicate that the microstructure between metal and semiconductor multilayer films can affect the phonon transmission characteristics.

4.2.4 Theoretical Analysis of Thermal Conductivity

4.2.4.1 Interface Effect of Heat Transfer of Multilayer Thermoelectric Film

The thermal conductivity is the most important parameters in our studying to understand the thermoelectric conversion mechanism of multilayer films. At the nanoscale, the main factor affecting the thermal conductivity of multilayer films is the scattering mechanism of phonons by structural defects, as shown in Fig. 4.18.

In addition to the traditional analytical methods, direct numerical simulation techniques have also been applied to the study of the thermal conductivity of multilayer films. The molecular dynamics (MD) method solves the equations of motion of particles that interact with each other to obtain the evolution of the spatial position and motion state of individual particles over time, so as to calculate the macroscopic behavior of the material. In the nanoscale range, as an effective supplement to the experiment, the MD method is widely used in research work on heat conduction.

4.2.4.2 Electro-Phonon Coupling Effect at the Interface of Multilayer Thermoelectric Films

For semiconductor/metal multilayer film systems, in addition to the interface effect, another effect that requires careful consideration is the electro-phonon coupling effect. As we all know, electrons are the main energy-carrying carriers in metals, while phonons dominate energy transfer in semiconductors. When heat transfer occurs between the metal-semiconductor interface, energy transfer between electrons and phonons is bound to occur. The thermal resistance at the interface of a metal-semiconductor structure is caused by two parts. The electrons in the first metal interact with phonons, then the phonons in the metal interact with phonons in the semiconductor; the second part is the electrons in the metal and semiconductor phonons at the metal-semiconductor interface.

In order to systematically study the coupling effects of electrons and phonons at the interface of metal semiconductor multilayer films, the thermal conductivity of Sb_2Te_3/Au and Sb_2Te_3/Ag multilayer films were calculated and simulated using a dual temperature model. The comparison of dual-temperature model and the experimental results are shown in Figs. 4.19. It can be seen from the figure that the simulation results of the two-temperature model are in good agreement with the experimental results, especially when the thickness of the metal semiconductor multilayer film is relatively large. The simulation results and the experimental results are in good agreement. The research shows that for metal-semiconductor multilayer thin-film systems, the interface has a lower coupling factor value, and its value range is generally 10^{16}–10^{17} W m^{-3} K^{-1}. Therefore, from the perspective of engineering applications, through reasonable optimization of the metal layer thickness, it can effectively reduce the effective thermal conductivity of the multilayer thin film, and at the same time, it does not weaken the electron transport ability, thereby enhancing the thermoelectric conversion merit of low-dimensional thermoelectric materials and improving the thermoelectric conversion efficiency.

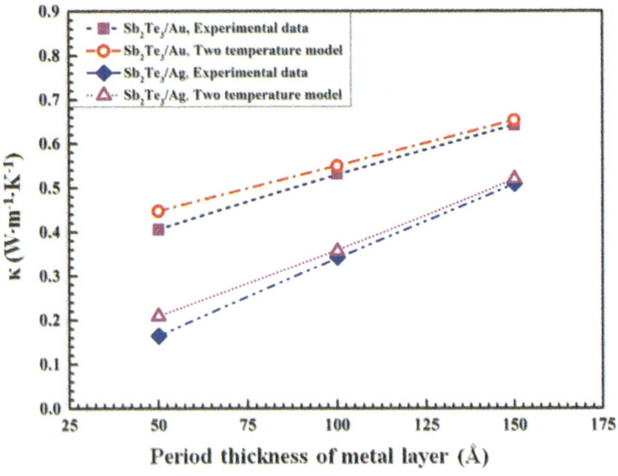

Fig. 4.19 Comparison of the dual-temperature model and experimental results of multilayer films

4.3 Cu/Sb_2Te_3 Multilayer Film

4.3.1 Film Preparation and Structure Analysis

The specific process of the experiment is as follows: The experimental instrument is a molecular beam epitaxy equipment (MBE) to prepare Sb_2Te_3/Cu periodic nano-multilayer films with different Cu thicknesses by co-evaporation of Sb, Te and Cu elements. The substrate selection was the same as in the previous section. When the vacuum reaches 10^{-8} Pa, the PID controller was turned on, and the Sb and Te beam source furnaces were separately heated, and the beam source furnace temperature is kept constant after the specified evaporation rate was reached. The specific evaporation temperature and deposition rate were designed as follows: the beam source furnace temperature of Te was constant at 589 K, and the corresponding deposition rate was 0.95 Å s^{-1}; the beam source furnace temperature of Cu was kept constant at 1323 K, and the corresponding deposition rate was 0.2 Å s^{-1}; and the deposition rate of the Sb beam source furnace temperature at 698 K was 0.45 Å s^{-1}. The Sb_2Te_3 layer was prepared using the scheme designed in the previous section to ensure the thickness of each layer being of 20 nm. After the Sb and Te beam source furnace baffle was closed, and at the same time the Cu beam source furnace baffle was opened to evaporate the Cu element. When the evaporation rate of Cu was stabilized, the sample baffle was opened. Repeated 10 times from time to time, Sb_2Te_3/Cu periodic nano-multilayer films with different Cu thicknesses were prepared by adjusting the time of copper evaporating, where the thickness of each layer of Cu was 0, 0.1 nm and 0.3 nm, which were named Sb_2Te_3/Cu (0 nm), Sb_2Te_3/Cu (0.1 nm), and Sb_2Te_3/Cu (0.3 nm), respectively. The samples were then annealed at 423 and 473 K for 6 h. TEM, EDS and other analytical methods were used to characterize the structure, morphology and composition, and the Seebeck coefficient and electrical conductivity were tested using homemade equipment.

Fig. 4.20 Surface morphology of Sb_2Te_3/Cu (0.3 nm) thin film by SEM and TEM

Figure 4.20a is the surface SEM of the Sb_2Te_3/Cu (0.3 nm) periodic nanofilm with an annealing temperature of 423 K. It can be seen from the picture that a large amount of particulate matter exists on the surface of the film, and the particles are Cu particles after energy spectrum analysis. Figure 4.20b is the result of TEM observation. It is also found that there is a large amount of granular Cu. Based on the analysis of the above two characterization methods, it can be considered that the copper in the sample does not form a layer, but exists in a granular state.

4.3.2 *Thin Film Electrical Properties*

It can be found from Tables 4.6 and 4.7 that the electrical properties have changed significantly after the addition of copper. The conductivity of 423 K annealed Sb_2Te_3/Cu periodic nano-films is 5.6 times (0.3 nm Cu) higher than that of samples without copper. The reason is that according to the conductivity formula $\sigma = ne\mu$, it can be known that the carrier concentration and the carrier mobility determine the conductivity, and the addition of metallic copper effectively increases the carrier concentration and thereby increases its conductivity. Therefore, it can be seen from the table that the higher the copper content, the greater the electrical conductivity (Table 4.6).

The Seebeck coefficient of Sb_2Te_3/Cu (0.3 nm Cu) periodic nano-films annealed at 423 K is 1.36 times higher than the samples without copper. The reason is: according to the Mott formula $\alpha = \frac{\pi^2 k_B^2 T}{3e}\left\{\frac{dn(E)}{ndE} + \frac{d\mu(E)}{\mu dE}\right\}_{E=E_F}$ (where n (E) and μ (E) are the relationship between carrier concentration and mobility and energy, and E_F is the Fermi level) [1, 2], the Seebeck coefficient can increase by increasing the carrier mobility energy derivative and decreasing the carrier concentration. Studies have shown that if there is a difference in work function between materials [3], the energy level barriers of carriers can be formed in the material body, thereby generating the effect of filtering carriers and changing the Seebeck coefficient. As shown in Fig. 4.21, the energy band diagram of Cu–Sb_2Te_3 metal-semiconductor contact. Generally, Sb_2Te_3 is a degenerate semiconductor, and the Fermi level is in the valence band [4]. The band gap width (E_g), electron affinity (E_A), and work function (Φ) of Sb_2Te_3 in Fig. 4.21 are all taken from the bulk material [5]. The energy band diagram of

Table 4.6 Electrical properties of 423 K annealed Sb_2Te_3/Cu

Sample	σ (S cm^{-1})	α (μV K^{-1})	PF (μW m^{-1} K^{-2})
Sb_2Te_3/Cu (0 nm)	200	90	162
Sb_2Te_3/Cu (0.1 nm)	800	65	340
Sb_2Te_3/Cu (0.3 nm)	1120	66	480

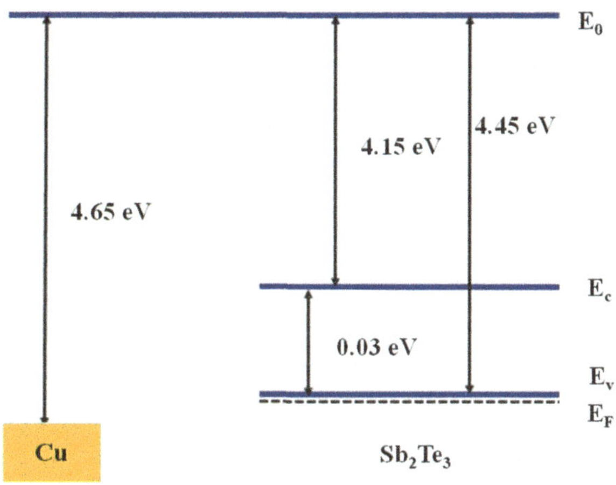

Fig. 4.21 Band diagram of Sb_2Te_3–Cu metal-semiconductor contact

Cu–Sb_2Te_3 metal-semiconductor contact at equilibrium is shown in Fig. 4.21. It can be seen from the figure that the work functions of Cu and Sb_2Te_3 are 4.65 eV and 4.45 eV, respectively, and the barrier height difference is about 0.2 eV. The numerical barrier height can theoretically filter low-energy carriers effectively without affecting the passage of high-energy carriers. Due to the existence of energy level barriers, at the interface between Cu and Sb_2Te_3, low-energy carriers (cold carriers) are strongly scattered, while high-energy carriers (hot carriers) are weakly scattered. Therefore, although the increase of the carrier concentration reduces the Seebeck coefficient, the Seebeck coefficient does not decrease much because of the carrier filtering effect. The schematic diagram is shown in Fig. 4.22.

As shown in Table 4.7, when the sample is annealed at 473 K, the crystallinity becomes better and better, resulting in an increase in carrier concentration and an increase in electrical conductivity. At the same time, as the annealing temperature

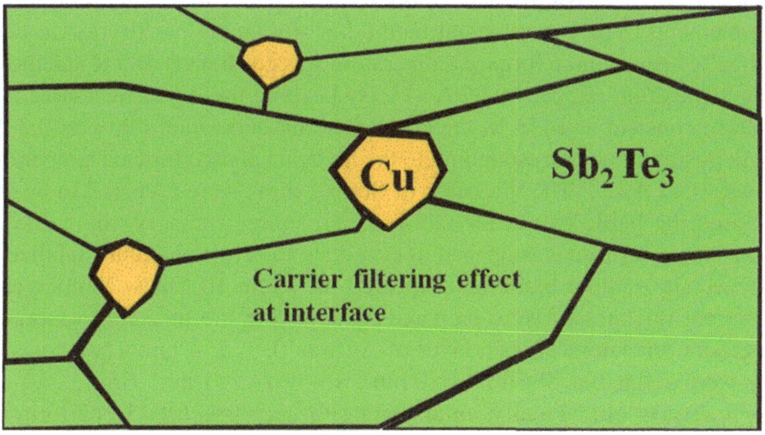

Fig. 4.22 Cu and Sb_2Te_3 interface carrier filtration mechanism

Table 4.7 Electrical properties of 473 K annealed Sb_2Te_3/Cu

Sample	σ (S cm^{-1})	α (μV K^{-1})	PF (μW m^{-1} K^{-2})
Sb_2Te_3/Cu (0 nm)	300	92	240
Sb_2Te_3/Cu (0.1 nm)	1587	43	290
Sb_2Te_3/Cu (0.3 nm)	1250	58	420

increases, Cu particles are incorporated into the Sb_2Te_3 material, which results in a reduction of copper particles, a decrease in scattering, a reduction in filtering effects, and a decrease in Seebeck. The integrated power factor remains at a large value of 420 μW m^{-1} K^{-2}. The Sb_2Te_3/Cu (0.3 nm Cu) sample has an optimal power factor after 423 K annealing. At this time, the power factor reached 480 μW m^{-1} K^{-2}, which was 3 times higher than that of Sb_2Te_3 film (160 μW m^{-1} K^{-2}).

4.4 Ag/Sb_2Te_3 Multilayer Film

In this section, Ag/Sb_2Te_3 periodic multilayer films are prepared by molecular beam epitaxy. By inserting Ag layers of different thicknesses in Sb_2Te_3, the effects of different metal thicknesses on the thermoelectric properties of multilayer thin films were investigated.

4.4.1 Film Preparation and Structure Analysis

Different Sb_2Te_3/Ag periodic nano-films were prepared by evaporation of Sb, Te and Ag elements. The experimental method is similar to the Sb_2Te_3/Cu periodic nano-film. Te beam source furnace temperature was constant at 589 K and the corresponding deposition rate was 0.95 Å s^{-1}. By keeping the Ag beam source furnace temperature constant at 1423 K, the corresponding deposition rate was 0.2 Å s^{-1}, and the deposition rate corresponding to the Sb beam source furnace temperature of 698 K was 0.45 Å s^{-1}. The thickness of Sb_2Te_3 films were prepared to be 20 nm. After closed the baffle of the Sb and Te beam source furnaces, the baffle of the Ag beam source furnace was opened to evaporate the Ag element at stabilized rate, then opened the sample baffle and repeated the cycle 10 times. Multilayer films with different thicknesses were prepared by adjusting the time for evaporation of Ag, where the thickness of each layer of Ag was 0, 1, 2, 4 nm. The names of the samples were defined as Sb_2Te_3/Ag (0 nm), Sb_2Te_3/Ag (1 nm), Sb_2Te_3/Ag (2 nm) and Sb_2Te_3/Ag (4 nm). Finally, an annealing furnace was used to perform 423 K annealing. The annealing time was 6 h, and then XRD, TEM, EDS and other analytical methods were used to analyze the structure, morphology and composition, and the Seebeck coefficient and conductivity are tested by using homemade equipment.

First, the samples were analyzed by phase analysis, and a wide-angle XRD test was performed using an X-ray diffractometer (D\max-2200, Japanese science) equipment, as shown in Fig. 4.23. The scanning angle is 0-65 degree and the scanning speed is 0.5 degree/min. The XRD test shows that the main component of the sample without silver is Sb_2Te_3, which indicates that the thin film prepared by MBE is a single compound of Sb_2Te_3. With the addition of Ag, new diffraction peaks appear in the XRD pattern. Comparing with the PDF card, it can be seen that corresponding to the (200), (220), and (222) crystal planes of $AgSb_2Te_3$, which indicates that the $AgSb_2Te_3$ phase appears in the thin film after adding Ag. With the increase of Ag content, the corresponding peak shape of $AgSb_2Te$ becomes sharper and the corresponding peak shape of Sb_2Te_3 gradually decreases, which indicates that with the increase of silver, the $AgSbTe_2$ phase gradually increases and the Sb_2Te_3, phase gradually decreases. At the same time, no Ag element was detected in the silver-added samples. This indicates that after annealing, Ag is extremely active and basically dissolves in the Sb_2Te_3 matrix to form intermetallic compounds. The composition of the entire film is $AgSbTe_2$ and Sb_2Te_3.

The structure of the film was further analyzed using a transmission electron microscope, as shown in Fig. 4.24. Figure 4.24a is a high-magnification TEM picture of the Sb_2Te_3/Ag periodic nano-film sample. It can be seen from the figure that there is no layered structure in the film, but there are two kinds of irregular lattice structures and more clearer grain boundaries. The electron diffraction pattern results in Fig. 4.24b show that the sample is polycrystalline and has good crystallinity, which is also consistent with the result of XRD test. In order to explore these two crystals, high-resolution TEM was used to analyze and explore their crystal types as shown

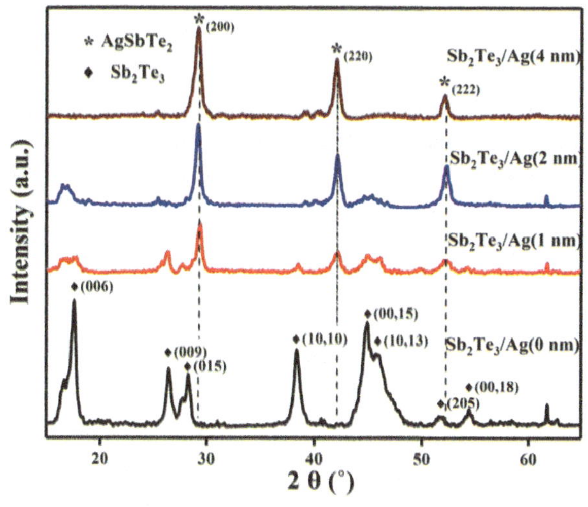

Fig. 4.23 XRD patterns of Sb_2Te_3/Ag periodic nano-film samples with different Ag layer thicknesses

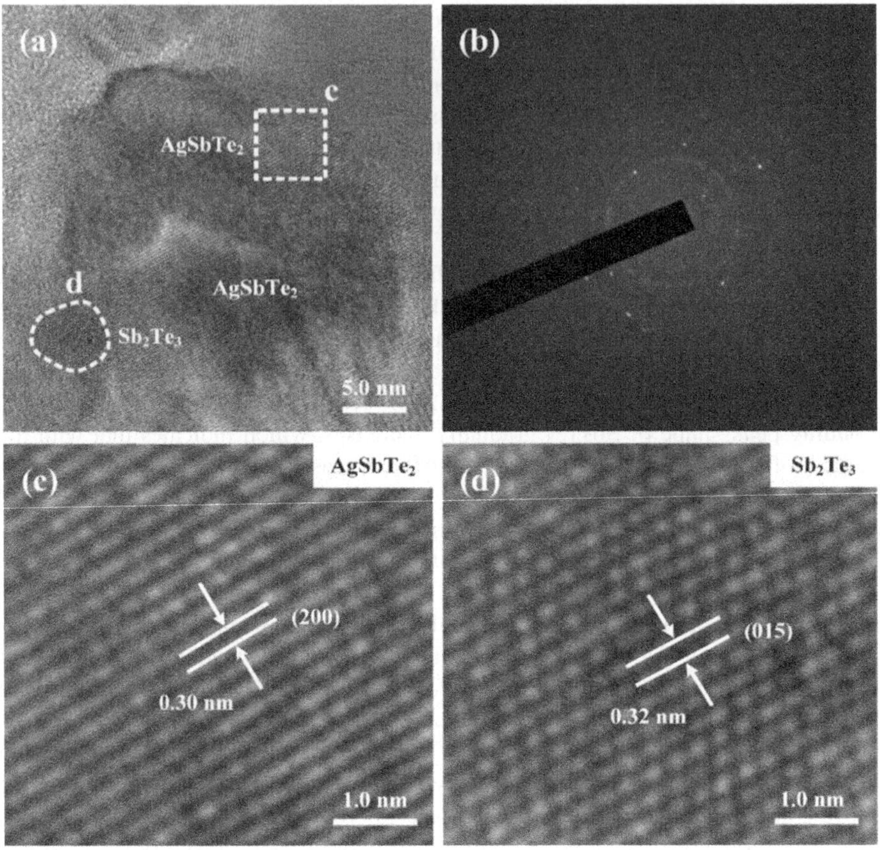

Fig. 4.24 TEM analysis of Sb_2Te_3/Ag periodic nano-film samples: **a** High power TEM; **b** Electron diffraction pattern; **c** and **d** High resolution TEM

in Fig. 4.24c, d. By calibrating the lattice constant, we can get the lattice constant of 0.3 nm in Fig. 4.24c, which corresponds to the (200) crystal plane of $AgSbTe_2$, and the lattice constant of 0.32 nm in Fig. 4.24d, which is the same as that of Sb_2Te_3. The lattice spacing of the (015) crystal planes is uniform. This is consistent with the XRD test results, there is no Ag simple substance in the sample but only two substances Sb_2Te_3 and $AgSbTe_2$.

4.4.2 Thin Film Thermoelectric Properties

First, the carrier mobility and carrier concentration of the samples were tested. Table 4.8 shows the changes in the conductivity, carrier mobility, and carrier concentration of Sb_2Te_3/Ag periodic nano-films with the thickness of Ag. It can be seen that

Table 4.8 Conductivity, carrier mobility, and carrier concentration of Sb_2Te_3/Ag periodic nano-film samples

Sample	σ (S cm^{-1})	μ (cm^2 V^{-1} s^{-1})	n (10^{20} cm^{-3})
Sb_2Te_3/Ag (4 nm)	454.45	5.86	4.85
Sb_2Te_3/Ag (2 nm)	75.66	2.28	2.07
Sb_2Te_3/Ag (1 nm)	163.94	5.53	1.85
Sb_2Te_3/Ag (0 nm)	138.99	32.13	0.27

with the addition of Ag, the carrier concentration of the Sb_2Te_3/Ag periodic nano-films rapidly increases, and the mobility decreases first and then increases. Due to the difference between the two, the electrical conductivity of the sample eventually increases with the thickness of the Ag layer increase first then decrease and finally increase again.

Samples of Sb_2Te_3 films with different Ag thicknesses were tested using the variable temperature Seebeck coefficient and conductivity tester. The test results are shown in Fig. 4.25. From Fig. 4.25a, it can be seen that the conductivity of Sb_2Te_3/Ag (0 nm), Sb_2Te_3/Ag (1 nm), and Sb_2Te_3/Ag (2 nm) increases with temperature showing semiconductor characteristics, while the conductivity of the Sb_2Te_3/Ag (4 nm) sample decreases with increasing temperature and shows metallicity, which may be caused by excessive Ag. With the increase of Ag content, the electrical conductivity increased from 130 S cm^{-1} of Sb_2Te_3 film to 590 S cm^{-1} (210 K) of Sb_2Te_3/Ag (4 nm) film, an increase of 4.5 times. Interestingly, the conductivity of the Sb_2Te_3/Ag (2 nm) sample film has decreased. The reason for this analysis is that the rate of increase of the carrier concentration at this time is lower than the rate of decrease of the carrier mobility, which causes the conductivity. The decline can

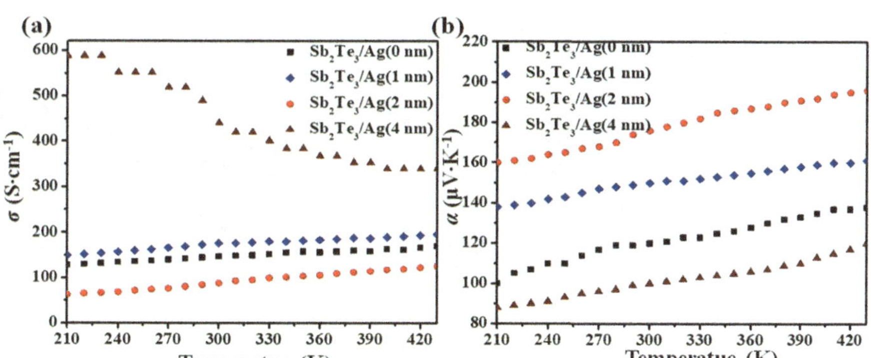

Fig. 4.25 Sb_2Te_3/Ag periodic nano-films with different Ag thicknesses **a** temperature-variable conductance and **b** seebeck coefficient

be proved by Table 4.8. First, the carrier mobility and carrier concentration of the sample were tested. It is the relationship between the conductivity, carrier mobility, and carrier concentration of Sb_2Te_3/Ag periodic nano-films as a function of Ag thickness. As can be seen from the table, with the addition of Ag, the carrier concentration of Sb_2Te_3/Ag periodic nano-films increased rapidly, and the mobility decreased first and then increased. Due to the different amplitudes of the two, the electrical conductivity of the sample increased first with the thickness of the Ag layer Lower it and increase it again.

It can be seen from Fig. 4.25b that the Seebeck coefficients of all samples are positive values, which proves that the Sb_2Te_3/Ag films grown by molecular beam epitaxy are all P-type semiconductor thermoelectric films. As the Ag content increases, the Seebeck coefficient increases rapidly. The Seebeck coefficient has increased from 130 $\mu V\ K^{-1}$ of pure Sb_2Te_3 film to 193 $\mu V\ K^{-1}$ (420 K) of Sb_2Te_3/Ag (2 nm) film, an increase of 1.5 times. However, the Seebeck coefficient of Sb_2Te_3/Ag (4 nm) films has decreased. The reasons for the analysis may be as follows:

(1) According to the Mott formula, it can be seen that an increase in the carrier concentration will lead to a decrease in Seebeck. The addition of excess Ag leads to an increase in carrier concentration, which leads to a decrease in Seebeck.
(2) The carriers of Ag are electrons, while the carriers of Sb2Te3 are holes. The increase of Ag causes the positive and negative Seebeck coefficients to cancel out.

The power factor can be calculated from Seebeck coefficient and conductivity. As shown in Fig. 4.26, the power factor increases with the increase of the thickness of Ag. The power factor of the Sb_2Te_3/Ag (4 nm) film reaches the maximum, with a value of 500 $\mu W\ m^{-1}\ K^{-2}$, which is 3.84 times higher than that of pure Sb_2Te_3 film (130 $\mu W\ m^{-1}\ K^{-2}$). Therefore, the addition of Ag can effectively improve the thermoelectric performance of Sb_2Te_3.

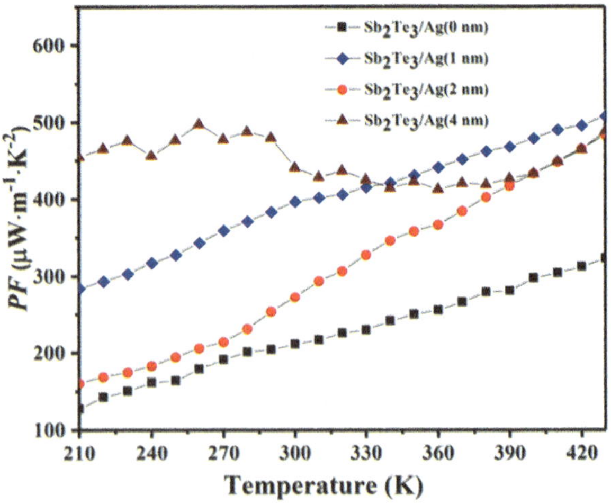

Fig. 4.26 Power factor of Sb_2Te_3/Ag periodic nano-films with different Ag thicknesses

In addition to the increase of the power factor, the thermal conductivity of periodic nano-multilayer films will also be significantly reduced due to phonon scattering, which will further increase the ZT value. The introduction of a discontinuous layered structure with a certain size and distribution can selectively scatter and transmit a large number of medium- and long-wavelength phonons [6].

4.5 Chapter Summary

(1) A series of Au/Sb_2Te_3 multilayer films with gold thicknesses of 0, 1, 3, 5, 10, and 20 nm were prepared by magnetron sputtering. The thermal conductivity of the films was tested using the 3ω method. The thermal conductivity decreases first and then increases. The minimum value appears when the thickness of the gold layer is 5 nm, and the thermal conductivity is 0.44 W m^{-1} K^{-1}, which is 44% of the pure Sb_2Te_3 sample. The classic thermal conduction model and the improved dual temperature model were used to analyze the thermal conductivity of the sample. It was found that when the gold layer is larger than 5 nm, the thermal conductivity trend that the dual temperature model conforms to is the electrophonon on the metal-semiconductor contact surface. Coupling plays a key role in heat transfer, and when the thickness of gold is less than 5 nm, the film may be discontinuous due to the agglomeration of the gold layer, and a sieve model is introduced for analysis, that is, the electrophonon coupling plays partial effect, and the effect becomes weaker as the thickness of the gold layer decreases.

(2) Magnetron sputtering $(Au/Ag/Cu/Pt/Cr/Mo/W/Ta)/Sb_2Te_3$ multilayer film was prepared, and its thermal conductivity was systematically experimentally and theoretically studied. Studies have shown that for metal-semiconductor thermoelectric multilayer film systems, the electron-phonon coupling effect at the interface significantly affects its cross-section thermal conductivity. By rationally adjusting the thickness of the metal layer or selecting a gold semiconductor multilayer film, the system can effectively reduce the thermal conductivity of its cross section, and at the same time does not weaken the electron transport characteristics in the thermoelectric material system to a certain extent. Furthermore, the thermoelectric conversion figure of merit of the thermoelectric material is improved as a whole, and the total energy conversion efficiency of the low-dimensional thermoelectric device is finally improved. In addition, the research also shows that for the metal-semiconductor multilayer thin film structure, the thermal conductivity simulated by the dual temperature model is in good agreement with the experimental results. It can be determined that the overall interface thermal resistance of the metal-semiconductor periodic system, The interfacial thermal resistance caused by electron-phonon coupling is dominant.

(3) Using MBE to alternately grow the Sb_2Te_3 film layer and the Cu layer. When the Cu layer is thin, the film does not form a layer, but forms discontinuous nanoparticles, and Sb_2Te_3 films with different contents of Cu nanoparticles

are obtained. The study found that Cu particles can effectively improve the electrical properties. After the film was annealed at 423 K, the power factor reached 480 $\mu W\ m^{-1}\ K^{-2}$.

MBE was used to prepare Sb_2Te_3 and Ag periodic nano-films. Because Ag is extremely active, the film does not form a multilayer structure but forms a mixture of $AgSbTe_2$ and Sb_2Te_3. The addition of Ag can greatly improve the electrical properties. The sample Sb_2Te_3/Ag (4 nm) has the largest power factor, its value reaches 500 $\mu W\ m^{-1}\ K^{-2}$, which is 3.84 times higher than the Sb_2Te_3 film (130 $\mu W\ m^{-1}\ K^{-2}$) without Ag.

References

1. Rowe IToECToECM (1995) CRC handbook of thermoelectrics. CRC Press
2. Chen L, Liu R, Shi X (2018) Thermoelectric materials and devices. Science Press, Beijing
3. Faleev SV, Léonard F (2008) Theory of enhancement of thermoelectric properties of materials with nanoinclusions. Phys Rev B 77(21):214304
4. Rose DH, Hasoon FS, Dhere RG, Albin DS, Ribelin RM, Li XS, Mahathongdy Y, Gessert TA, Sheldon P (1999) Fabrication procedures and process sensitivities for CdS/CdTe solar cells. Prog Photovoltaics Res Appl 7(5):331–340. https://doi.org/10.1002/(SICI)1099-159X(199909/10)7:5%3c331:AID-PIP257%3e3.0.CO;2-P
5. Matin M, Aliyu MM, Quadery AH, Amin N (2010) Prospects of novel front and back contacts for high efficiency cadmium telluride thin film solar cells from numerical analysis. Sol Energy Mater Sol Cells 94(9):1496–1500
6. Minnich AJ, Dresselhaus MS, Ren ZF, Chen G (2009) Bulk nanostructured thermoelectric materials: current research and future prospects. Energy Environ Sci 2(5):466–479. https://doi.org/10.1039/B822664B

Chapter 5
Preparation of Sb_2Te_3/Bi_2Te_3 Thin Films by Magnetron Sputtering

Sb_2Te_3 and Bi_2Te_3 are the semiconductor materials with the best thermoelectric properties near room temperature. The current research is mainly bulk. The adopted methods include solution method, solid phase synthesis method, hot pressing method, microwave method, ion exchange method, melting method and spark plasma sintering (SPS). Various means to enhance the performance of thermoelectric materials mainly increase the Seebeck coefficient (including energy filtering effect) under the premise of ensuring electrical conductivity, so that the power factor increases. In addition, the introduction of interfacial scattering phonons reduces the thermal conductivity. However, the research on thermoelectric thin film is not enough. The research on thermoelectric thin film will be beneficial to the utilization of micro-scale heat and sensors. This chapter uses the magnetron sputtering process in the micro-nano machining process to adjust the structure and thermoelectric properties of Sb_2Te_3 and Bi_2Te_3.

5.1 Controlling Sputtering Power, Annealing, Thickness Deposition of Sb_2Te_3 Film

Sb_2Te_3 film has many excellent physical and chemical properties, and has been widely concerned by people. The structure and performance of Sb_2Te_3 thin film are inseparable from the preparation technology and process. The thin films prepared by different technologies differ greatly in morphology and performance. Even if the same method adopts different preparation processes, thin films with very different properties will be obtained. In this section, magnetron sputtering is used to control the sputtering power, annealing conditions and film thickness to deposit Sb_2Te_3 film. By changing the conditions of magnetron sputtering, adjusting the structure of the film, exploring the factors affecting its thermoelectric performance, and providing a basis for improving its thermoelectric performance.

Z. Hu and Z. Wu, *Nanostructured Thermoelectric Films*,
https://doi.org/10.1007/978-981-15-6518-2_5

5.1.1 Controlling Sputtering Power

5.1.1.1 Film Preparation

Sb_2Te_3 thin film has many excellent physical and chemical properties, that has a wide range of applications in high-tech fields such as optics and microelectronics, and has been receiving widespread attention. The structure and performance of Sb_2Te_3 film are inseparable from the preparation technology and process. The morphology and properties of films prepared by different technologies are very different. Even if the same method is used for different preparation processes, films with very different properties could be obtained. In this section, RF magnetron sputtering is used to deposit Sb_2Te_3 films on silicon substrates and glass substrates, respectively. The effects of target sputtering power on the film structure and thermoelectric properties are explored. The substrate used in the experiment was immersed in a mixed solution of concentrated sulfuric acid and hydrogen peroxide, acetone, and alcohol for 10 min before the film was deposited, and then washed with deionized water and blow dried with high-purity N_2. The experimental background vacuum is less than 2×10^{-6} Torr, the argon flow rate is stable at 13.5 sccm; the target base distance is 20 cm, and the substrate table rotation speed is 20 rpm, so that a uniform Sb_2Te_3 film can be obtained. During the experiment, the sputtering powers of the antimony telluride alloy targets were adjusted to 15, 20, and 25 W, respectively. The thickness of the film was monitored in real time through a quartz crystal plate. Finally, the thickness of the film was measured by a step meter to about 240 nm. In order to optimize the properties of the film, the prepared samples were post-annealed at an annealing temperature of 100, 150, 200, 250, 300 °C, and an annealing time of 6 h. At the same time, XRD, SEM, EDS and other analytical methods were used to characterize the prepared Sb_2Te_3 thin film, and its electrical properties were tested using homemade equipment.

5.1.1.2 Thin Film Structure

Figure 5.1a is the relationship between the deposition rate of Sb_2Te_3 film and the RF power. It can be seen from the figure (a) that as the RF power increases, the deposition rate increases; and there is a good linear relationship between the deposition rate and the RF power. Radio frequency power not only determines whether the sputtering phenomenon can be generated, but also the power will affect the energy of the ions that bombard the target, and then affect the sputtering rate of the target. With the increase of RF power, the ionization rate of the gas increases, the plasma density increases, the number of ions that bombard the target surface increases, and the energy increases, so the growth rate of the film increases with the increase of sputtering power. Figure 5.1b shows the relationship between the composition of Te and RF power in the Sb_2Te_3 film. It can be seen that as the RF power increases, the Te content in the film decreases. When the RF power of the antimony telluride target is 15 W, the Te content in the

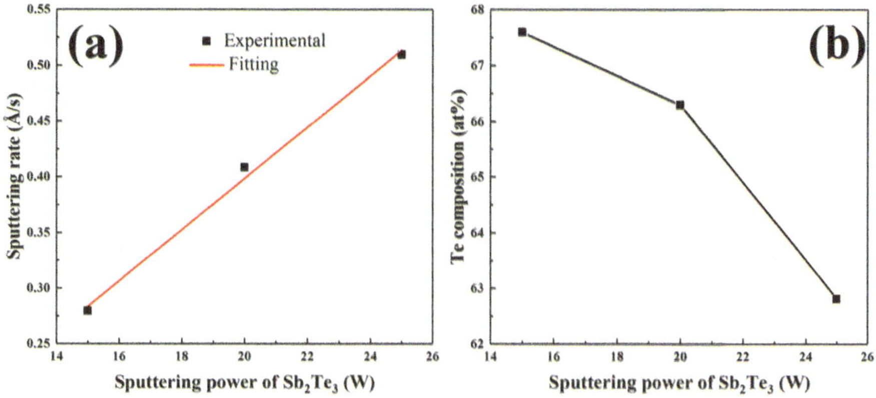

Fig. 5.1 Relationship between the deposition rate of Sb_2Te_3 film and the Te composition in the film and the sputtering power

film is 67.6%; and when the RF power is increased to 25 W, the Te content in the film is 62.8%. At the same time, it was found that Te-rich Sb_2Te_3 films deposited at room temperature deviated from the chemical composition ratio, which may be due to the higher sputtering yield of Te than that of Sb.

As shown in Fig. 5.2, these are the SEM images of the Sb_2Te_3 film under different sputtering powers. It can be seen that the surface morphology of the film deposited under different sputtering powers does not change much, and the film surface is more

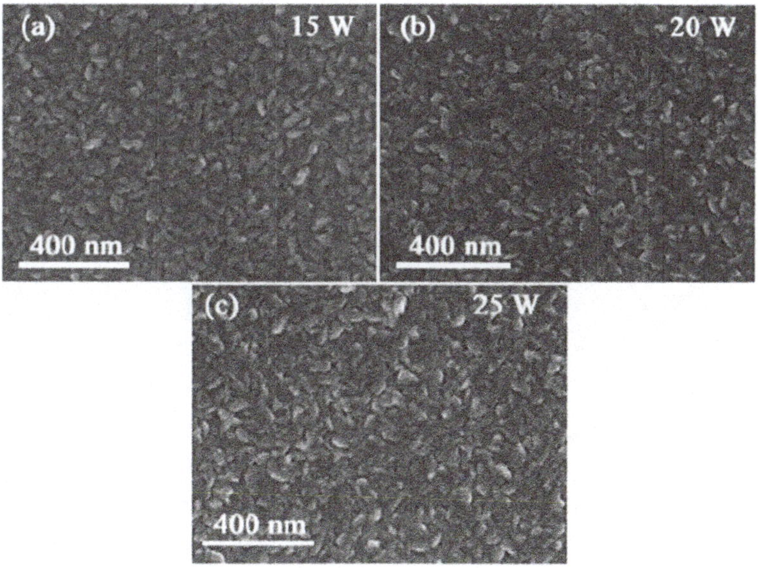

Fig. 5.2 SEM images of Sb_2Te_3 thin films at different powers **a** 15 W, **b** 20 W, **c** 25 W

uniform and denser. With the increase of sputtering power, the crystal grain size of the film increases slightly. This is because when the sputtering power is increased, the energy of the atoms flying from the target surface to the substrate is increased, the atom diffusion ability is enhanced, the crystal nuclei fuse together and grow, and the film grain size increases.

5.1.1.3 Effect of Sputtering Power on Its Thermoelectric Performance

Figure 5.3 shows the electrical properties of Sb_2Te_3 thin films at different sputtering powers. It can be seen from Fig. 5.3a that when the RF sputtering power is 15 W, the resistivity of the film is 12.14 mΩ cm; as the RF power increases, the resistivity of the film gradually decreases; the RF power increases to 25 W. At this time, the resistivity of the thin film decreased to 7.97 mΩ cm. The resistivity of the Te material itself is relatively high, and the sample with a high Te content in the sample has a high resistivity. Therefore, the decrease in resistivity is mainly due to the decrease in the Te composition in the thin film. It can be seen from the figure that the Seebeck coefficients of all samples are positive, which indicates that the prepared Sb_2Te_3 thin film samples are P-type. When the sputtering power is increased from 15 W to 20 W, the Seebeck coefficient of the film increases from 136.43 to 147.24 $\mu V\ K^{-1}$; further increase of the sputtering power reduces the Seebeck coefficient of the film to 140.26 $\mu V\ K^{-1}$. It can be seen that the Seebeck coefficient of the film increases first and then decreases as the RF power increases.

According to the resistivity and Seebeck coefficient, the power factor of the Sb_2Te_3 film can be calculated. The calculation results are shown in Fig. 5.3b. It can be seen that as the sputtering power increases, the power factor of the film increases first and then decreases slightly. This is because the power factor is the result of combining both resistivity and Seebeck coefficient. Although the resistivity of the thin film decreases with increasing sputtering power, the Seebeck coefficient increases and

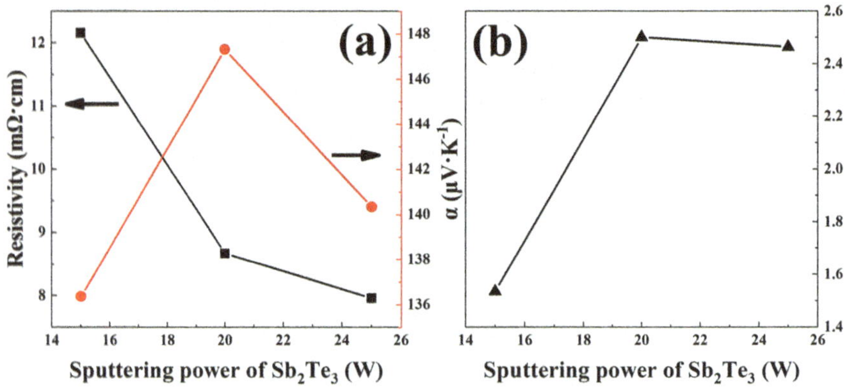

Fig. 5.3 Electrical properties of Sb_2Te_3 thin films at different sputtering powers

then decreases with increasing power. When the power is 20 W, the maximum power factor of the Sb_2Te_3 film deposited at room temperature is 2.5 μW cm^{-1} K^{-2}.

5.1.2 Controlling the Annealing Process

In order to further optimize the properties of the thin film, a Sb_2Te_3 alloy target with a RF power of 20 W was selected and placed in a well annealing furnace for post-annealing. The annealing temperature was 100, 150, 200, 250, 300 °C, and the annealing time was 6 h. The effect of annealing treatment on the morphology and properties of the film was studied.

Figure 5.4 is an SEM image of the Sb_2Te_3 film before and after annealing, where the original sample at room temperature is labeled As-deposited. It can be seen from Fig. 5.4a and b that the surface of the deposited film is relatively uniform and dense at room temperature, and the grain size of the film gradually grows with the increase of annealing temperature. When the annealing temperature was increased to 250 °C, massive nanocrystals precipitated on the surface of the film, and there were a few tiny nanopores on the surface. The appearance of nanopores may be due to the re-evaporation of Te during the annealing process, resulting in voids on the film surface.

At the same time, it can also be seen from Fig. 5.5 that as the annealing temperature increases, Te atoms tend to re-evaporate, and the composition of Te in the film gradually decreases and gradually approaches the stoichiometric ratio. EDS composition

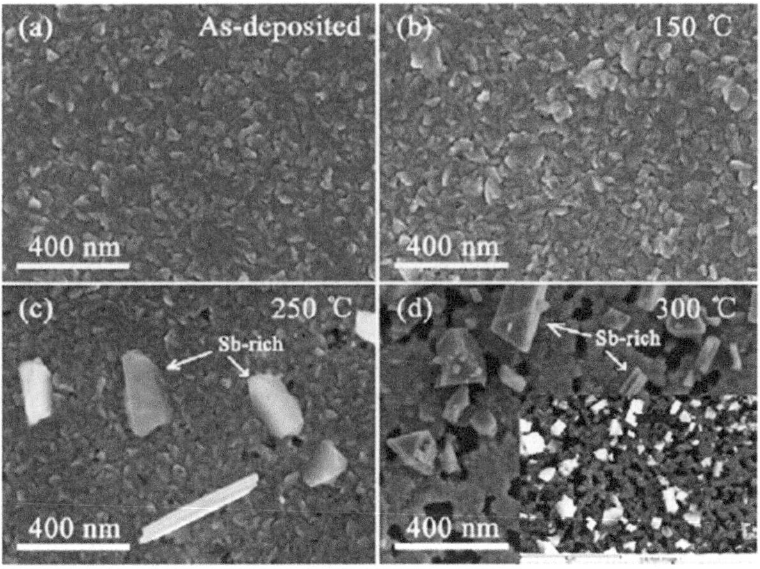

Fig. 5.4 SEM images of Sb_2Te_3 film before and after annealing **a** As-deposited, **b** 150 °C, **c** 250°C, **d** 300 °C

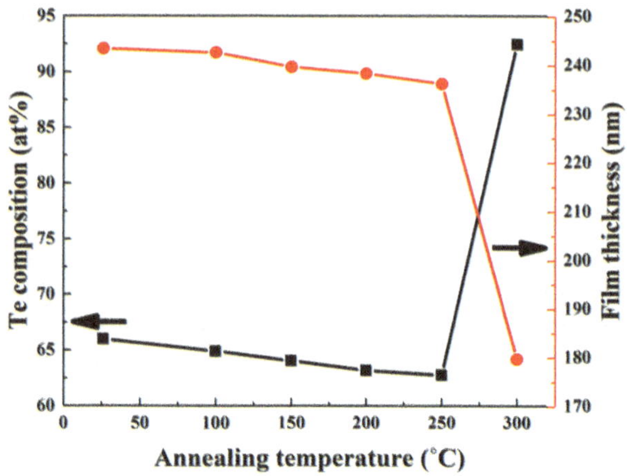

Fig. 5.5 Relationship between the content of Te in the Sb_2Te_3 film and the thickness and the annealing temperature

analysis shows that the precipitated bulk nanocrystals are rich in Sb, and the composition ratio of Sb and Te in the crystals is close to 82:18, which may be due to the fact that Sb atoms tend to diffuse at the grain boundaries during the annealing process condensation. As the annealing temperature and time increase, the precipitated lumps gradually increase and the size gradually increases; when the annealing temperature increases to 300 °C, a large amount of massive Sb-rich substances precipitate on the surface of the film, and part of the film on the surface of the film Shedding, the surface of the film was motley. It is worth noting that when the annealing temperature is increased to 300 °C, the composition of Te in the film increases sharply to about 93%. This is mainly due to the relatively increased composition of Te in the thin film as the Sb bulk precipitates. At the same time, it can be seen from the figure that as the annealing temperature increases, the thickness of the film gradually decreases, which may be caused by the re-evaporation of the Te film during the annealing process.

The XRD patterns of Sb_2Te_3 film before and after annealing are shown in Fig. 5.5 is. It can be seen that the diffraction peaks of the thin films deposited at room temperature are weaker and the peak envelope is larger, which may be due to the lack of sufficient energy and time for the films deposited at room temperature to migrate and condense. When the annealing temperature is 150 °C, the three strongest peaks at 2θ of 28.26°, 38.29°, and 45.86° are in agreement with the standard spectra of Sb_2Te_3, corresponding to the (015), (1010) and (1013) crystal planes, respectively. It is indicated that a polycrystal having a hexagonal dense structure has been formed. With the further increase of the annealing temperature, the diffraction peak in the XRD pattern further strengthened and the peak envelope gradually weakened indicating that the crystallinity of the film gradually increased with the increase of the annealing temperature. It is worth noting that when the annealing temperature is increased to 300 °C, a new diffraction peak appears at 2θ of 32.30°, 52.00°, 67.82°, respectively. Compared with the PDF (26-0101) card of Sb, the diffraction peaks at 2θ of 32.30° and 67.82° correspond to the (200) and (103) crystal planes of Sb, respectively; while the diffraction peak at 2θ of 52.00° corresponds to Te The (103) crystal plane (Te, PDF (36-1452)). At the same time, it can be seen from Fig. 5.4d that at a high annealing temperature, a large amount of Sb crystals is precipitated from the film, which indicates that when the annealing temperature is 300 °C, phase separation begins to occur in the film. A large amount of excessive Te existing in the thin film in the film may enter the crystal lattice of Sb_2Te_3 to recrystallize, and the degree of crystallization is intensified, but it also affects the structure of the film. Therefore, excessively high temperature annealing may increase the defects of the film, increase the influence of factors such as thermal expansion and lattice matching, and cause the quality of the Sb_2Te_3 film to decrease (Fig. 5.6).

The grain size can be obtained according to the Scherrer formula [1]: $D = K\lambda / FWHM \cos\theta$, where (D is the grain size (nm), K is a constant, the value is 0.89, λ is the X-ray wavelength (nm), and Cu is XRD during the experiment The wavelength of Kα ray (λ = 1.5406 Å), FWHM is the half-width of the peak (rad), and θ is the diffraction angle (rad).

The relationship between the annealing temperature shown in Table 5.1 and the grain size of the Sb_2Te_3 film on the (1010) plane was obtained according to the Scherrer formula. It can be seen from Table 5.1 that the crystallinity of the film

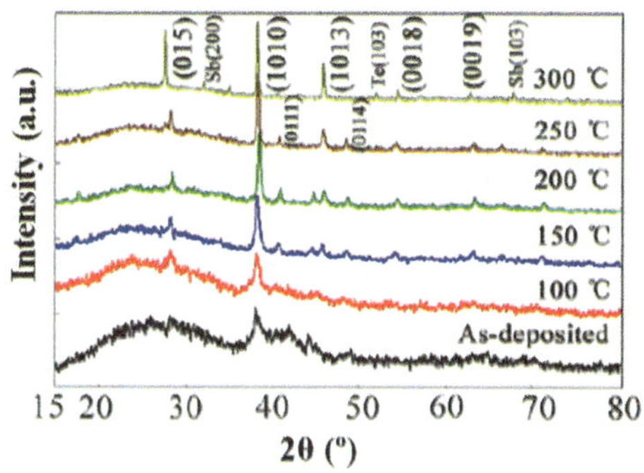

Fig. 5.6 XRD patterns of Sb_2Te_3 films before and after annealing

Table 5.1 The relationship between the grain size of the film (1010) surface and the annealing temperature

Annealing Temp. (°C)	FWHM *B*(°)	*B*cos*θ*	Average grain size *D* (nm)	Carrier mean *Free Path* *l* (nm)
–	1.413	0.02316	5.9	0.57
100	1.307	0.02142	6.4	1.51
150	0.448	0.00710	19.3	4.47
200	0.310	0.00479	28.6	6.26
250	0.271	0.00410	33.5	8.24
300	0.199	0.00280	49.0	

deposited at room temperature is weak, and the grain size is about 5.9 nm. When the annealing temperature is 300 °C, the grain size increases to about 49.0 nm. The increase in annealing temperature gradually increases.

Figure 5.7 shows the relationship between the carrier concentration and mobility of the Sb_2Te_3 film and the annealing temperature. Due to the deterioration of the film quality at higher temperatures, the carrier concentration and mobility of the samples were not tested at the annealing temperature of 300 °C. It can be seen from Fig. 5.8 that the carrier concentration of the thin film deposited at room temperature is 13×10^{19} cm^{-3}, and the mobility is 5.53 cm^2 V^{-1} s^{-1}. When the annealing temperature is 250 °C, the carrier concentration of the film decreases to 7.71×10^{19} cm^{-3}, and the mobility increases to about 95 cm^2 V^{-1} s^{-1}. The carrier concentration of the film decreases as the annealing temperature increases, while the carrier mobility of the film increases as the annealing temperature increases. The carrier mobility increases with the increase of annealing temperature, which can be mainly considered from two aspects, namely defect scattering and grain boundary scattering. After annealing

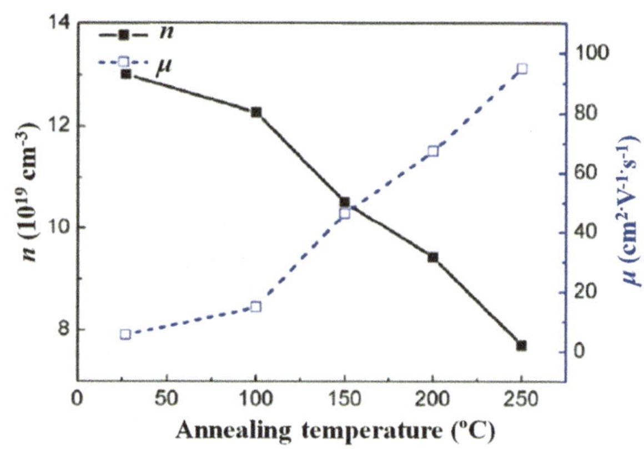

Fig. 5.7 Relationship between the carrier concentration and mobility of the Sb_2Te_3 film and the annealing temperature

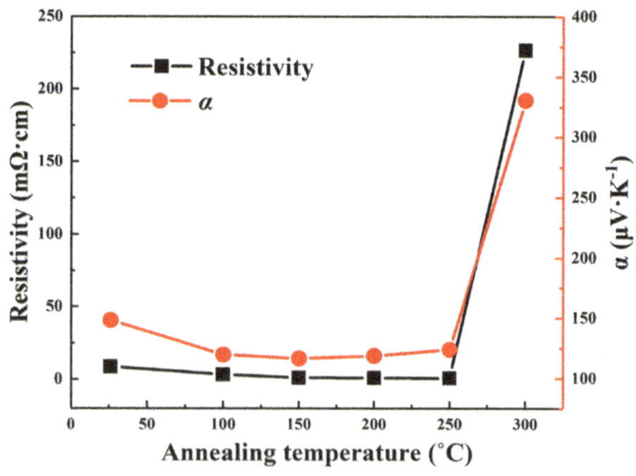

Fig. 5.8 Relationship between electrical properties and annealing temperature of Sb_2Te_3 thin films

treatment, defects in the film are reduced, and carriers are less likely to be scattered by defects during transport. In addition, the increase in grain size reduces the density of grain boundaries in the film, and carriers are trapped at the grain boundaries during transport. The probability of scattering is reduced, which contributes to an increase in carrier mobility.

The relationship between the mean free path of the Sb_2Te_3 film carriers and the annealing temperature is shown in Table 5.1. As can be seen from the table, the average free path of the thin-film carrier deposited at room temperature is only 0.57 nm. As the annealing temperature increases, the average free path of the carriers increases, which is consistent with expectations. On the other hand, one may note that the mean free path of the carriers is smaller than the grain size of the film. This may be because some defects and ionized impurities still exist in the annealed film, and their scattering of carriers makes the mean free path of the carriers of the film smaller than the grain size of the film.

Figure 5.8 shows the relationship between the electrical properties of Sb_2Te_3 films and the annealing temperature. It can be seen from Fig. 5.9a that the resistivity of the Sb_2Te_3 film deposited at room temperature is about 8.66 mΩ cm, and the resistivity is significantly reduced after annealing treatment. When the annealing temperature is further increased from 150 °C, the tendency of the resistivity to decrease is slowed down, and the minimum value is 0.85 mΩ cm when the annealing temperature is 250 °C. The decrease in resistivity is mainly due to the increase in crystallinity of the thin film and the increase in grain size. When the annealing temperature is further increased from 250 °C, the resistivity of the film increases significantly. This may be due to the significant change in the composition of the thin film after high temperature annealing, which deteriorates the quality of the Sb^2Te^3 thin film, which in turn leads to a sharp increase in the resistivity of the thin film. It can be seen from Fig. 5.9b that the Seebeck coefficients of all the samples are positive values, indicating that the Sb_2Te_3 films prepared are p-type. The Seebeck coefficient (148.02 μV K^{-1}) of the films deposited at room temperature is high because the

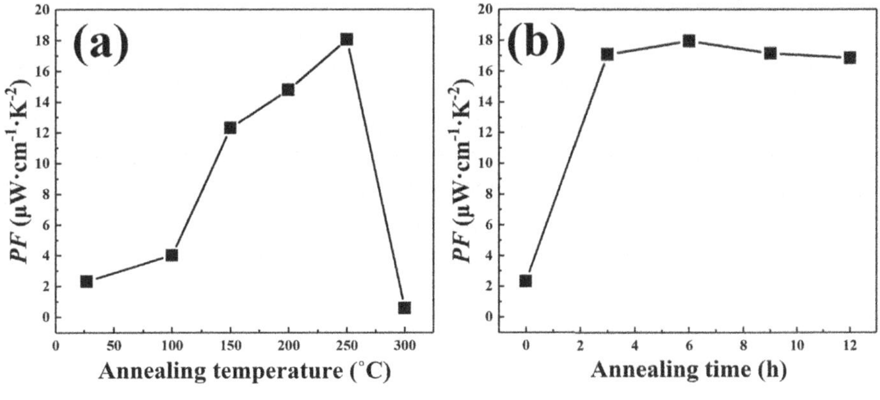

Fig. 5.9 Power factor of Sb_2Te_3 film before and after annealing

crystallinity of the films deposited at room temperature is not high and is almost amorphous. The amorphous Sb_2Te_3 film has a high Seebeck coefficient, and the Seebeck coefficient of the thin film is significantly reduced after annealing, mainly because the film changes from amorphous to polycrystalline. When the annealing temperature was further increased from 150 °C, the Seebeck coefficient of the film increased slightly. For P-type thermoelectric materials, the Seebeck coefficient is inversely proportional to the logarithm of the carrier concentration. Therefore, the increase in the Seebeck coefficient is due to the decrease in the carrier concentration. When the annealing temperature is increased to 300 °C, the Seebeck coefficient of the film increases sharply to about 330.7 $\mu V\ K^{-1}$, which is equivalent to that of the Te material. It can be seen from Fig. 5.4d that a large amount of Sb-rich crystals is precipitated, but the crystals are randomly distributed on the surface of the film and do not form a continuous structure. At this time, the main role may be Te in the film, so the samples under this condition have higher resistivity and Seebeck coefficient.

It can be seen from Fig. 5.9a that the power factor of the Sb_2Te_3 film deposited at room temperature is about 2.5 $\mu W\ cm^{-1}\ K^{-2}$, and the performance of the film is poor. After annealing, the thermoelectric properties of the film were significantly enhanced. When the annealing temperature is 250 °C, the power factor of the film reaches a maximum of 18.09 $\mu W\ cm^{-1}\ K^{-2}$. It can be seen from Fig. 5.9b that the annealing time has no significant effect on the properties of the film. Therefore, under the experimental conditions, the annealing temperature is considered to be the main factor affecting the thermoelectric properties of the thin film.

5.1.3 Regulating Film Thickness

Thickness is an important parameter that affects the quality of thin films and has a certain impact on the properties of thin film materials. In addition, when manufacturing thin film devices using micro-machining processes, the thickness of functional materials has a significant impact on device performance. In this section, the effect of thickness on the microstructure and thermoelectric properties of antimony telluride films is mainly studied. Radio frequency magnetron sputtering was used to deposit Sb_2Te_3 films on silicon substrates and glass substrates, respectively, to explore the effect of thickness on the structure and thermoelectric properties of the films. During the experiment, ensure that other process parameters are unchanged. During the deposition process, the time for changing the film deposition is 40, 60, 85, 100, and 120 min, respectively. In order to optimize the properties of the film, the prepared samples were subjected to post-annealing. When the annealing temperature is 250 °C, the Sb_2Te_3 film has good thermoelectric properties. When Sb_2Te_3 thin film samples with different thicknesses were post-annealed, the annealing temperature was selected to be 250 °C, and the annealing time was 6 h. Finally, the effect of thickness on its properties was studied.

Table 5.2 Thickness and composition of Sb_2Te_3 films at room temperature as a function of deposition time

Sample	Sputtering time (min)	Thickness (nm)	Te content (at.%)	Grain size (nm)
1#	40	90	66.2	3.1
2#	60	140	65.9	3.6
3#	85	190	66.3	4.4
4#	100	230	66.1	5.9
5#	120	270	65.8	6.2

5.1.3.1 Effect of Thickness on Its Structure and Surface Topography

The thickness of the film is controlled by adjusting the film deposition time. Table 5.2 shows the relationship between the thickness and composition of the obtained film and the deposition time. As can be seen from the table, the thickness of the film and the deposition time are approximately linear. In addition, there is no obvious rule between the content of Te in the film and the deposition time. Considering the test error, the Te content of the film deposited at room temperature is about 66%, and the film composition is rich in Te, which deviates from the chemical composition ratio. This may be due to the higher sputtering yield of Te than that of Sb at room temperature. It can also be seen that as the thickness of the film increases the grain size of the film gradually increases.

As shown in Fig. 5.10, Sb_2Te_3 films of different thicknesses deposited at room temperature can be seen that as the thickness of the film increases the grains on the surface of the film increase and the size increases. At the beginning of thin film deposition, since the substrate is at a relatively low temperature (room temperature), the atom diffusion ability from the target surface to the substrate is weak, forming a large number of flat and dense microcrystals as shown in Fig. 5.10a. With the extension of the sputtering time, the relative temperature of the substrate increases under the irradiation of the sputtering glow, the film is heated the atoms get enough energy and time to diffuse and recrystallize and the small crystal grains in the film fuse with each other. As the film grows, the film grain size increases slightly.

5.1.3.2 Effect of Thickness on Its Thermoelectric Performance

The relationship between the electrical properties and film thickness of Sb_2Te_3 films before and after annealing is shown in Fig. 5.11. It can be seen from Fig. 5.11a that when the film thickness increases from 90 to 230 nm, the resistivity of the film decreases from 27.25 to 8.66 mΩ cm; as the thickness increases, the resistivity of the film gradually decreases. When the thickness is further increased, the tendency of the resistivity to decrease is slowed and the minimum value is 8.38 mΩ cm at a

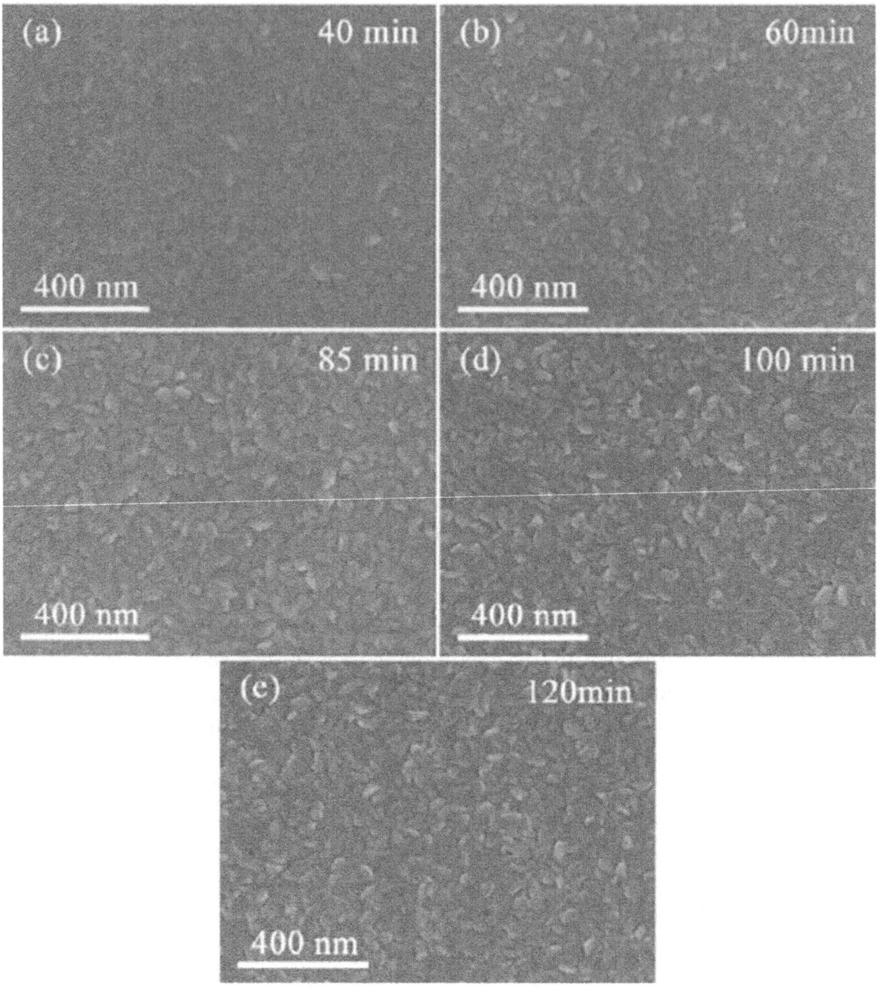

Fig. 5.10 SEM images of Sb_2Te_3 films of different thicknesses deposited at room temperature

thickness of 270 nm. On the other hand, after post-annealing (250 °C), the resistivity of Sb_2Te_3 films with different thicknesses decreased significantly and the resistivity of the annealed Sb_2Te_3 films still decreased with increasing thickness. After the Sb_2Te_3 film with a thickness of 270 nm is annealed, the resistivity of the film has a minimum value of 0.77 mΩ cm. The resistivity of the film deposited at room temperature is higher than that after annealing. There are three main factors, namely surface scattering and internal microcrystalline grain boundary scattering and defects (point defects, line defects). When the thickness of the film is thin, in addition to the strong scattering of carriers by the microcrystalline grain boundaries and defects inside the film, the surface scattering of the film is strong, so the resistivity of the film

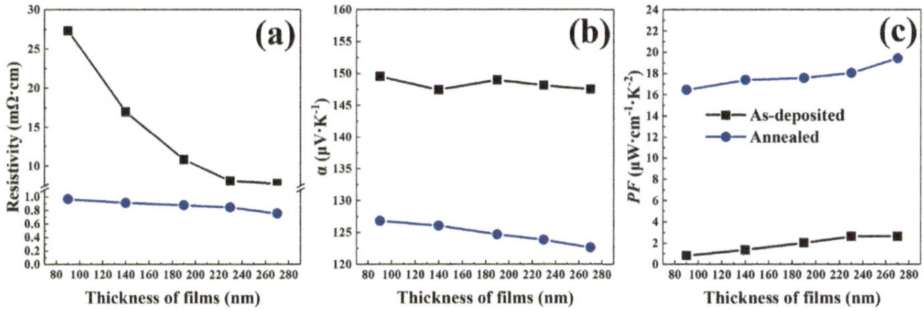

Fig. 5.11 The relationship between the thermoelectric properties and thickness of Sb_2Te_3 film before and after annealing: **a** resistivity, **b** seebeck coefficient, **c** power factor

is high. With the increase of thickness, under the irradiation of sputtering glow for a long time, the atoms get enough energy and time to diffuse, the density of the film increases, the grains grow and the density of grain boundaries decreases. Surface scattering is reduced, which reduces resistivity. During the annealing process, the diffusion and recrystallization of atoms and the growth of grains are significantly enhanced. At the same time, the degree of defects (vacancies, dislocations, etc.) in the film is significantly reduced, and the factors that cause carrier scattering are reduced. The resistivity is significantly reduced after annealing. Figure 5.11b shows the relationship between Seebeck coefficient and film thickness of Sb_2Te_3 films before and after annealing. It can be seen from the figure that the Seebeck coefficients of the films under different thicknesses are basically the same, and there is no obvious change rule. After annealing (250 °C), the Seebeck coefficient of the film decreases slightly with increasing film thickness. It is worth noting that the Seebeck coefficient of the annealed film is relatively lower than that deposited at room temperature. After annealing, the Seebeck coefficient of the films decreased significantly, mainly due to the transition from amorphous to polycrystalline Sb_2Te_3 films.

Figure 5.11c shows the relationship between the power factor and thickness of the Sb_2Te_3 film before and after annealing. It can be seen that the power factor of the film deposited at room temperature gradually increases with increasing thickness. When the film thickness increases from 90 nm to 270 nm, the power factor of the film increases from 0.82 to 2.60 $\mu W\ cm^{-1}\ K^{-2}$. After annealing (250 °C) optimization, the power factor of Sb_2Te_3 films with different thicknesses increased significantly. The maximum power factor of the 270 nm-thin film after annealing at 250 °C is 19.55 $\mu W\ cm^{-1}\ K^{-2}$. It can be seen that, compared with the annealing treatment, the influence of the film thickness on the thermoelectric properties of the material is weak.

5.2 Controlling Sputtering Power, Annealing, Thickness, Substrate Temperature to Deposit Bi–Sb–Te-Based Films

Bi–Sb–Te ternary thin films can be formed by substituting Bi atoms in Sb_2Te_3 thin films, and the thermoelectric properties of such thin films at room temperature are also excellent. In this section, magnetron sputtering is used to prepare Bi–Sb–Te series thin films, and the power, annealing conditions, film thickness and substrate temperature of sputtering are controlled to deposit Bi–Sb–Te series ternary films. By changing the conditions of magnetron sputtering, adjusting the structure of the film, exploring the factors that affect its thermoelectric performance, so as to improve the process and improve the film thermoelectric performance.

5.2.1 Controlling Sputtering Power

5.2.1.1 Film Preparation

In this section, radio frequency magnetron co-sputtering was used to deposit bismuth-tellurium-antimony alloy thin films on silicon and glass substrates to explore the effects of target sputtering power on the structure and thermoelectric properties of the films. During the experiment, the sputtering power of the Sb_2Te_3 alloy target was fixed at 20 W, and the sputtering power of the Bi alloy target was adjusted to 3, 4, and 5 W, respectively. The thickness of the film was monitored in real time through a quartz crystal resonator plate. Finally, the thickness of the film was measured by a step scanner to about 240 nm. In order to optimize the properties of the film, the prepared samples were post-annealed at an annealing temperature of 100 °C, 150 °C, 200 °C, 250 °C, and 300 °C, respectively, and an annealing time of 6 h for lateral comparison with the prepared Sb_2Te_3 film. At the same time, XRD, SEM, EDS and other analytical methods were used to characterize the prepared films, and their electrical properties were tested using homemade equipment.

5.2.1.2 Thin Film Structure

Figure 5.12 is the distribution of each element in the Bi–Sb–Te alloy thin film when the Bi target sputtering power is 3 W. It can be seen from Fig. 5.12b and c that the Sb and Te elements in the prepared alloy film have a relatively uniform distribution. It is worth noting that in Fig. 5.12a, it can be found that some bismuth nanocrystals are randomly distributed in the film, and the macroscopic distribution is more uniform, which indicates that co-sputtering is used to prepare a more uniform doping at room temperature alloy film.

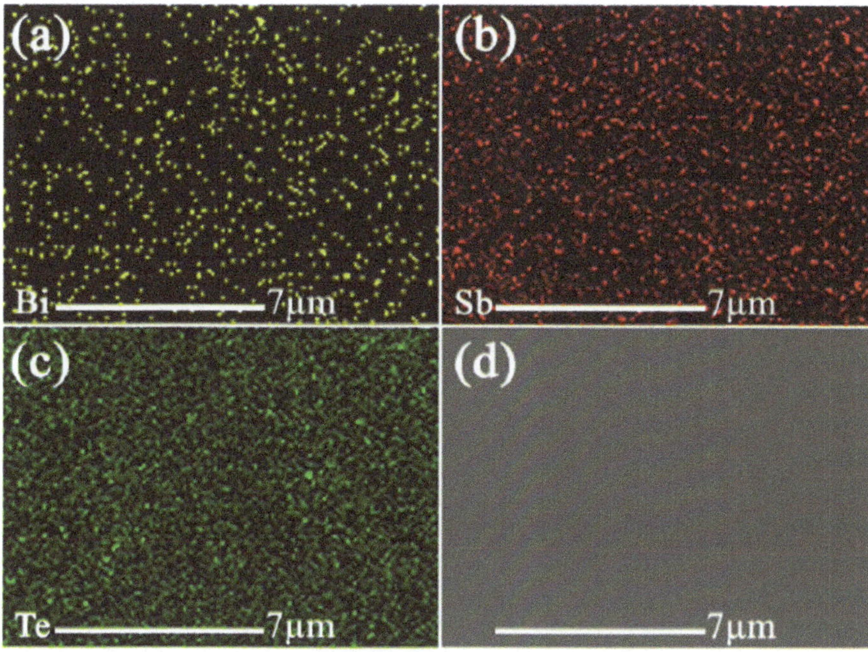

Fig. 5.12 Distribution diagram of each element in a Bi–Sb–Te alloy film with a Bi target power of 3 W **a** Bi, **b** Sb, **c** Te, **d** Bi–Sb–Te alloy film SEM

Figure 5.13 are SEM images of Bi–Sb–Te alloy films at different Bi target powers. It can be seen that the surface morphology of the films deposited under different Bi target powers does not change much, the film surface is flat and dense and some micro-grains are randomly distributed. With reference to Fig. 5.12, it is considered that the main component of these microcrystals may be elemental bismuth. With the increase of the sputtering power of the Bi target, the energy of the bismuth atom sputtered from the target surface to the substrate increases, the atom diffusion ability is enhanced, the nucleus fuses with each other and grows up, and the size of the nano-crystal grains increases slightly.

Figure 5.13 shows the effect of different Bi target powers on the composition of Bi–Sb–Te alloy films. From Fig. 5.13, when the Bi target sputtering power is 3 W, the composition of Bi in the alloy thin film is about 8.9%, the composition of Sb is about 30.2%, and the composition of Te is about 60.9%. With the increase of the Bi target power, the Bi content in the alloy film increases. This is because when the power of the Bi target is 3–5 W, the sputtering yield and the target power increase linearly due to the lower target power; as the Bi target power increases, the output of the Bi atom increases, so in the alloy film the composition of Bi increases (Fig. 5.14).

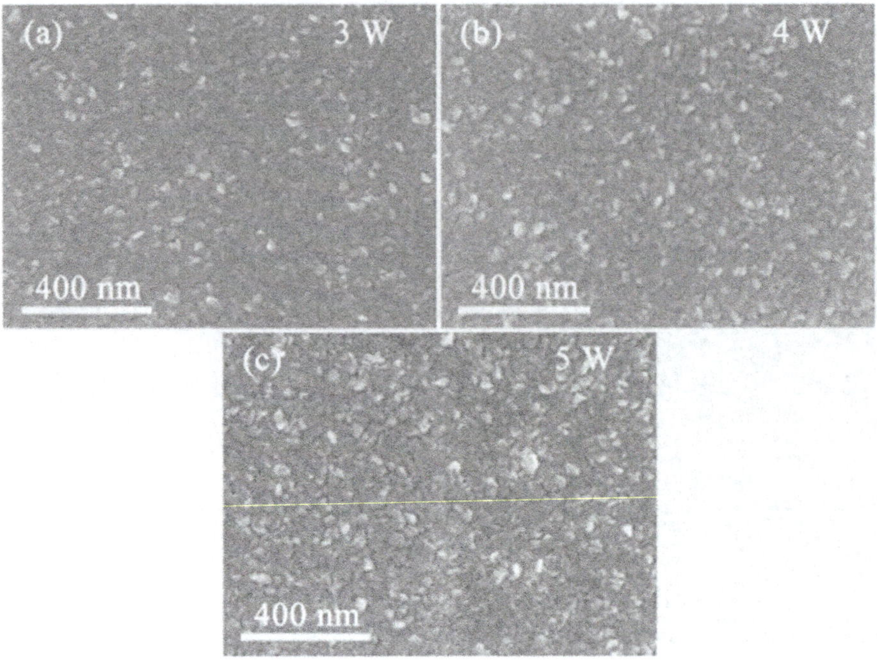

Fig. 5.13 SEM images of Bi–Sb–Te alloy films with different Bi target powers **a** 3 W, **b** 4 W, and **c** 5 W

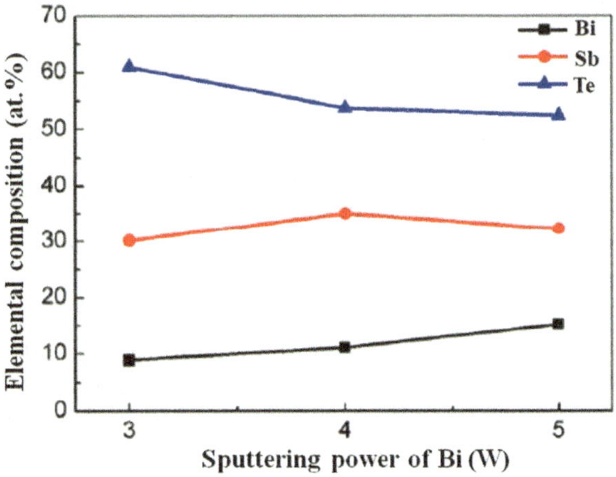

Fig. 5.14 Relationship between the composition of each element in the Bi–Sb–Te alloy film and the sputtering power of the Bi target

5.2.1.3 Thermoelectric Performance

Figure 5.15 shows the electrical properties of Bi–Sb–Te alloy films at different Bi target sputtering powers. It can be seen from Fig. 5.15a that with the increase of RF

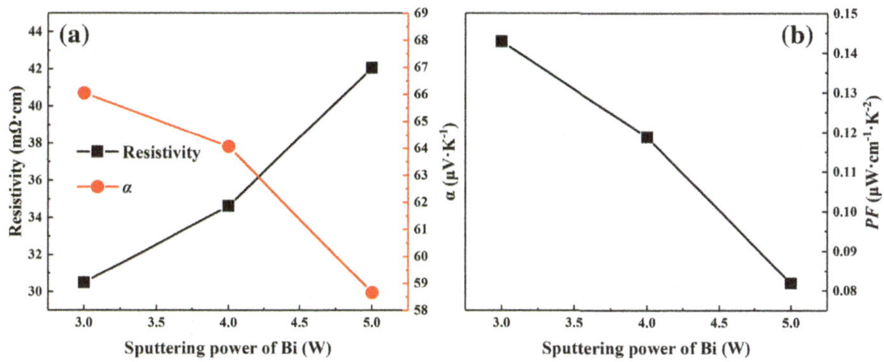

Fig. 5.15 Thermoelectric properties of Bi–Sb–Te alloy thin films at different Bi target sputtering powers: **a** resistivity and Seebeck coefficient, **b** power factor

power, the resistivity of the film gradually increases. When the Bi target sputtering power is increased from 3 to 5 W, the resistivity of the film increases from 30.47 to 42.02 mΩ cm. Although with the increase of the sputtering power of the Bi target, the content of the Te component in the alloy thin film is reduced, which is beneficial to the reduction of the film resistivity; but at the same time, the doping amount of Bi is also increased, which increases the grain boundary interface in the alloy thin film and increases. This reduces the interface scattering of the carriers, which increases the resistivity. On the other hand, the element Bi itself has poor conductivity, and the addition of elements with poor conductivity may also lead to an increase in the resistivity of the film. It can be seen from Fig. 5.15b that the Seebeck coefficients of all samples are positive, indicating that the Bi–Sb–Te alloy films prepared are all p-type. When the Bi target sputtering power is increased from 3 W to 5 W, the Seebeck coefficient of the alloy thin film is reduced from 66.1 to 58.73 μV K^{-1}. The Seebeck coefficient of the thin film increases with the Bi target sputtering power gradually decreases.

The power factors of Bi–Sb–Te alloy films at different Bi target sputtering powers are shown in Fig. 5.15b. It can be seen that the power factor of the thin film gradually decreases as the sputtering power of the Bi target increases. When the Bi target sputtering power is increased from 3 W to 5 W, the power factor of the alloy thin film decreases from 0.14 to 0.08 μW cm^{-1} K^{-2}. It is worth noting that, compared with the Sb_2Te_3 film, the Bi–Sb–Te alloy film prepared by co-sputtering at room temperature has poor thermoelectric properties. This is mainly because the alloy film has a higher resistivity than the Sb_2Te_3 film, which makes the alloy film's thermoelectric performance worse.

5.2.2 Controlling the Annealing Process

In order to further optimize the properties of the film, a sample with a Bi target RF power of 3 W was selected and placed in a well annealing furnace for post-annealing. The annealing temperature was 100, 150, 200, 250, 300, 350 °C, and the annealing time was 6 h. The effects of annealing on the morphology and properties of the films were studied.

Figure 5.16 is the XRD pattern of Bi–Sb–Te alloy film before and after annealing. It can be seen from the figure that the alloy film deposited at room temperature has a weak diffraction peak at 2θ of 41.61°, and the peak envelope is larger. Comparing Bi's PDF (65.6203) card, it can be seen that the diffraction peak is the (220) crystal plane of Bi. This may be due to the strong diffusion ability of bismuth atoms during the deposition of the thin film, and the atoms diffused and condensed to form bismuth nanocrystals. When the annealing temperature is 150 °C, the main diffraction peak at 2θ is 38.29° coincides with the (1010) crystal plane of $Bi_{0.5}Sb_{1.5}Te_{3.0}$, indicating that the bismuth-tellurium-antimony alloy thin film begins to form. This is because the annealing process intensifies the interaction between atoms, and the atoms diffuse to condense and form an alloy thin film. When the annealing temperature is 250 °C, the diffraction peak in the XRD pattern is further enhanced, and the peak envelope is almost disappeared, indicating that the crystallinity of the film is further enhanced. It is worth noting that a new diffraction peak appears at 2θ of 17.47°, 44.61°, and 54.15°, corresponding to the (006), (0015), and (0018) crystal planes of $Bi_{0.5}Sb_{1.5}Te_{3.0}$, respectively. It shows that the alloy thin film begins to form a layered polycrystalline structure. With the further increase of the annealing temperature, the diffraction peaks of the film along the (001) direction are significantly enhanced, indicating that the films preferentially grow along the (001) direction to form a layered structure. In addition, the average size of the crystal grain size of the film can be calculated according to the Scherrer formula, and the results are shown in Table 5.3 from which the grain size of the film gradually increases with the increase of the annealing temperature.

Figure 5.17 shows the surface morphology and cross section of the Bi–Sb–Te film before and after annealing. It can be seen from Fig. 5.17a that the surface of

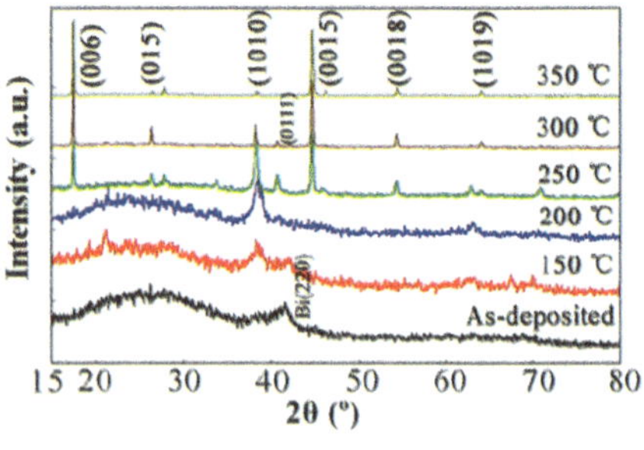

Fig. 5.16 XRD patterns of Bi–Sb–Te alloy films before and after annealing

Table 5.3 The relationship between the grain size of the film and the annealing temperature

Annealing temperature (°C)	FWHM *B* (°)		*B*sos*θ*		Average grain size (nm)	
	(1010)	(006)	(1010)	(006)	(1010)	(006)
As-deposited	–					
150	1.012	–	0.01668	–	8.2	–
200	0.865	–	0.01426	–	9.6	–
250	0.391	0.144	0.00644	0.00248	21.3	54
300	–	0.107	–	0.00184	–	75.3

the film deposited at room temperature is relatively flat and dense, and some nanocrystal grains are randomly distributed on the surface. As the annealing temperature increases, the grain size of the alloy film increases. When the annealing temperature is increased to 300 °C, the grain size increases to about 80 nm, which is consistent with the XRD test results. It is worth noting that, compared with Sb_2Te_3 film, Bi–Sb–Te alloy film does not significantly deteriorate the surface of the film after high temperature annealing at 300 °C, which indicates that the proper amount of Bi is beneficial to the thermal stability of the film. After further increasing the annealing temperature, the film quality began to deteriorate, as shown in Fig. 5.17e. Nanocrystals were deposited on the film surface, and the film surface was spotted. Analysis of the composition of EDS shows that the precipitated nanocrystals are rich in Sb, and the composition ratio of Sb to Te in the precipitated crystals is close to 89:11, which may be due to the re-evaporation of Te atoms in the annealing process, and the Sb atoms in the Grain boundaries tend to diffuse and condense, so that they precipitate from the film surface at grain boundaries. Figure 5.18 is a graph showing the content of each element in the alloy film before and after annealing. It can be seen from the figure that when the sputtering power of the Bi target is 3 W, the content of Bi element in the sample deposited at room temperature is 8.9%, the content of Sb element is 30.2%, and the content of Te element is 60.9%. As the annealing temperature increased, the content of each element in the film did not change significantly, indicating that the prepared alloy film has better thermal stability. When the annealing temperature was increased to 350 °C, with the precipitation of Sb-rich crystals, the composition of Sb and Te in the film changed significantly, and the composition of Te in the film increased significantly.

Figure 5.19 shows the relationship between the carrier concentration and mobility of the Bi–Sb–Te alloy film and the annealing temperature. Since the film quality deteriorated at higher temperatures, the carrier concentration and mobility of the samples were not tested at the annealing temperature of 350 °C. It can be seen from Fig. 5.19 that the carrier concentration of the Bi–Sb–Te alloy film deposited at room temperature is 18.79×10^{19} cm^{-3}, and the mobility is 1.14 cm^2 V^{-1} s^{-1}. Compared with Sb_2Te_3 films deposited at room temperature, alloy films have higher carrier concentration and lower mobility. This is mainly because the doped bismuth nanocrystals greatly increase the grain boundary interface in the film, and the probability of carriers

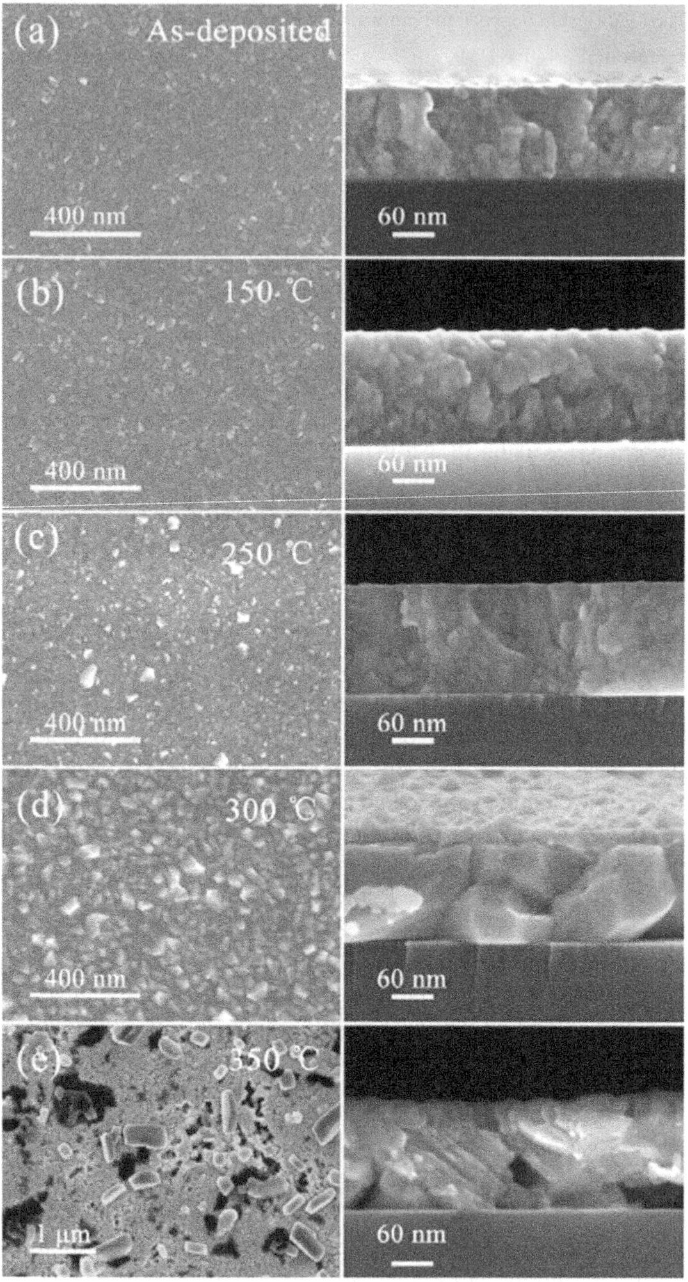

Fig. 5.17 SEM surface and cross-sectional views of Bi–Sb–Te alloy films before and after 20 annealed **a** As-deposited, **b** 150 °C, **c** 250 °C, **d** 300 °C, **e** 350 °C

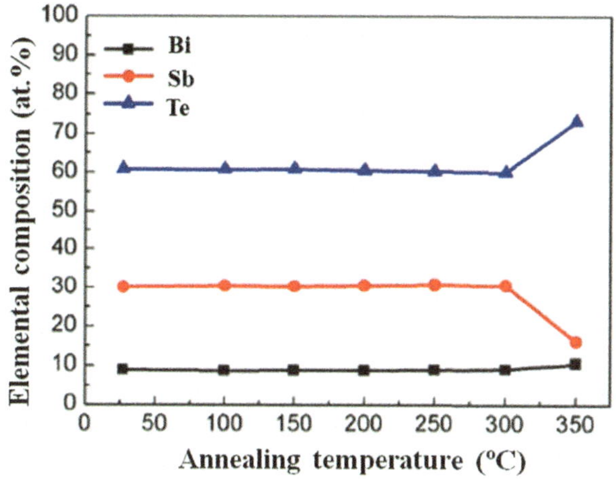

Fig. 5.18 Relationship between the composition of each element in the Bi–Sb–Te alloy film and the annealing temperature

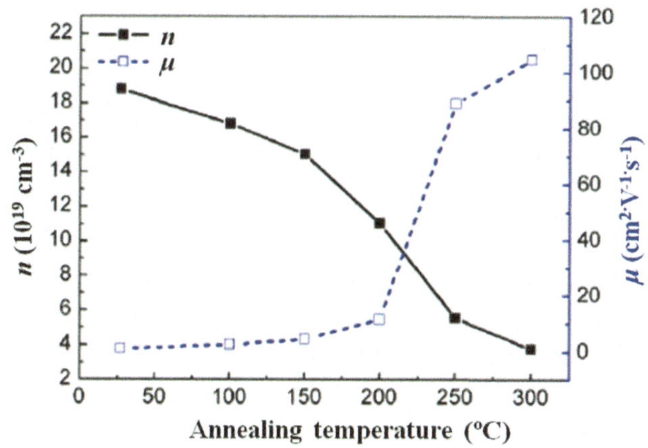

Fig. 5.19 Relationship between carrier concentration and mobility and annealing temperature of Bi–Sb–Te alloy thin films

being scattered at the grain boundaries during transport increases. Therefore, the Bi–Sb–Te alloy film deposited at room temperature has a low carrier mobility. As the annealing temperature increases, the carrier concentration of the alloy film decreases and the carrier mobility increases. When the annealing temperature is 300 °C, the carrier concentration of the film is reduced to approximately 3.76×10^{19} cm^{-3}, and the mobility is increased to approximately 104.8 cm^2 V^{-1} s^{-1}. The decrease in the carrier concentration is mainly due to the reduction of defects in the film by the annealing treatment. The significant increase in carrier mobility is mainly due to the

Table 5.4 Relationship between the mean free path of the film and the annealing temperature

Annealing temperature (°C)	Mean free path of carrier *l* (nm)	
	Sb_2Te_3 薄膜	Bi–Sb–Te film
As-deposited	0.57	0.13
100	1.51	0.29
150	4.47	0.48
200	6.26	1.12
250	8.24	6.91
300	–	7.15

increase in grain size and the decrease in grain boundary density; the reduction of defects in the film also contributes to the improvement of mobility.

The average free path of the carriers was calculated from the carrier concentration and mobility of the film. Table 5.4 shows the relationship between the mean free path of the carriers of Sb_2Te_3 film and Bi–Sb–Te alloy film and the annealing temperature. It can be seen from the table that the average free path of the carrier of the Bi–Sb–Te alloy film deposited at room temperature is only 0.13 nm, which is much smaller than the 0.57 nm of the Sb_2Te_3 film. This is mainly because the doped bismuth nanocrystals increase the carrier concentration of the thin film and also increase the interface in the thin film, so that the probability of collision and scattering of the carriers during transportation is increased, so that for Bi–Sb–Te the average free path of the carrier of the alloy film is low, which is in line with expectations.

Figure 5.20 shows the relationship between the electrical properties of Bi–Sb–Te alloy films and the annealing temperature. It can be seen from Fig. 5.20a that the resistivity of the Bi–Sb–Te alloy film deposited at room temperature is 30.5 mΩ cm, which is significantly higher than that of the Sb_2Te_3 film, which is mainly due to the incorporation of the alloy film for bismuth nanocrystals, the probability of carriers being scattered during transportation is greatly increased, which causes the resistivity

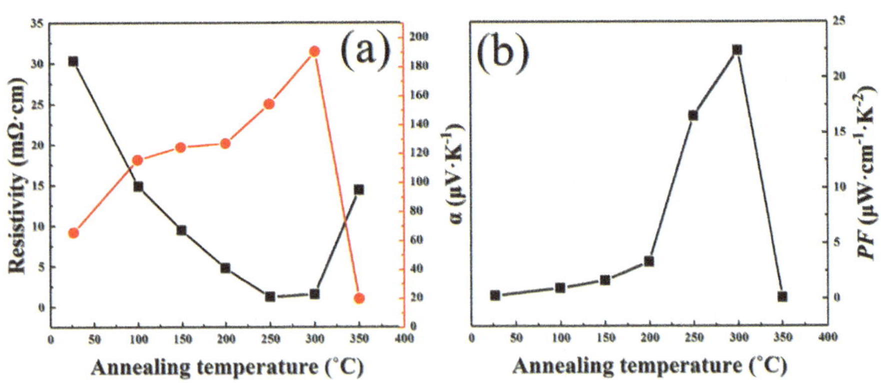

Fig. 5.20 The relationship between the thermoelectric properties of Bi–Sb–Te alloy film and the annealing temperature: **a** resistivity and Seebeck coefficient, **b** power factor

of the alloy film to increase. After annealing treatment, the resistivity of the alloy film is significantly reduced, which is mainly due to the increase of the crystallinity of the film. As the annealing temperature increases further from 250 °C, the resistivity increases. When the annealing temperature is 250 °C, the minimum resistivity of the Bi–Sb–Te alloy film is 1.4 mΩ cm, which is higher than that of the Sb_2Te_3 film (0.85 mΩ cm). When the annealing temperature is 350 °C, the resistivity of the film increases significantly. This phenomenon may be caused by the significant change in the composition of the alloy film after high temperature annealing. It can be seen from Fig. 5.20a that the Seebeck coefficient of the film deposited at room temperature is 66.1 $\mu V\ K^{-1}$, and the Seebeck coefficient of the film after annealing increases, and it gradually increases with the increase of the annealing temperature. This is because there is a close relationship between the Seebeck coefficient and the carrier concentration of thermoelectric materials. From Fig. 5.19, it can be seen that as the annealing temperature increases, the carrier concentration in the film decreases, and the Seebeck coefficient of the film varies with the annealing temperature increases. When the annealing temperature is 300 °C, the Seebeck coefficient of the film has a maximum value of 190.6 $\mu V\ K^{-1}$. Compared with the Sb_2Te_3 film, the Seebeck coefficient of the annealed Bi–Sb–Te alloy film increased significantly. When the annealing temperature is increased to 350 °C, the Seebeck coefficient of the alloy film decreases sharply to about 19.9 $\mu V\ K^{-1}$, which may be related to the composition change of the alloy film.

The power factor of the film can be calculated based on the resistivity and Seebeck coefficient. The calculation results are shown in Fig. 5.20b. The power factor of Bi–Sb–Te alloy films deposited at room temperature is about 0.14 $\mu W\ cm^{-1}\ K^{-2}$, which is lower than that of Sb_2Te_3 films. After annealing, the thermoelectric properties of the Bi–Sb–Te alloy film were significantly enhanced. When the annealing temperature is 300 °C, the power factor of the film reaches a maximum of 22.54 $\mu W\ cm^{-1}\ K^{-2}$. It can be seen that by optimizing the annealing temperature, Bi–Sb–Te alloy films doped with a proper number of Bi nanocrystals have better thermoelectric properties and thermal stability than Sb_2Te_3 films.

5.2.3 Control Thickness

In this section, the effects of thickness on the microstructure and thermoelectric properties of Bi–Sb–Te alloy films are mainly studied. During the experiments, the sputtering powers of the fixed Sb_2Te_3 alloy and Bi alloy targets were 20 W and 3 W, respectively. Bi–Sb–Te alloy films were deposited on silicon substrates and glass substrates by radio frequency magnetron co-sputtering. Other process parameters remain unchanged. By adjusting the film deposition time to be 40, 60, 80, 105, 125 min, respectively, alloy films of different thicknesses can be prepared. In order to optimize the properties of the film, the prepared samples were subjected to post-annealing. According to the previous section, the alloy film has good thermoelectric properties when the annealing temperature is 300 °C. When alloy thin film samples

Table 5.5 Relationship between thickness and composition of Bi–Sb–Te alloy films and deposition time

Sample	Sputtering time (min)	Thickness (nm)	Elemental composition (at.%)		
			Bi	Sb	Te
1#	40	90	8.4	29.0	62.6
2#	60	140	8.8	29.4	61.8
3#	80	180	8.8	29.9	61.3
4#	105	240	8.9	30.2	60.9
5#	125	280	9.1	30.1	60.8

with different thicknesses are post-annealed, the annealing temperature is selected to be 300 °C, and the annealing time is 6 h. At the same time, the prepared alloy thin films were characterized by XRD, SEM and other analytical methods, and their electrical properties were tested using homemade equipment.

5.2.3.1 Thin Film Structure

The thickness of the film is controlled by adjusting the film deposition time. The thickness and composition of the prepared Bi–Sb–Te alloy film are shown in Table 5.5. It can be seen from the table that as the thickness of the alloy film increases, the content of the element Bi in the film increases slightly, while the content of the element Te decreases. When the film thickness is 280 nm, the content of Bi in the film is 9.1%, the content of Sb is 30.1%, the content of Te is 60.8%, and the composition of the film is closest to the stoichiometric ratio $Bi_{0.5}Sb_{1.5}Te_{3.0}$.

Figure 5.21 are SEM images of Bi–Sb–Te alloy films with different thicknesses. It can be seen from the figure that the surface morphology of different thickness Bi–Sb–Te alloy films deposited at room temperature does not change much, the film surface is flat and dense and some bismuth nanocrystals are randomly distributed. As the thickness increases, the number of bismuth nanocrystals on the surface of the film increases and the size increases. With the increase of sputtering time, the relative temperature of the substrate increased under the irradiation of the sputtering glow, the film was heated, the atoms obtained sufficient energy and time to diffuse, and the small crystal grains in the film merged and grew up. Therefore, the number of nano crystal grains in the film increases and the size increases.

5.2.3.2 Thermoelectric Performance

Figure 5.22 shows the relationship between the electrical properties and film thickness of Bi–Sb–Te alloy films before and after annealing. From Fig. 5.22a, when the thickness of the film is 90 nm, the resistivity of the film is 61.42 mΩ cm; as the thickness increases, the crystallinity of the alloy film increases, defects decrease,

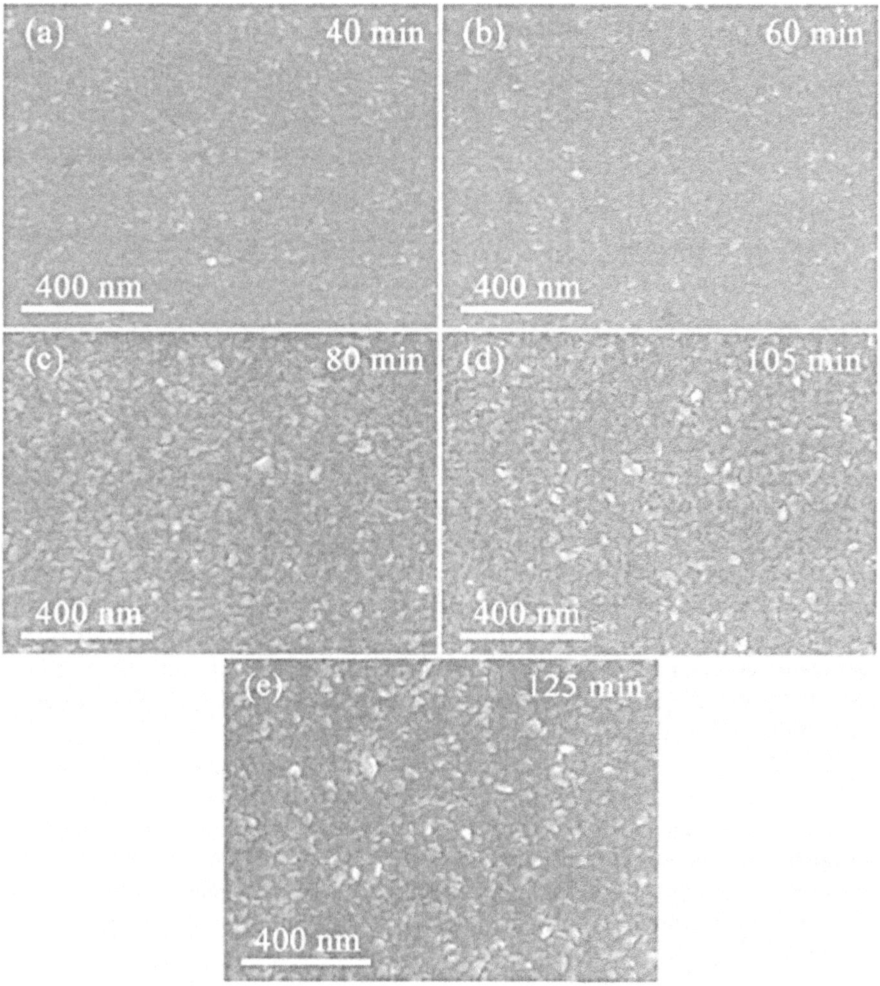

Fig. 5.21 SEM images of Bi–Sb–Te alloy films with different thicknesses

and the resistivity of the film gradually decreases. It is worth noting that the resistivity of Bi–Sb–Te alloy films is higher than that of Sb_2Te_3 films, which is consistent with expectations. Because the incorporation of bismuth nanocrystals increases the interface in the thin film, the probability of carriers being scattered during transport increases, and the average free path decreases, leading to an increase in resistivity. On the other hand, after annealing (300 °C) treatment, the resistivity of the film is significantly reduced, and the resistivity of the film is still reduced with increasing thickness. The main reason is that during the annealing process, the process of atom diffusion and recrystallization and grain growth is significantly enhanced, and the degree of defects inside the film is also significantly reduced, and the factors that

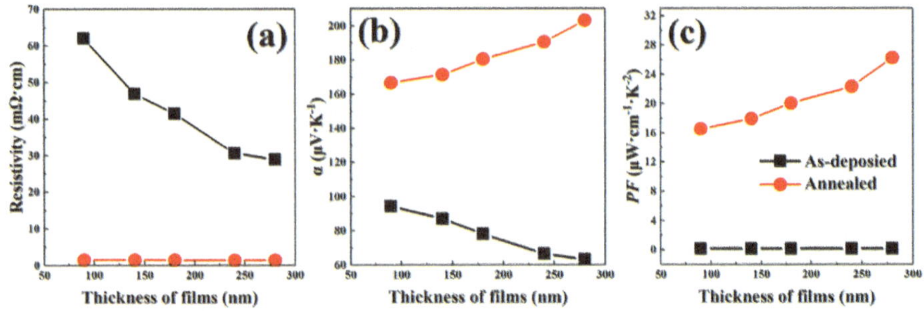

Fig. 5.22 The relationship between the thermoelectric properties and thickness of Bi–Sb–Te alloy film before and after annealing: **a** resistivity, **b** Seebeck coefficient, **c** power factor

cause carrier scattering are reduced, thereby reducing the resistivity of the film significantly reduced after annealing. After the film with a thickness of 280 nm is annealed at 300 °C, the resistivity of the film has a minimum value of 1.57 mΩ cm. It can be seen from Fig. 5.22b that when the thickness of the Bi–Sb–Te alloy film deposited at room temperature is 90 nm, the film's Seebeck coefficient is 93.9 $\mu V\ K^{-1}$. As the thickness increases, the Seebeck coefficient of the film gradually decreases. After annealing (300 °C), the Seebeck coefficient of the film increased significantly. When the thickness is 280 nm, the Seebeck coefficient has a maximum value of 203.82 $\mu V\ K^{-1}$. The increase in Seebeck coefficient is mainly due to the decrease in the carrier concentration of the film after annealing.

Figure 5.22c shows the relationship between the power factor and thickness of the Bi–Sb–Te alloy film before and after annealing. It can be seen that the Bi–Sb–Te alloy film deposited at room temperature has a low power factor and poor thermoelectric properties. After annealing (300 °C) optimization, the power factor of the film is significantly improved, and the thermoelectric performance is significantly improved. The maximum power factor of the 280 nm thick thin film after annealing is 26.41 $\mu W\ cm^{-1}\ K^{-2}$.

5.2.4 *Controlling Substrate Temperature*

The electrical properties of thermoelectric materials are greatly affected by temperature. Generally, there are mainly two kinds of temperatures that play an important role:post-annealing and in situ heating. In situ heating refers to heating the substrate to a set temperature and then depositing the film at this value. This temperature is also called the substrate temperature. During the film deposition process, the particles sputtered from the target surface will continue to move on the substrate surface with an initial kinetic energy after reaching the substrate surface. The proper substrate temperature will help the particles on the substrate surface to migrate to the optimal position, reduce the growth defects of the film, and improve the crystalline quality of the film. In addition, the temperature of the substrate has a great influence on the morphology and electrical properties of the thermoelectric film.

In this section, Bi–Sb–Te alloy films are deposited on silicon substrates and glass substrates by magnetron co-sputtering. During the sputtering process, the RF power of the fixed Sb_2Te_3 alloy target was 20 W and that of the Bi alloy target was 3 W. The substrate heating temperature range was 150–250 °C, while other process parameters were unchanged. In order to further optimize the properties of the film, the prepared samples were placed in a well annealing furnace for post-annealing. The annealing temperatures were 150, 200, 250, 300, and 350 °C, respectively, and the annealing time was 6 h. The effects of annealing on the morphology and properties of the films were studied.

5.2.4.1 Influence of Substrate Temperature on Its Structure and Surface Morphology

Figure 5.23 shows the XRD patterns of Bi–Sb–Te alloy films at different substrate temperatures. It can be seen from the figure that all the samples have a large steamed bread peak in the range of 2θ from 20° to 35°. For the Bi–Sb–Te alloy film deposited at room temperature, the diffraction peak at 2θ of 41.61° is consistent with the (220) crystal plane of Bi. For alloy films under substrate heating conditions, the (1010) crystal plane of the main diffraction peak $Bi_{0.5}Sb_{1.5}Te_{3.0}$ with 2θ of 38.29° coincides, which indicates that the crystal grains preferentially grow along the (1010) crystal plane and form Bi–Sb–Te alloy film. This is because the substrate heating has a similar annealing effect. The atoms flying from the target surface to the substrate have sufficient kinetic energy to diffuse and migrate under the action of the heated substrate. The atomic nuclei of bismuth, tellurium and antimony mutually grow up to form alloy thin films.

The relationship between the composition of each element in the Bi–Sb–Te alloy film and the substrate temperature is shown in Fig. 5.24. It can be seen that under the heating condition of the substrate, the content of Bi and Te in the alloy film is lower than that deposited at room temperature; as the substrate temperature increases, the content of Bi in the film increases and the content of Te gradually decreases.

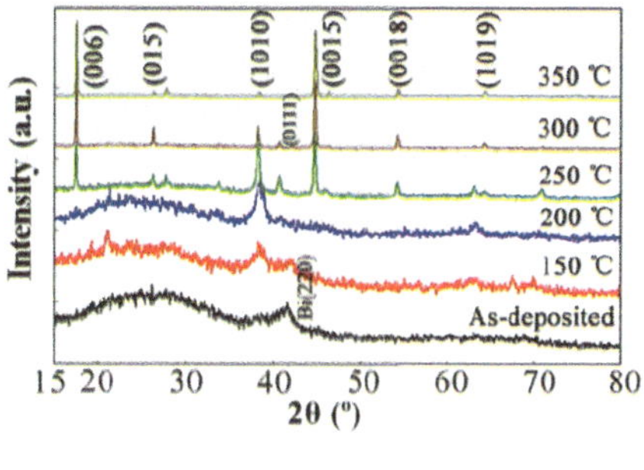

Fig. 5.23 XRD patterns of alloy films at different substrate temperatures

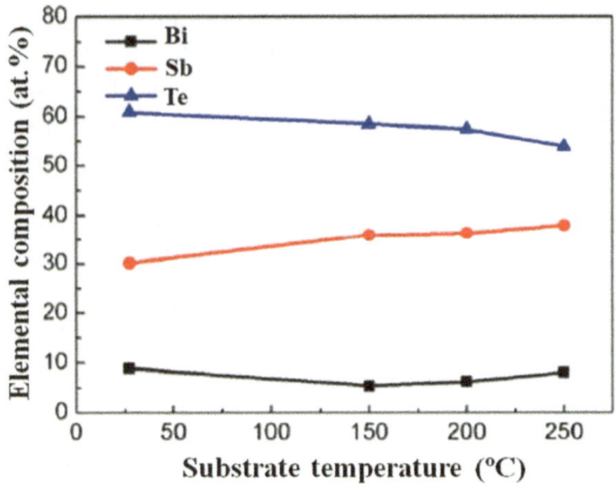

Fig. 5.24 Relationship between the composition of each element in the Bi–Sb–Te alloy film and the substrate temperature

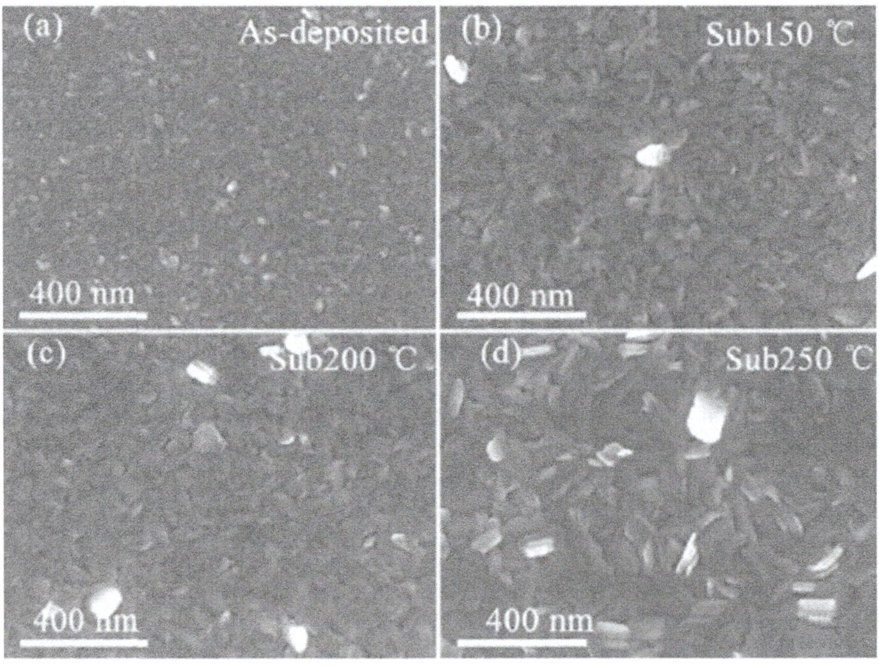

Fig. 5.25 Surface morphology of the alloy thin film at different substrate temperatures

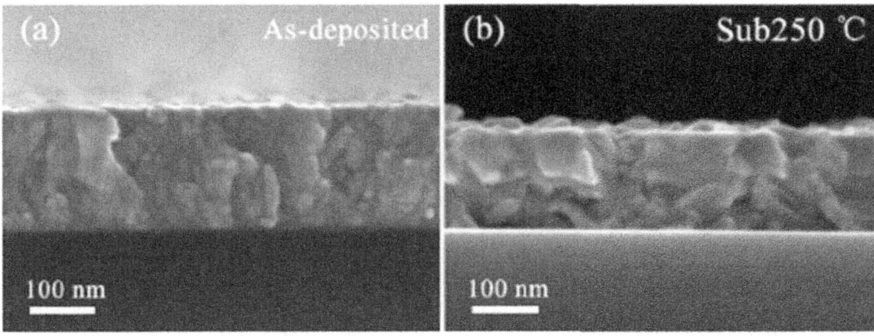

Fig. 5.26 Cross-sectional SEM image of Bi–Sb–Te alloy film **a** Room temperature (240 nm), **b** Substrate temperature 250 °C (200 nm)

Figures 5.25 and 5.26 are the surface and cross-section morphology of Bi–Sb–Te alloy films at different substrate temperatures. It can be seen that when the film is deposited at room temperature, the atoms flying from the target surface to the substrate nucleate on the substrate, and then grow and thicken based on the crystal nucleus. The film contains a large number of nano-grains. Under the condition that the substrate is heated, the atoms are heated to obtain sufficient energy and time to diffuse, and the crystal grains grow together and become flaky, which produces an effect similar to that of annealing. Substrate heating not only affects the morphology of the film, but also the roughness of the film, as shown in Fig. 5.27. It can be seen from the figure that the roughness (~1.3 nm) of the thin film deposited at room temperature is small, and the surface roughness of the thin film is larger under the condition of substrate heating, and it gradually increases with the increase of substrate temperature. When the substrate temperature is 250 °C, the roughness of the film is about 6.7 nm.

Fig. 5.27 Relationship between film roughness and substrate temperature

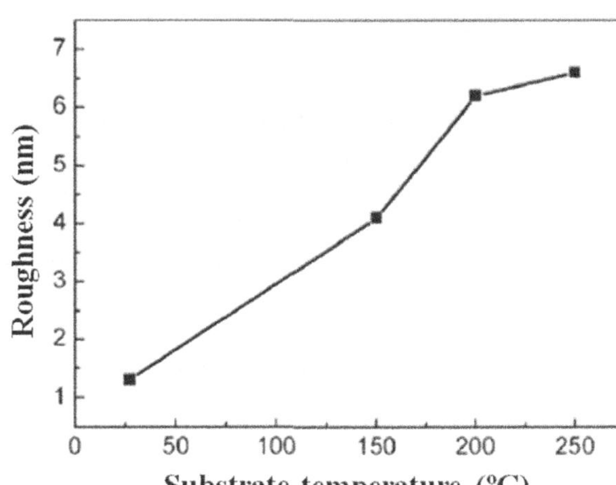

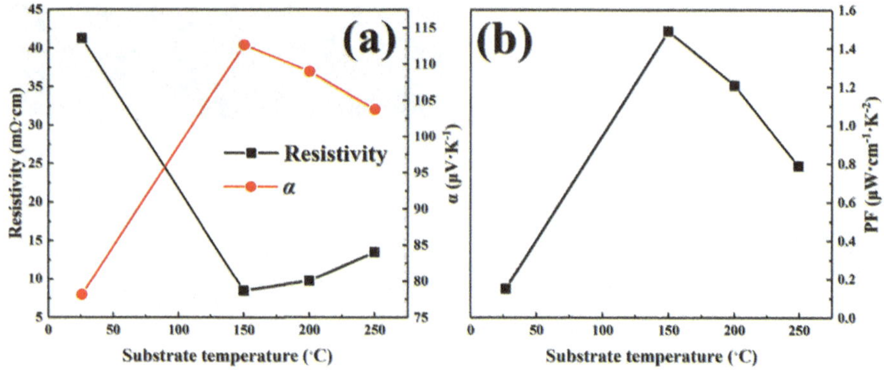

Fig. 5.28 The relationship between the thermoelectric properties of Bi–Sb–Te alloy film and the substrate temperature: **a** resistivity and Seebeck coefficient, **b** power factor

5.2.4.2 Influence of Substrate Temperature on Its Thermoelectric Performance

The power factor of Bi–Sb–Te alloy films at different substrate temperatures is shown in Fig. 5.28. It can be seen that the power factor of the film increases when the substrate is heated compared to the film deposited at room temperature, but it is not obvious. When the substrate temperature is 150 °C, the power factor of the alloy thin film increases from 0.15 $\mu W\ cm^{-1}\ K^{-2}$ at room temperature to 1.49 $\mu W\ cm^{-1}\ K^{-2}$. In addition, with the increase of temperature, the power factor of Bi–Sb–Te alloy films has a tendency to decrease.

5.2.4.3 Effect of Annealing on Microstructure and Thermoelectric Properties of Bi–Sb–Te Thin Films at Different Substrate Temperatures

In order to further optimize the properties of the thin film, the prepared thin film sample was placed in a well annealing furnace for post-annealing. The annealing temperatures were 150, 200, 250, 300, and 350 °C, respectively, and the annealing time was 6 h. The effects of annealing on the morphology and properties of the films were studied.

Figure 5.29 are XRD patterns of Bi–Sb–Te alloy films before and after annealing. It can be seen from the figure that the alloy film deposited at a substrate temperature of 250 °C preferentially grows along the (1010) crystal plane. With the increase of annealing temperature, the diffraction peak of the thin film increases and the crystallinity increases. When the annealing temperature is 250 °C, the diffraction peak in the XRD pattern is further enhanced, and the peak envelope is almost disappeared, indicating that the crystallinity of the film is further improved. In addition, a new diffraction peak appears at 2θ of 17.47°, 44.61°, and 54.15°, corresponding

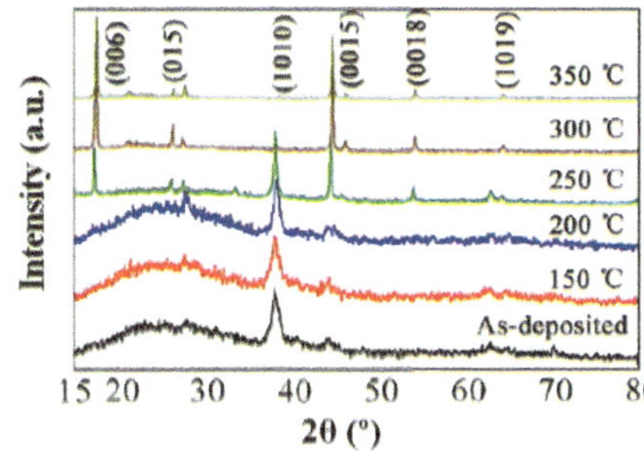

Fig. 5.29 XRD patterns of Bi–Sb–Te alloy film (Substrate temperature 250 °C) before and after annealing

to the (006), (0015), and (0018) crystal planes of $Bi_{0.5}Sb_{1.5}Te_{3.0}$, respectively and began to form a layered polycrystalline structure. With the further increase of the annealing temperature, the diffraction peaks of the film along the (001) direction are significantly enhanced, indicating that the films preferentially grow along the (001) direction to form a layered structure. Figure 5.30 is the surface morphology of the alloy thin film before and after annealing. The surface of the film deposited before annealing was flat and dense, and the grain size of the film surface increased after annealing. For samples with a substrate temperature of 250 °C, the grain size of

Fig. 5.30 SEM images of Bi–Sb–Te alloy films before and after annealing

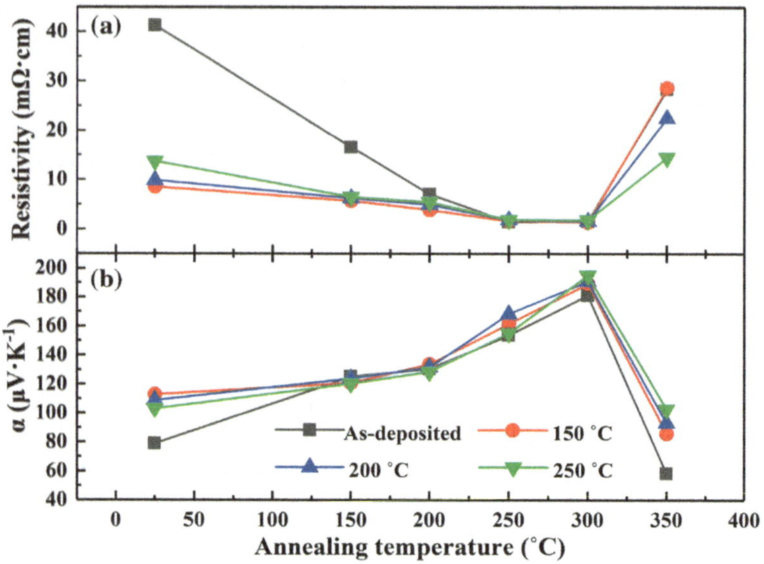

Fig. 5.31 Relationship between the thermoelectric properties and annealing temperature of Bi–Sb–Te alloy films at different substrate temperatures

the film surface increased significantly after annealing, and the surface was rough (Fig. 5.30).

Figure 5.31 shows the electrical properties of Bi–Sb–Te alloy films at different substrate temperatures before and after annealing. It can be seen from Fig. 5.31a that the resistivity of the film decreases significantly with the increase of annealing temperature. It is worth noting that when the annealing temperature is further increased from 250 °C, the resistivity of the film deposited at room temperature increases and the resistivity of the film heated by the substrate further decreases. The film with a substrate temperature of 150 °C has a minimum resistivity of 1.44 mΩ cm when the annealing temperature is 300 °C. It can be seen from Fig. 5.31b that the Seebeck coefficient of the Bi–Sb–Te alloy thin film increases under the condition of substrate heating compared with room temperature. After the annealing treatment, the Seebeck coefficient of the alloy film is further increased. The film with a substrate temperature of 250 °C has a Seebeck coefficient of 195.64 μV K^{-1} after annealing at 300 °C.

Figure 5.32 shows the relationship between the power factor and annealing temperature of Bi–Sb–Te alloy films at different substrate temperatures. It can be seen that the thermoelectric properties of the alloy films deposited at room temperature are poor. After annealing, the thermoelectric properties of the film were significantly enhanced. When the annealing temperature is 300 °C, the power factor of the film with a substrate temperature of 150 °C has a maximum value of 25.32 μW cm^{-1} K^{-2}. It can be seen that the proper substrate temperature can improve the thermoelectric properties of the film.

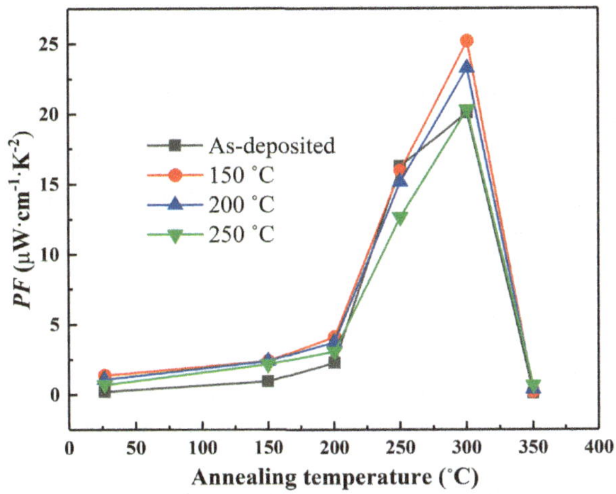

Fig. 5.32 Relationship between power factor and annealing temperature of Bi–Sb–Te alloy films at different substrate temperatures

5.3 Magnetron Sputtering Sb_2Te_3/Te Heterojunction Film

Electronic Topological Transition (ETT) is an electronic topological transition that can be caused by pressure or strain. Due to Van Huff's singularity, the topological morphology of the material changes under a certain pressure, and the density of electronic states near the Fermi level is strongly redistributed, eventually causing a series of discontinuous changes in electrical and thermal parameters [2, 3]. Sb_2Te_3 is reported as a topological insulator with a conductive surface state and an insulator state [4]. Topological insulators often have excellent thermoelectric properties. The small band gap of topological insulators and their metal states associated with ETTs are the main reasons for their superior thermoelectric performance [5]. In addition, their electronic structure's sensitivity to strain makes their thermoelectric properties highly tunable. In a large number of reports, the Raman shift of Sb_2Te_3/Bi_2Te_3/Sb_2Se_3/Bi_2Se_3 crystals has been studied using Raman spectroscopy. [6–9] Phonon softening means an increase in the intensity of phonon scattering, which is reflected in an increase in the full width at half maximum of the Raman spectrum. According to related research, ETT occurs when the Sb_2Te_3 crystal with a rhombic structure (R-3 m space group) is near an external pressure of 3 GPa [6, 10]. The thermal voltage, electrical conductivity, carrier concentration, and mobility will change abruptly at the ETT point, and the thermoelectric properties will be optimized at the ETT point [7, 11, 12]. In this section, Te is introduced into Sb_2Te_3 to construct a Sb_2Te_3/Te heterojunction. There is a lattice strain at the interface between Te and Sb_2Te_3. On the one hand, the lattice strain is conducive to reducing the lattice thermal conductivity. On the other hand, the crystallization and growth of Te nanoparticles in Sb_2Te_3 exerts internal pressure on Sb_2Te_3, and the possible ETT transformation is discussed.

5.3.1 Thin Film Preparation and Annealing Process

A Sb_2Te_3-Te composite film was prepared by co-sputtering using a magnetron sputtering device. The substrate was selected from a Si wafer with 250 nm SiO_2 on the surface, and the co-sputtering targets were a high-purity Sb_2Te_3 alloy target (purity 99.99%) and a Te simple substance target (purity 99.99%). Before the film is deposited, the substrate is cleaned using a silicon oxide wafer cleaning process. The background vacuum of the chamber is 10^{-6} Torr. High-purity argon gas is introduced during the film deposition process, the flow rate is 40 sccm, and the working vacuum is about 7 mTorr. Sb_2Te_3 uses radio frequency (RF) power sputtering with a power of 150 W, and Te uses a direct current (DC) power sputtering with a power of 10 W. To ensure the uniformity of the film, the sample holder abutment was rotated at a constant speed of 20 rpm when the film was deposited. After the deposition of the film was completed, the samples were annealed in a vacuum annealing furnace. The vacuum was maintained at about 1.5 Pa during the annealing process, the annealing temperature was 423 K, and the annealing times were 0 min, 30 min, and 180 min. The samples were named as A0, A1, and A2. Then the unannealed and annealed samples were subjected to structural characterization and performance testing to study the thermoelectric properties of Sb_2Te_3-Te heterojunction.

5.3.2 Film Morphology and Structure Characterization

Firstly, the surface and composition of the film were tested. Scanning electron microscopy could not effectively distinguish the details of the morphology, which was related to the flatness of the film surface. Randomly selected the area of the sample surface for EDS scanning, as shown in Fig. 5.33, the film composition is evenly distributed. For each sample, three regions were randomly selected for energy spectrum analysis. The atomic percentage content of the elements is shown in Table 5.6. By taking the average atomic ratio, the Te/Sb atomic percentages in the sample components are all about 1.8, indicating that the atomic ratio did not change significantly during the annealing process. That is, the annealing treatment does not affect the composition of the film.

Subsequently, as shown in Fig. 5.34 the cross section of the film was observed, and the thickness of the samples A0, A1, and A2 was determined to be 85, 80, and 75 nm by using a step meter and a scanning electron microscope, respectively. On the one hand, because the initial state of the film has a certain degree of porosity, as the annealing time increases, the film porosity decreases, the film becomes dense, and its thickness decreases. On the other hand, the reduction of this film is also related to the release of stress, the crystallization of the material and the rearrangement of atoms. In addition, it can also be seen from the cross section that the film structure is dense, but the roughness is different, which reflects the difference in the internal structure to a certain extent.

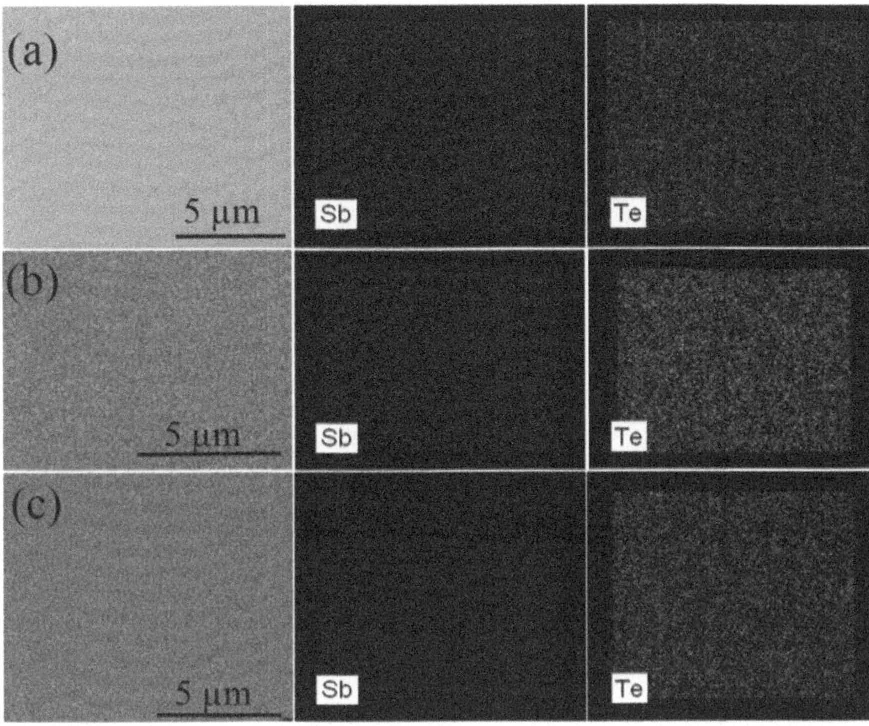

Fig. 5.33 Sample surface and EDS spectrum: **a** Sample A0; **b** Sample A1; **c** Sample A2

The roughness of the film is initially different from the cross-section of the film. To further confirm this, the surface of the film was characterized using AFM. The observation area is 2.5 × 2.5 μm, as shown in Fig. 5.35. The surface root-mean-square roughness of samples A0, A1 and A2 were 0.533 nm, 0.640 nm and 0.582 nm, respectively. The difference of the surface roughness of the film reflects the difference of its internal structure to a certain extent. At the same time, the surface roughness of the film is different, so the electron-phonon transmission on the film surface will also be different. The change of the surface roughness of the sample indicates that the internal structure of the film has changed with the progress of annealing. It is worth noting that the roughness of samples A0, A1 to A2 first increased and then decreased, this trend is closely related to the performance trend of the later test.

In order to facilitate comparison and analysis later, the basic parameters of the samples have been arranged in Table 5.7.

The structural changes of the Sb_2Te_3/Te heterojunction film during annealing were identified by XRD, as shown in Fig. 5.36.

The peak spectrum of A0 in the film can be indexed to the Te phase (JCPDS card number 85.0561, space group P3121), while A1 and A2 can be indexed to the Sb_2Te_3 phase (JCPDS card number 71-0393, space group R-3 m). In the unannealed sample A0, the thin films mainly consist of amorphous Sb_2Te_3 peaks and weaker

Table 5.6 Area scan element distribution of samples

Sample	Element	wt%	at.%	Average atomic ratio (Te/Sb)
A0	Sb-L	34.81	35.88	64.42:35.58
	Te-L	65.19	64.12	
	Sb-L	34.63	35.70	
	Te-L	65.37	64.30	
	Sb-L	34.11	35.17	
	Te-L	65.89	64.83	
A1	Sb-L	35.41	36.50	64.45:35.55
	Te-L	64.59	63.50	
	Sb-L	34.17	35.23	
	Te-L	65.83	64.77	
	Sb-L	33.85	34.91	
	Te-L	66.15	65.09	
A2	Sb-L	34.45	35.52	64.31:35.69
	Te-L	65.55	64.48	
	Sb-L	34.92	36.00	
	Te-L	65.08	64.00	
	Sb-L	34.50	35.56	
	Te-L	65.50	64.44	

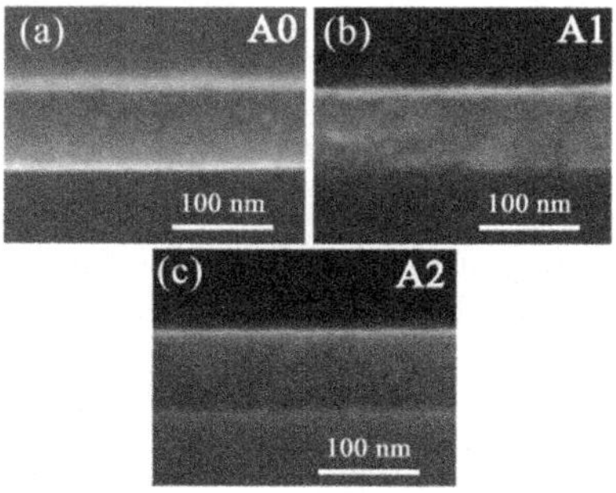

Fig. 5.34 SEM image of the cross section of the film: **a** A0; **b** A1; **c** A2

Te peaks. The characteristic of magnetron sputtering is the main reason for the amorphous Sb_2Te_3 peaks. With the increase of annealing time, the atoms rearranged and crystallized. Sb_2Te_3 was dominated by (001) orientation with (015) orientation peak. During the transition from amorphous to crystalline, the most obvious feature is the sharpening of peaks. However, an interesting phenomenon appeared in this XRD pattern. The peaks between 65°–75° first became flat and then sharpened, and moved to the right. Through comparison of the PDF card, it was finally determined that the peak in this range in the sample A0 was the Sb_2Te_3 peak, and the peak in this range in the samples A1 and A2 was the Te (211) peak. Another noteworthy

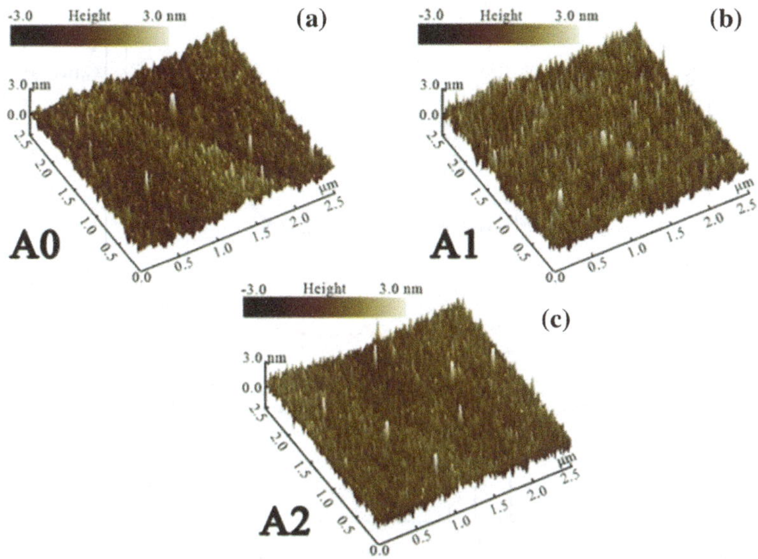

Fig. 5.35 Topography of the film surface before and after annealing: **a** A0; **b** A1; **c** A2

Table 5.7 Basic parameters of Sb_2Te_3/Te heterojunction by magnetron co-sputtering

Sample	Atomic percentage (Te/Sb)	Annealing time (min)	Thickness (nm)	Roughness (nm)
A0	64.42:35.58	0	85	0.533
A1	64.45:35.55	30	80	0.640
A2	64.31:35.69	180	75	0.582

phenomenon is that the occurrence of this process is accompanied by the preferential crystallization of Sb_2Te_3, while the recrystallization and growth of Te are relatively delayed. Compared to sample A2, the peak Te (211) in A1 is relatively flat, which means that the short-range order of the atomic arrangement and its grains are finer.

The grain size can be calculated by the Scherrer formula. The Te (211) grain sizes in A1 and A2 are 2.8 nm and 3.6 nm, respectively. In addition, the XRD pattern can also reflect the internal structure strain of the material to a certain extent. As the peak intensity decreases, the peak width increases and the strain increases. Taking the Te (211) peak with a particularly obvious change as a comparison, the strains here in A1 and A2 are 2.1% and 1.7%, respectively.

Since the Sb_2Te_3 peaks in A1 and A2 in XRD spectra are roughly equal, for the sake of comparison, it is considered that the Sb_2Te_3 phase crystals of the two have approached completeness. Since the Te nanoparticles in the A1 sample are smaller than the Te nanoparticles in A2, the interface density between the Sb_2Te_3 and Te phases in A1 is greater under the same Te content, and the lattice will bend at the

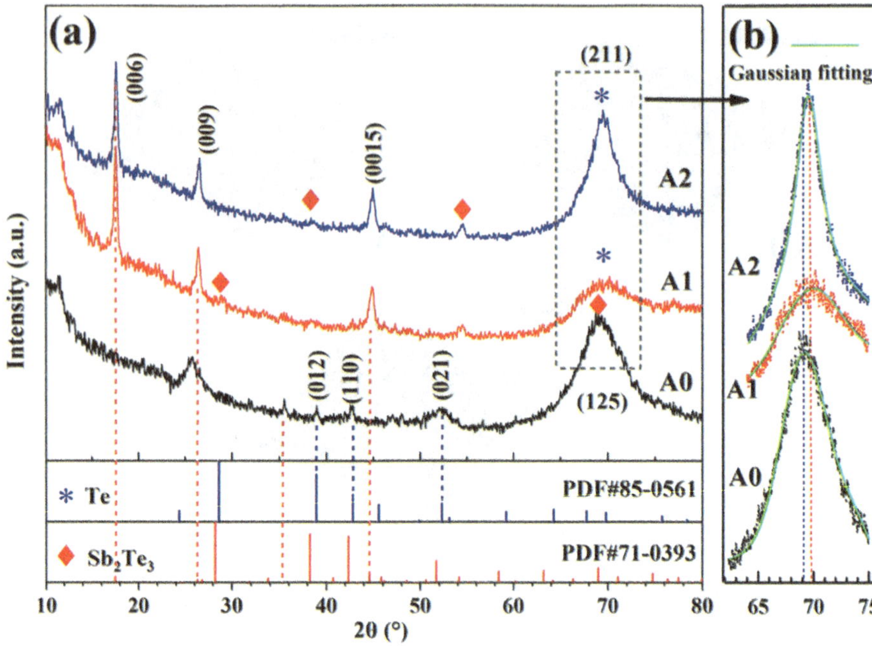

Fig. 5.36 XRD patterns of samples: **a** A0, A1, and A2 correspond to the annealing time of 0 min, 30 min, and 180 min, respectively; **b** Gaussian fitting curve

interface of Sb_2Te_3 and Te, resulting in a crystal lattice. strain. From this, it can be preliminarily judged that the lattice strain in the sample A1 is larger than A2, which is consistent with the strain result calculated by XRD here. Since the phonons are strongly scattered at the interface, the lattice thermal conductivity can be effectively reduced. The density of the interface between the nanophase and the main phase in A1 is large, and the corresponding lattice strain is large. Therefore, it can be predicted that the lattice thermal conductivity is smaller than that of A2 [5], which is confirmed by the performance test results later.

In addition, the chemical states of the elements on the surface of the film were characterized using XPS, as shown in Fig. 5.37. The peak centers of A0 in Fig. 5.37a are −573.1 and −583.4 eV, which correspond to the peaks of zero valence in elemental Te. As the annealing progresses, the center of the peak moves to a lower energy, reflecting the Te shift to the −2 valence state [13]. In the corresponding XRD pattern, Sb_2Te_3 in A0 is mainly amorphous, and the Te peak is more obvious. With the progress of annealing, Sb_2Te_3 is crystallized, so the Te of −2 valence is more obvious in the XPS pattern. The XRD pattern is consistent with the XPS pattern. The center of the peak in Fig. 5.37b is −528.8 and −538.2 eV, corresponding to the +3 valence Sb element in Sb_2Te_3. Peaks of oxidation states of Sb and Te appear in the XPS spectrum, while no peaks of related element oxidation states appear in the XRD spectrum. This is because during the experimental test, the XRD spectrum and

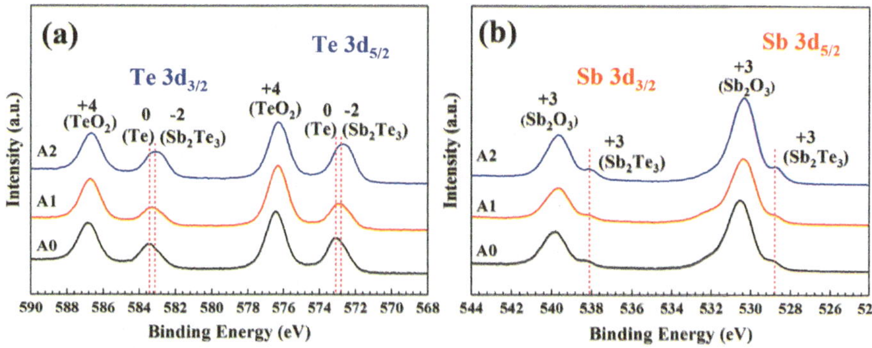

Fig. 5.37 XPS spectrum of the sample: **a** Te 3d; **b** Sb 3d

the Raman spectrum test are preferentially performed, and the XPS supplementary experiment is performed to further prove the existence of Te. Due to the sample being left for too long, the surface has an oxidation state. Since XPS is the final supplementary experiment, it does not affect the relevant parameters of the subsequent performance tests.

5.3.3 Heterogeneous Lattice Strain Characterization

Raman spectroscopy is usually used to detect phonon vibration modes in materials. One application is to detect the electronic topological conversion phenomenon that occurs in materials when they are subjected to external pressure. Specifically, as the pressure changes, the phonon mode of the material changes with the pressure. When a certain pressure value is reached, the Raman shift and the full width at half maximum of the corresponding phonon mode in the material will change abruptly. Or turning changes, the corresponding electrical and thermal characteristics will change at this point, and the thermoelectric performance at this point will also be improved. In order to characterize the effect of Te crystallization and growth on the phonon vibration mode in Sb_2Te_3/Te, Raman spectroscopy was performed on the samples, as shown in Fig. 5.38.

The Raman active phonon vibration modes of Sb_2Te_3 crystal are, and, the corresponding Raman shifts are 68.4 cm^{-1}, 110.9 cm^{-1}, and 165 cm^{-1}, respectively. [14] The Raman active phonon vibration modes of Te crystals are, and, the corresponding Raman shifts are 92.2 cm^{-1}, 120.4 cm^{-1}, and 140.7 cm^{-1}, respectively. [15] The atomic vibration modes of the two are shown in Fig. 5.38f. The spectrum of the sample obtained by Raman detection is a cumulative curve, which contains the composite information of Sb_2Te_3 and Te. Therefore, the Gaussian fit is used to understand the coupling of the Raman spectrum. The decoupled peaks of the three samples can be indexed to Sb_2Te_3 And Te. The Raman shift and the FWHM of the

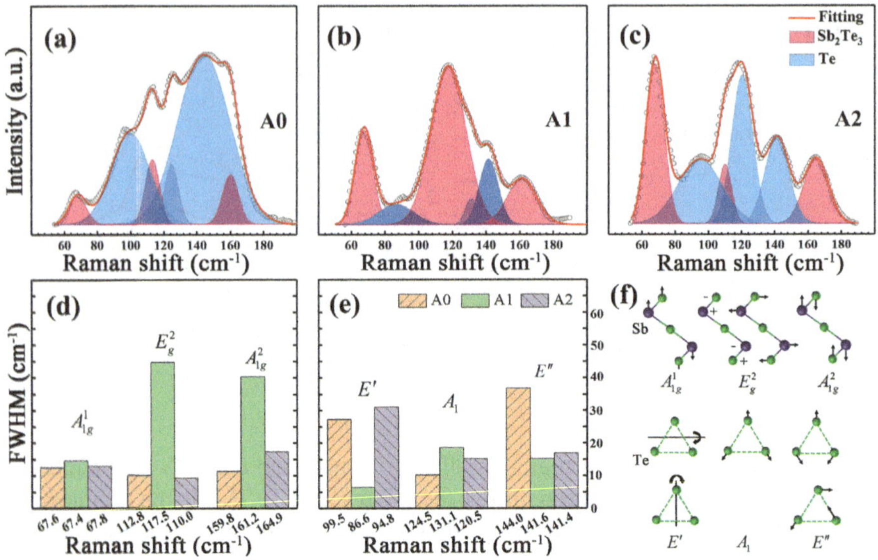

Fig. 5.38 Raman spectra of samples: **ac** represents the Raman spectra of samples A0, A1, and A2 and their corresponding Gaussian decoupling curves; **d** and **e** the Raman shifts and half values of Sb_2Te_3 and Te respectively height and width; **f** Raman phonon vibration modes of Sb_2Te_3 and Te [28]

two phases are statistically shown in Fig. 5.38d, e. As the annealing time increases, the Raman shift and the FWHM of the two phases change significantly.

This section focuses on the changes in Sb_2Te_3 phonon vibration. Studies have shown that in Sb_2Te_3 [16], as the annealing progresses, the full width at half maximum of the peak in the Raman spectrum decreases during the transition from amorphous to crystalline Sb_2Te_3, while the peak displacement remains substantially unchanged. The Raman shift to some extent reflects the change in stress during annealing. Studies show that [6, 17], with the increase of applied pressure, the Raman shift of all phonon modes in Sb_2Te_3 (R-3 m) increases with the increase of pressure, and the Raman shift of all phonon modes in Te with the pressure increases and decreases. Different from the application of uniform pressure in the report, inducing the stress by precipitating nanoparticles internally will cause the Raman shift of the phonon mode in Sb_2Te_3 to be preferentially oriented, but not all Raman shifts will change simultaneously. The Sb_2Te_3 mode corresponds to the (001)-oriented Sb_2Te_3 in this article, and the Sb_2Te_3 mode and the Te mode in the figure have the most obvious changes. For simple comparison, select Sb_2Te_3 mode and Te mode for strain comparison analysis. The coefficient of variation of Raman shift in Sb_2Te_3 mode with increasing pressure is 2.11 cm^{-1}/GPa, and the coefficient of variation of Raman shift in Te mode is -0.7 cm^{-1}/GPa [6, 17]. The Raman shifts of the Sb_2Te_3 modes in A0, A1, and A2 samples are 112.8, 117.5, and 110.0 cm^{-1}, and the Raman shifts of the Te mode in A0, A1, and A2 samples are 99.5, 86.6 and 94.8 cm^{-1}. Compared with the

Raman shifts in Sb_2Te_3 and Te crystals, the pressures on Sb_2Te_3 and Te in sample A1 are 3.1 GPa and 8.0 GPa, respectively. Compared to elemental crystals, the lower Raman shift in Sb_2Te_3 and the higher Raman shift in Te are under tension, so the corresponding phonon modes in A0 and A2 are under tension. It is also related to the area of the Raman test being the surface of the film, which is often subjected to tension. Compared with the reported ETT of Sb_2Te_3 near 3 GPa [6], the sample A1 is preliminarily considered to be close to ETT. Therefore, it can be predicted that the thermoelectric performance of A1 will be much better than A0 and A2. At the same time, the full width at half maximum of the Raman peak reflects the degree of phonon softening and the magnitude of the lattice thermal conductivity [18]. The mode of Sb_2Te_3 in A1 has the largest FWHM, so it can be predicted that its lattice thermal conductivity is the smallest. In the subsequent experiments, the above two points were confirmed.

In order to analyze the lattice strain in the Sb_2Te_3-Te heterojunction, a sample A1 was selected for HRTEM analysis. When preparing HRTEM samples, the film was scraped into absolute ethanol using a blade, dissolved by ultrasonic vibration, and then dropped on an ultra-thin carbon film to be dried for TEM observation. As shown in Fig. 5.39, the (001) and (015) planes of Sb_2Te_3 and the (110) plane of Te can be observed, and these peaks are consistent with the peaks observed in the XRD pattern. As shown in Fig. 5.39b, c, a large area of Sb_2Te_3 crystal with a (0012) crystal plane was observed, and the lattice fringe spacing was 0.253 nm. Dislocations were also observed in Sb_2Te_3. The inverse Fourier transform is shown in Fig. 5.39a.

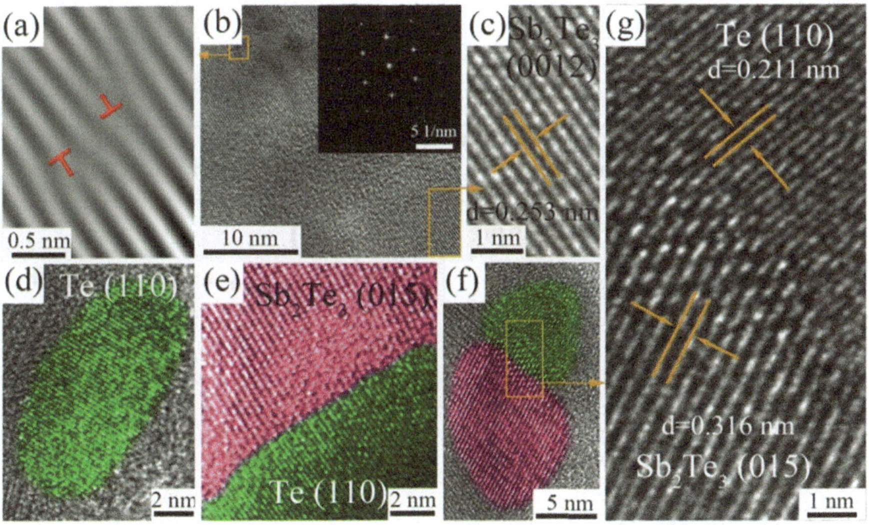

Fig. 5.39 TEM image of sample A1: **a** Inverse Fourier transform of selected area in (**b**); **b** TEM image and Fourier transform of Sb_2Te_3; **c** Figure (**b**) High resolution of selected area Figures; **d** Te nanoparticles in Sb_2Te_3; **e**, **f**, and **g** HRTEM images of Sb_2Te_3 and Te interfaces and partial enlarged view

Dislocations cause strain, lattice lines, and mass fluctuations, which are caused by dislocation nuclei, while the interface is strained by the rotation of the grains in pairs. [18–20] Te nanoparticles with a (110) crystal plane were observed in Sb_2Te_3 with a diameter of about 8.5 nm, which is different from the results calculated by the Scherrer formula in XRD. The orientation of the grains is different. On the other hand, there is an error in calculating the grains by Scherrer formula. In addition, the nano interface of the Sb_2Te_3 (015) plane and the Te (110) plane was also observed, in which the bending of the crystal lattice at the interface is clearly visible, as shown in Fig. 5.39f, g.

5.3.4 *Carrier Transport and Thermoelectric Performance*

The thermoelectric properties of Sb_2Te_3/Te at different temperatures are shown in Fig. 5.40. The thermoelectric properties of the films varied significantly from 360

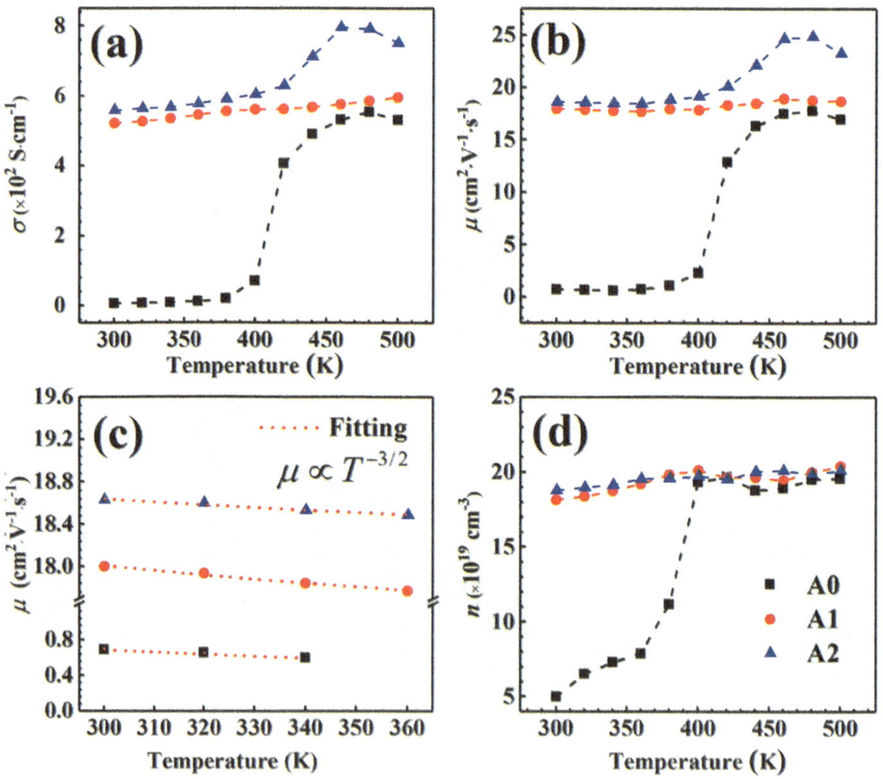

Fig. 5.40 Thermoelectric performance-temperature curve: **a** conductivity; **b** and **c** carrier mobility; **d** carrier concentration

to 440 K, especially for sample A0. The change trend of carrier concentration and conductivity with temperature is basically the same. The interval where the carrier concentration and mobility change most significantly with temperature is concentrated at 400–440 K, while the interval where the carrier concentration changes most significantly is 360–400. K.

As the temperature increases, the electrical conductivity of all samples continues to increase first, and then gradually decreases after higher temperatures, as shown in Fig. 5.40a. The change in conductivity reflects the comprehensive result of changes in carrier concentration and mobility. The carrier scattering mechanism can be inferred from the change in carrier concentration with temperature. Generally, at low temperature, the carrier mobility increases with increasing temperature, and impurity scattering is the main scattering mechanism at this time. As the temperature increases, the carrier mobility decreases with increasing temperature after the impurities are completely ionized. At this time, the phonon scattering of the lattice vibrations plays a dominant role [21]. From Fig. 5.40b, the main scattering mechanism with temperature can be divided into three parts:(1) Acoustic scattering is the main (temperature range of A0 is 300–340 K, temperature range of A1 and A2 is 300–360 K); (2) impurity scattering is dominant (temperature range of A0 is 360–460 K, temperature range of A1 and A2 is 380–460 K), the contribution of impurity scattering in this paper is mainly reflected in the change of defects with temperature; (3) Acoustic scattering is dominant (temperature range of all samples is higher than 480 K). The fitting relationship of carrier mobility with temperature change (below 360 K) is proportional to $T^{-3/2}$, which shows that this temperature range is dominated by acoustic scattering and the scattering factor is $-1/2$, as shown in Fig. 5.40c.

The trend of carrier concentration as a function of temperature is shown in Fig. 5.40d. As the temperature increased during the test (above 360 K), the amorphous phase in A0 began to crystallize. When the concentration of Sb'_{Te} defects exceeds the concentration of Te vacancies, the thin film exhibits P-type conductivity. In the presence of Sb vacancies, Sb atoms diffuse from the Te site back to its original sublattice site, generating additional Te vacancies and electrons. Long-term annealing at high temperature can alleviate this effect, reduce Te vacancies, and increase the concentration of P-type carriers [22]. When the annealing time is longer, this effect tends to be constant, so that the carrier concentration in A1 and A2 after the annealing treatment changes little with temperature. In the end, the carrier concentration of all the samples approaches a level (all are P-type), because the impurity ionization saturation in the Sb_2Te_3/Te system of the same composition is equivalent. Therefore, through the annealing temperature interval and time, a suitable conductivity can be obtained.

Seebeck coefficient as a function of temperature is shown in Fig. 5.41a. With increasing temperature, the Seebeck coefficient decreases, which is opposite to the change trend of carrier concentration and carrier mobility. The most obvious temperature range for Seebeck coefficient change is 360–400 K, which is consistent with the interval where the carrier concentration changes with temperature. The change

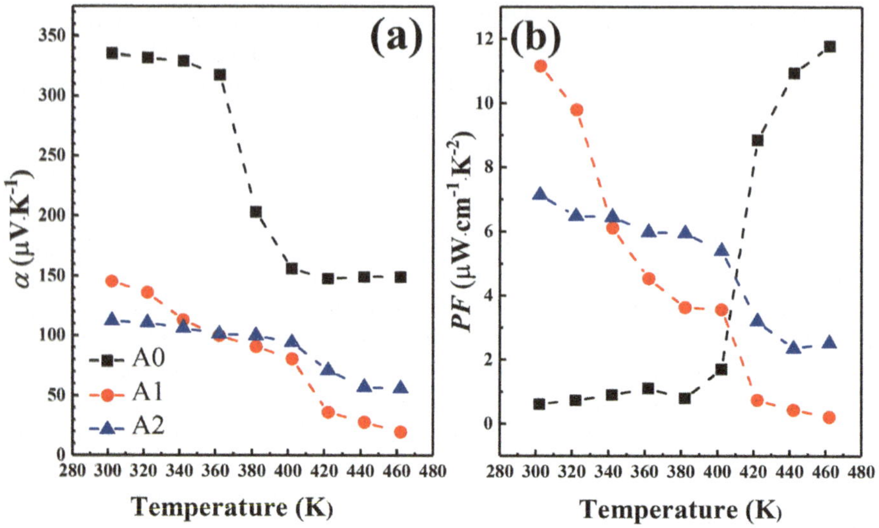

Fig. 5.41 Thermoelectric performance as a function of temperature: **a** Seebeck coefficient; **b** power factor

of Seebeck coefficient can be explained by Mott's formula. Higher carrier concentration and mobility will lead to a decrease in Seebeck coefficient. Here, the Seebeck coefficient depends more on the carrier concentration.

The variation of the power factor of each sample in the Sb_2Te_3/Te system with temperature is shown in Fig. 5.41b. The power factor of sample A0 increases with increasing temperature, while the power factors of samples A1 and A2 decrease with increasing temperature. The power factors of A1 and A2 samples near room temperature are higher than A0. It is worth noting that during the temperature change test, the carrier concentration in the A0 sample changed greatly in the range of 360–440 K. In order to confirm whether this test process is reversible, the test was repeated on the sample A0 and labeled as sample SA0, As shown in Fig. 5.42. The room temperature conductivity, Seebeck coefficient, and power factor of the second test were 253.2 S cm^{-1}, 215.2 μV K^{-1}, and 11.7 μW cm^{-1} K^{-2}, respectively. The test results show that the change in performance of A0 during the temperature change test is irreversible, because the process of temperature change test can also be regarded as an annealing treatment to a certain extent. The above results show that the power factor can be optimized by controlling the annealing time of the composite film.

According to the sign of Seebeck coefficient and the Hall test result, the carrier type is determined to be p-type, which indicates that the defects in the Sb_2Te_3/Te heterojunction are mainly Sb_{Te} inversion defects at the Te position. Combining the defects, interfaces and possible ETT in the Sb_2Te_3/Te heterojunction, the typical defects and structures in the film are shown in Fig. 5.43a and b, while the electrons and phonons at the heterojunction interface The transmission diagram is shown in Fig. 5.43c. In order to explore the effects of Te nanoparticles, dislocations, nano-interfaces, and

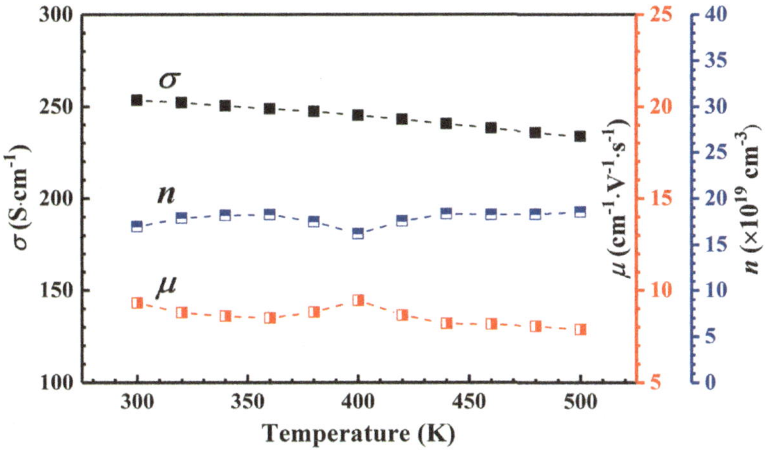

Fig. 5.42 The second thermoelectric performance test of the conductivity of sample A0

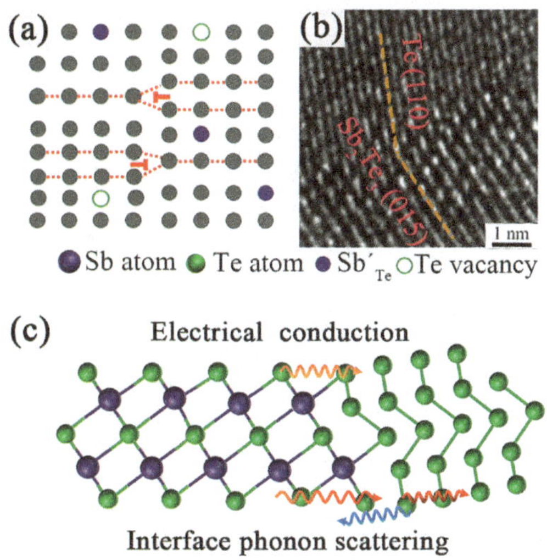

Fig. 5.43 Defects and interfaces: **a** Point defects and dislocations in Sb_2Te_3; **b**, **c** Sb_2Te_3 (015) and Te (110) interfaces and electron/phonon transmission

possible ETTs on the thermoelectric properties of Sb_2Te_3-Te heterojunctions, the relevant thermoelectric performance parameters of the films at room temperature are shown in Table 5.8.

The room temperature conductivities of samples A0, A1, and A2 increased from 5.5 S cm^{-1}, 524.4 S cm^{-1} to 560.5 S cm^{-1}, respectively. The carrier concentration and mobility in the annealed samples were higher than those in the unannealed sample A0. Assuming that the scattering mechanisms do not affect each other, the relevant parameters can be calculated by the following formulas [23, 24]:

Table 5.8 Thermoelectric properties of samples at 300 K

Sample	Te Size (nm)	n (10^{19} cm^{-3})	μ (cm^2 V^{-1} s^{-1})	l (nm)	L (V-2 K^{-2})	m^*/m_0	η	A (μV K^{-1})	Ke (W m^{-1} K^{-1})	PF (μW cm^{-1} K^{-2})
A0	–	5.0	0.7	0.05	1.51E–8	5.5	⁻1.81	335.4	0.003	0.6
A1	2.8	18.2	18.0	2.0	1.74E-8	2.6	1.11	145.9	0.273	11.2
A2	3.6	18.8	18.6	2.4	1.85E–8	1.9	1.98	112.9	0.312	7.1

$$\sigma = ne\mu_T, \tag{5.1}$$

$$\mu_{in} = De\left(1/2\pi m^* k_B T\right)^{1/2} \exp(-\phi_b/k_B T), \tag{5.2}$$

$$l = h(3n/\pi)^{1/3}\mu_T/2e \tag{5.3}$$

$$1/\mu_T = 1/\mu_m + 1/\mu_{in}. \tag{5.4}$$

where n is the carrier concentration, e is the amount of electronic charge, and μ_T, μ_m, and μ_{in} are the total carrier mobility of the sample, the mobility of the material body, and the mobility related to interfacial potential scattering. D, m, k_B, T, l, and ϕ_b are the grain size, effective mass, Boltzmann constant, absolute temperature, mean free path of the carrier, and barrier height, respectively. h is Planck's constant.

Annealing is beneficial to reduce defects and release stress, which in turn increases μ_m. According to the Te (211) peak in the XRD pattern, the grain size becomes larger as the annealing time increases. According to Eq. 5.2, a large grain size is conducive to the increase of μ_{in}. According to Eq. 5.3, the average free path of the carriers in A0, A1, and A2 can be 0.05 nm, 2.0 nm, and 2.4 nm, respectively. According to Eqs. 5.3 and 5.4, the total mobilities in A0, A1, and A2 increase sequentially, which is consistent with the Hall test results.

In order to compare the effect of Te on the Seebeck coefficient of the thin film, the relationship between the Seebeck coefficient of the thin film and the carrier concentration at room temperature (Pisarenko) is shown in Fig. 5.44. The figure also lists the relevant Sb_2Te_3 systems doped with two phases (Te, Pt, Cu, Ag_xTe_y, γ-Sb_2Te_3, I^-) and the bulk Sb_xTe_y for comparison. It can be seen from the figure that in the same system, the Seebeck coefficient generally decreases with the increase of the carrier concentration. The relevant literature data in the figure show that the Seebeck coefficient of Sb_2Te_3 can be increased by adjusting the potential barrier between the main phase and the secondary phase, which is the energy filtering effect. In addition, the Seebeck coefficient can also be adjusted by controlling the interface density between the primary and secondary phases. In this paper, the Seebeck coefficients in A0, A1, and A2 decrease in sequence from 335.4, 145.9, and 112.9 $\mu V\ K^{-1}$. At the same time, the corresponding m are 5.5 m_O, 2.6 m_O, and 1.9 m_O, where m_O is the free electron mass. The Fermi level changed from negative to positive, indicating that the Fermi level of the film has entered the valence band from the forbidden band. With the increase of annealing time, Sb_2Te_3 crystallizes preferentially, and then Te crystals crystallize and grow, accompanied by a decrease in grain boundary density and a decrease in Seebeck coefficient. The Te grains in the A1 sample are smaller than those in the sample A2, which means that there is a higher interface density and lattice strain in A1. The m in the sample A1 is larger than A2. This is because the ETT may occur in A1, and the energy band near the Fermi surface becomes sharp. To sum up, the appropriate Seebeck coefficient (145.9 $\mu V\ K^{-1}$) and conductivity (524.4 $S\ cm^{-1}$) in A1 make its power factor higher (11.2 $\mu W\ cm^{-1}\ K^{-2}$).

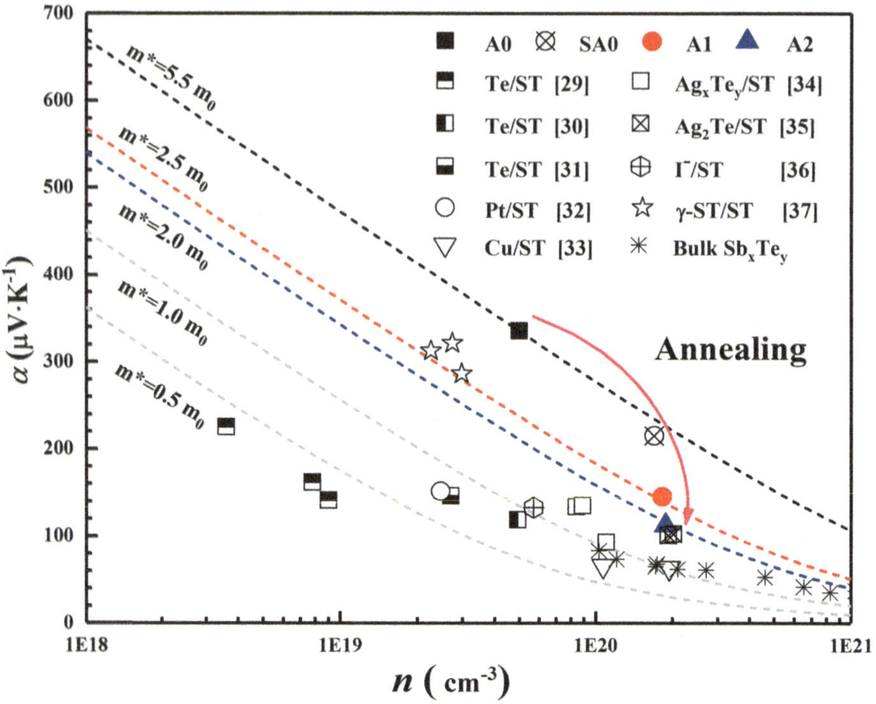

Fig. 5.44 The relationship between the Seebeck coefficient and the carrier concentration at 300 K (ST stands for Sb_2Te_3) [29–37, 49]

The power factors of samples A0, A1, and A2 increased from 0.6to 11.2 μW cm^{-1} K^{-2}, and then decreased to 6.9 μW cm^{-1} K^{-2}. This trend is the same as the previously observed tendency of the roughness of the film surface. The change of the surface roughness reflects the change of the internal structure of the film to a certain extent. With the crystallization and growth of Te, the roughness increases first and then decreases. Appropriate Te nanoparticles can induce appropriate stress and may cause ETT in Sb_2Te_3, which ultimately optimizes the power factor.

On the one hand, Te may induce the ETT of Sb_2Te_3 to help optimize the power factor. On the other hand, the lattice strain induced by Te grains will help reduce the thermal conductivity. The thermal conductivity data measured using the TDTR method are shown in Fig. 5.45. The thermal conductivity in samples A0, A1, and A2 is 0.41, 0.55, and 0.69 W m^{-1} K^{-1} in this order. Through Widermann-Franz law and Lorentz constant obtained by calculation, the electronic thermal conductivity can be calculated. The electronic thermal conductivity in samples A0, A1, and A2 was 0.003 W m^{-1} K^{-1}, 0.27 W m^{-1} K^{-1}, and 0.31 W m^{-1} K^{-1}, respectively. Lattice thermal conductivity in samples A0, A1, and A2 were 0.41 W m^{-1} K^{-1}, 0.28 W m^{-1} K^{-1}, and 0.38 W m^{-1} K^{-1}, respectively.

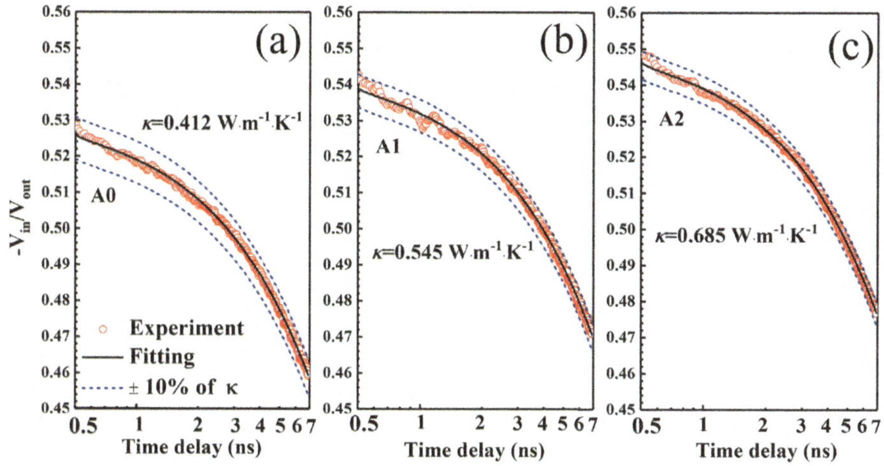

Fig. 5.45 Thermal conductivity data for the TDTR test

The magnitude of the strain can reflect the intensity of phonon scattering to a certain extent. A large number of fine Te particles induce lattice strain, increase the interface density and strongly scatter phonons. The lattice thermal conductivity of the sample decreases from A0, A2, and A1 in sequence, which is consistent with the results predicted by XRD and Raman. Considering that the film is a mixed film prepared by a co-sputtering method, and the composition is uniform, the ZT calculation can be roughly performed from the power factor in the film plane and the longitudinal thermal conductivity. The ZT values in samples A0, A1 and A2 increased from 0.05 to 0.61 and then decreased to 0.31. The corresponding relationship between the changes of Te particles and related parameters is shown in Fig. 5.46. Table 5.9 compares the results of this article with the thermoelectric properties of Sb_2Te_3 based heterojunctions reported in related literatures, showing strong competitiveness.

5.4 Preparation of Sb_2Te_3 and Bi_2Te_3 Films by Magnetron Co-sputtering

In this experiment, the sputter deposition of Sb_2Te_3 and Bi_2Te_3 thermoelectric thin films was co-sputtered. The target is a three-inch (Φ76.2 mm) sputtering target with a single target thickness of 5 mm, of which the thickness of the sputtering material is 3 mm. In order to prevent the target from being thermally cracked during the sputtering process, the target was bound on to 2 mm thick copper disk. The purity of elemental Sb target, Bi target and Te target is 99.99%, and they are prepared by hot pressing method. Of these three materials, the Te target has the worst conductivity and is sputtered by radio frequency (RF) method. The other two targets are sputtered

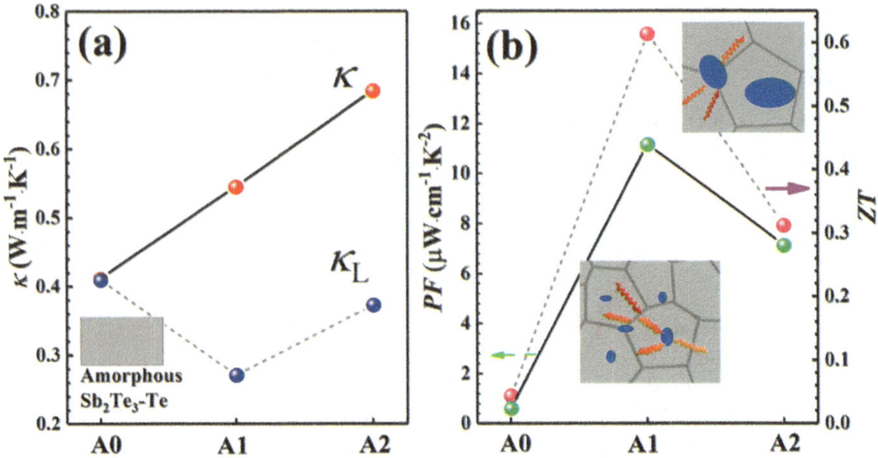

Fig. 5.46 Thermoelectric properties of the sample: **a** Total thermal conductivity and lattice thermal conductivity of the sample; **b** ZT value and power factor of the sample

by direct current (DC) method. That is, Sb_2Te_3 thermoelectric thin film is deposited by means of Te (RF) + Sb (DC) co-sputtering, and Bi_2Te_3 thermoelectric thin film is deposited by means of Te (RF) + Bi (DC) co-sputtering. Te (RF) sputtering power was set to 10 W, Sb (DC) sputtering power was set to 15 W, and Bi (DC) sputtering power was set to 13 W. The Sb_2Te_3 film and Bi_2Te_3 film were all sputter-deposited under the sputtering pressures of 0.2 Pa, 0.5 Pa, 0.8 Pa, and 1.1 Pa, respectively. The sputter deposition time of each sample was 30 min. The high purity argon gas used for sputtering has a purity of 99.999%. The sputtering working pressure is controlled by adjusting the flow rate of argon gas, and the control accuracy is 1 sccm.

5.4.1 Film Preparation and Technology

The substrate used in this experiment is a single-sided polished single crystal silicon wafer covered with a 1 μm thick SiO_2 oxide layer. The total thickness of the silicon wafer is 550 μm and the diameter of the silicon wafer is 100 mm. Each complete substrate is cut into four pieces. In order to eliminate the adverse effects of impurities and stains on the surface of the substrate on the thermoelectric film, the substrate needs to be cleaned. First, ultrasonically clean the silicon wafer in acetone for 5 min, quickly remove it with tweezers and rinse with deionized water, and then ultrasonically clean with ethanol for 5 min, quickly remove it with tweezers, rinse with deionized water, and then put in isopropyl ultrasonic wash in alcohol for 5 min, quickly clip out with tweezers and rinse with deionized water. Then put it into a mixed solution of concentrated sulfuric acid and hydrogen peroxide with a volume ratio of 3:1 and soak it for 10 min. Remove with tweezers, rinse with deionized water, and blow dry with

Table 5.9 Comparison of Sb_2Te_3-based heterojunction thermoelectric properties with related literatures

	Method	Temp. (K)	σ (S cm^{-1})	α (μV K^{-1})	PF (μW cm^{-1} K^{-2})	κ (W m^{-1} K^{-1})	ZT	Ref.
Sb_2Te_3/Te	MS	300	524.4	145.9	11.2	0.545	0.61	This work
S-doping Sb_2Te_3	PVD	RT	1575	186	54.4	3.03	0.54	[38]
$Sb_2Te_3/(I^-)$	Microwave synthesis	300	869	135	15.84	0.89	0.53	[36]
Sb_2Te_3	MS	RT	282.1	135.5	5.2	0.37	0.42	[39]
Sb_2Te_3/Cu_2Te	Solid-state synthesie	370	3841.7	71.9	19.9	1.98	0.37	[40]
Sb_2Te_3/Te	Co-evaporation	300	666.7	160	17.1	1.8	0.3	[41]
Sb_2Te_3/Te	Sintering	340	852.1	117.4	11.7	1.48	0.27	[42]
Sb_2Te_3/Te	Fusion method	300	3534.7	80.9	23.1	2.8	0.25	[43]
Sb_2Te_3/Te	SS	323	811	118.8	11.3	1.54	0.24	[30]
Sb_2Te_3/Te	SS	323	358.2	170.3	10.4	1.52	0.22	[13]
Sb_2Te_3	PVD	RT	4000	80	25.6	3.5	0.22	[38]
Sb_2Te_3/Te	SS	325	233.1	189.3	8.4	1.67	0.16	[44]
Sb_2Te_3/Te	Solution Casting	300	146	249	5.3	1.04	0.15	[31]
Sb_2Te_3/Te	MBE	300	1289.3	110	15.63			[45]
Sb_2Te_3/Te	Co-evaporation	RT	277.5	190.8	10.1			[46]
Sb_2Te_3/Te	Electrodeposition	RT	302	168	8.5			[47]
Sb_2Te_3/Te	Electrodeposition	300	335.4	145.7	7.1			[29]
$Sb_2Te_3/\gamma\text{-}Sb_2Te_3$	Electrodeposition	RT	56.7	322	5.9			[37]
Sb_2Te_3/Te	SS	300	150.2	148.1	3.3			[48]
Sb_2Te_3/Cu	MS	313	307.4	80.5	2.0			[33]
Sb_2Te_3/Pt	SS	300	73	152	1.02			[32]
Sb_2Te_3/Ag_xTe_y	Ligand exchange	RT	56	135	1.02			[34]

Note In the table, MS indicates magnetron sputtering, SS indicates solution synthesis method, and RT indicates room temperature

dry compressed N_2. Before putting the film on the magnetron sputtering film coater, use an Ar ion source to perform back-sputter cleaning on the coated substrate, so that the surface of the substrate can be cleaned at the atomic level, thereby improving the bonding quality between the deposited film and the substrate. In order to ensure that the substrate is protected from secondary contamination, the sample is always held by clean tweezers during the operation. Sb_2Te_3 and Bi_2Te_3 thermoelectric films were prepared under different sputtering working pressures.

In the experiment, Sb_2Te_3 and Bi_2Te_3 films were annealed. Annealing can not only improve the crystalline state of the thermoelectric film and stabilize the structure of the film, but also adjust the thermoelectric properties of the film. Sb_2Te_3 and Bi_2Te_3 thin film samples were put into an annealing furnace for annealing treatment. The annealing furnace is heated by the bottom resistance wire. The sample is placed on a stainless-steel disc on the resistance heating wire. The stainless-steel tray is insulated from the heating resistance wire by a heat-resistant ceramic pad. First, the furnace cavity is evacuated. After the vacuum degree is lowered below 0.5 Pa, high purity nitrogen is introduced to atmospheric pressure. Then re-evacuate and fill with high purity nitrogen. Repeat three times to ensure nitrogen purity in the chamber. The samples were annealed with nitrogen at 200, 250, 300, 350, and 400 °C for 2 h, and then cooled to room temperature before use.

5.4.2 Topography of the Film

Figure 5.53 is the surface morphology of Sb_2Te_3 film under different sputtering working pressures. It can be seen from the figure that when the sputtering working pressure is 0.2 Pa, the surface of the Sb_2Te_3 thin film is smooth and dense, and only agglomerated particles appear in some locations. When the sputtering pressure is increased to 0.5 Pa, particles of uneven size begin to appear on the film surface. When the sputtering working pressure is increased to 0.8 Pa, the surface of the film is uniformly distributed with particles, and some holes are also distributed between the particles. When the working pressure reaches 1.1 Pa, the particle size of the film surface becomes larger and the uniformity becomes worse. When the sputtering working pressure is low, that is, 0.2 Pa, the collision probability between the sputtered target atoms and the working gas is low, the energy loss is the least, the rate of reaching the substrate surface is high, the energy is high, and the formed film is the densest. No obvious agglomeration of particles, as shown in Fig. 5.47a. When the gas pressure rises, the probability of collision between the sputtered target atoms and the working gas increases, and the rate at which the atoms reach the substrate decreases. Sb and Te atoms have a certain time to move and diffuse and reunite on the substrate. Agglomerated particles began to appear on the bottom film surface, as shown in Fig. 5.47b. The working pressure continues to increase, which not only causes the rate of Sb and Te atoms to reach the substrate to continue to decrease, but also the working gas atoms may be adsorbed on the surface of the substrate and form pore inclusions in the film. The rate at which Sb and Te atoms reach the substrate surface decreases, and the arriving atoms migrate and agglomerate for a longer time, which is more conducive to the migration and agglomeration of atoms when the film is formed, forming agglomerated particles on the surface of the film, and holes generated, as shown in Fig. 5.47c. If the Sb and Te atoms have sufficient migration time, the agglomerated particles will continue to grow, as shown in Fig. 5.47d.

Figure 5.48 is the surface morphology of Bi_2Te_3 film under different sputtering working pressures. It can be seen from the figure that when the sputtering working

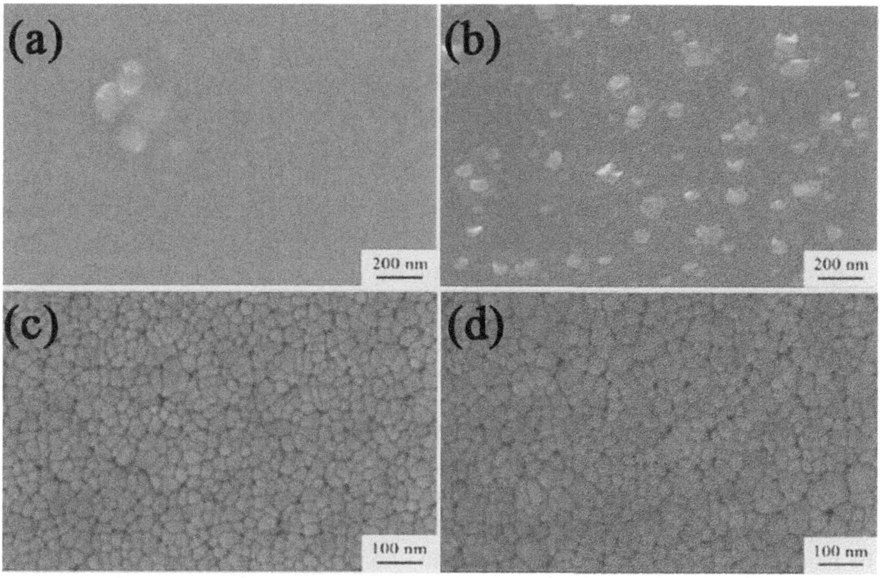

Fig. 5.47 Surface morphology of Sb_2Te_3 films under different sputtering working pressures:The working pressures are **a** 0.2 Pa, **b** 0.5 Pa, **c** 0.8 Pa, and **d** 1.1 Pa

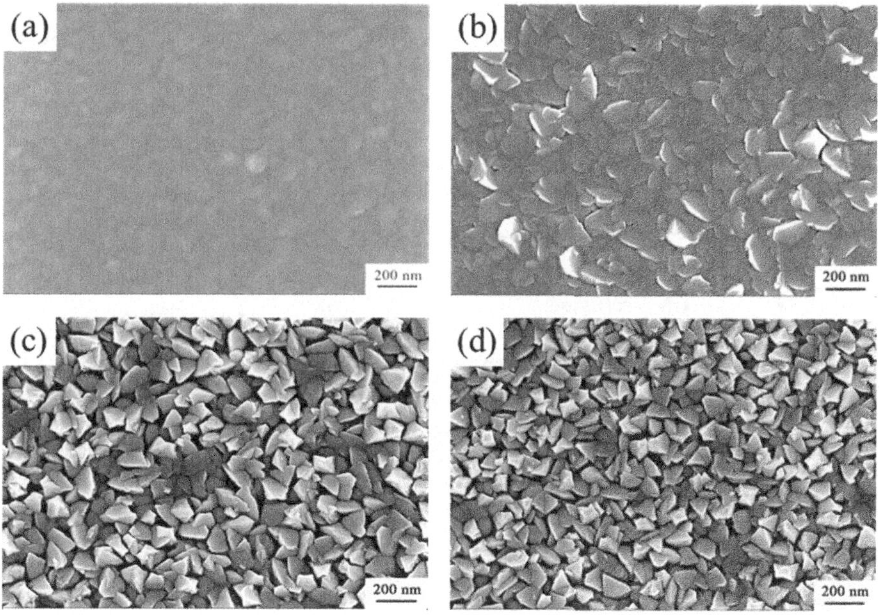

Fig. 5.48 Surface morphology of Bi_2Te_3 films under different sputtering working pressures: the working pressures are **a** 0.2 Pa, **b** 0.5 Pa, **c** 0.8 Pa, and **d** 1.1 Pa

pressure is 0.2 Pa, the surface of the Bi_2Te_3 film has a lamellar contour, but the contour is not clear, as shown in Fig. 5.48a. When the working pressure of the sputtering is increased to 0.5 Pa, the surface of the film shows a clear sheet-like structure, and the polyhedron structure is included between the lamellar structures, as shown in Fig. 5.48b. With the increase of the working pressure, the layered structure disappeared and completely transformed into a polyhedral structure, as shown in Fig. 5.48c. The working pressure continued to increase, and the size of the polyhedron structure on the surface of the film decreased, as shown in Fig. 5.48d. The main reason for the change in the surface structure of the Bi_2Te_3 film is the increase in working pressure, which affects the rate at which Bi atoms and Te atoms reach the substrate, thereby affecting the migration and diffusion agglomeration and growth of Bi atoms and Te atoms on the substrate surface [24], which causes the surface structure changes in appearance.

The XRD patterns of Sb_2Te_3 film and Bi_2Te_3 film under different working pressures are shown in Fig. 5.49. According to XRD diffraction pattern analysis results, the Sb_2Te_3 film and Bi_2Te_3 film obtained under different sputtering working pressures contained a large number of amorphous phases and other non-equilibrium phases. The internal structure of the films was in an unbalanced state, and no good crystals Sb_2Te_3 phase and Bi_2Te_3 phase were formed in the films.

The XRD patterns of Sb_2Te_3 film and Bi_2Te_3 film are fitted to calculate the crystallinity. The crystallinity of Sb_2Te_3 film and Bi_2Te_3 film are shown in Fig. 5.50. It can be seen from the figure that there are a large number of amorphous phases in the Sb_2Te_3 film and Bi_2Te_3 film obtained by different sputtering working pressures, and the film structure is in an unstable state. At the same time, it can be seen that

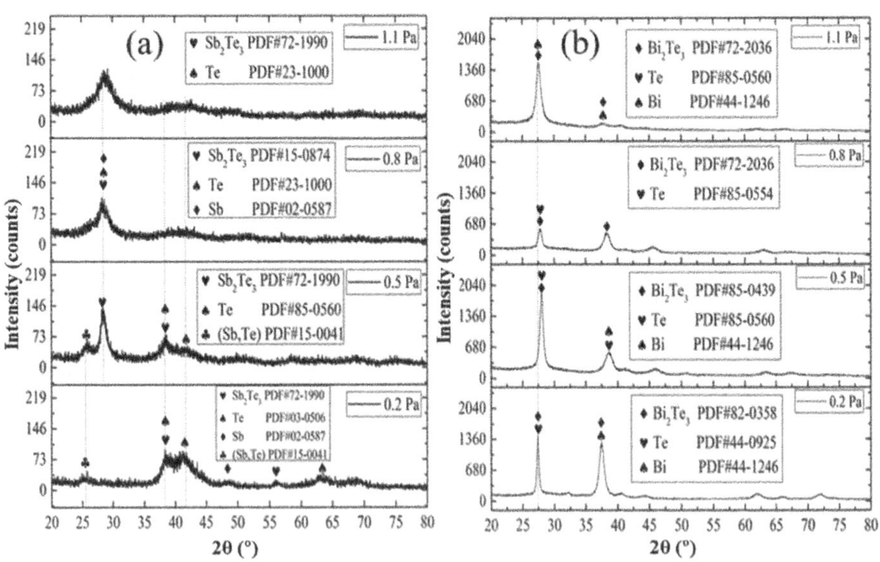

Fig. 5.49 XRD patterns of films under different working pressures: **a** Sb_2Te_3 film **b** Bi_2Te_3 film

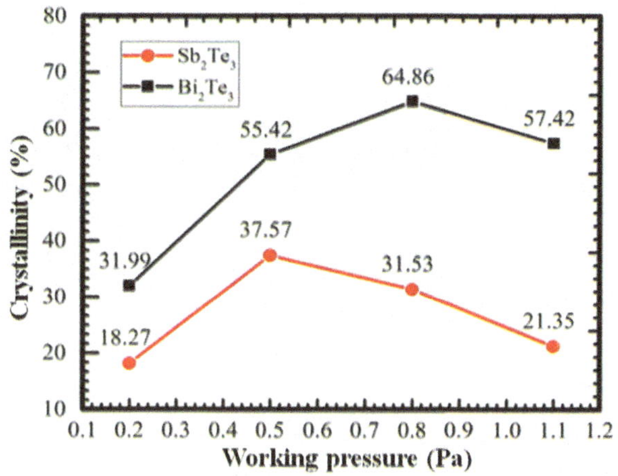

Fig. 5.50 Crystallinity of Sb_2Te_3 films and Bi_2Te_3 films

the crystallinity of Bi_2Te_3 film is higher than that of Sb_2Te_3 film, which is consistent with the results observed by SEM.

Therefore, in order to obtain Sb_2Te_3 film and Bi_2Te_3 film with stable structural properties, the film needs to be recrystallized and annealed, and the recrystallized annealing is used to improve the crystalline state of the thermoelectric film so that the structural properties of the film reach a stable state.

Figure 5.51 shows the SEM images of the surface and cross-section morphology of the Sb_2Te_3 film under different annealing conditions. The small picture in the upper right corner of each picture is the cross-sectional morphology of the film under the corresponding conditions. It can be seen from the figure that with the increase of the annealing temperature, the surface morphology and cross-sectional morphology of the Sb_2Te_3 film change significantly. Before annealing, the surface of the Sb_2Te_3 film was smooth, and the internal structure was dense from the cross section. When the Sb_2Te_3 film was annealed at 200 °C for 2 h, particles precipitated on the surface of the film. From the cross-sectional morphology, it can be seen that crystallization occurred inside the film, which is consistent with the XRD diffraction pattern analysis test results of Sb_2Te_3 film. As the annealing temperature continues to increase, the size of the precipitate particles on the surface of the Sb_2Te_3 film increases. The pores in the film are generated due to the volatilization of the Te element, and the size of the pores increases with the increase of the annealing temperature. Even the holes are connected to form large holes. The large holes penetrate the film surface and are exposed on the film surface. Therefore, the excessively high annealing temperature not only affects the stoichiometric ratio of Te and Sb elements in the Sb_2Te_3 film, but also changes in the stoichiometric ratio of Te and Sb elements in the film affects the performance, especially the thermoelectric properties, of the Sb_2Te_3 film. The high annealing temperature makes large continuous or even holes on the surface of the Sb_2Te_3 film, which will not only seriously affect the thermoelectric properties of the film, but also seriously affect the mechanical stability of the film and reduce its mechanical strength, which is not good for its application in device processing.

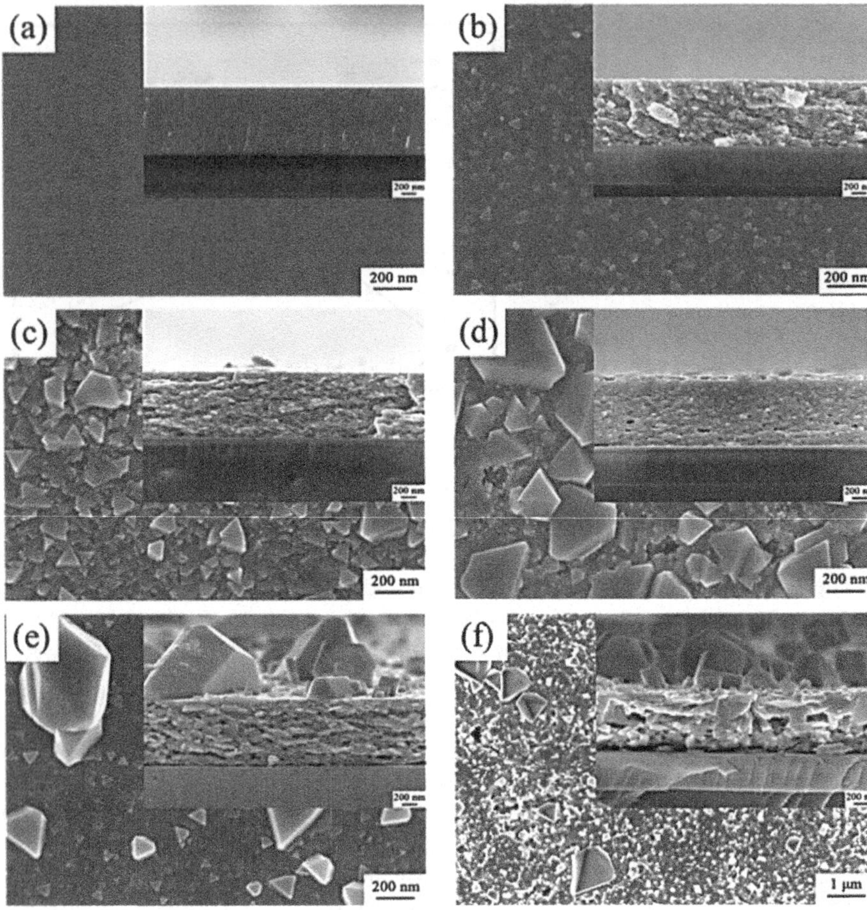

Fig. 5.51 The surface of the Sb_2Te_3 thin film at different annealing temperatures, that is, the cross-sectional topography structure. The small picture in the upper right corner of the figure is the corresponding section: **a** unannealed, **b** 200 °C annealed for 2 h, **c** 250 °C Anneal for 2 h, **d** 300 °C for 2 h, **e** 350 °C for 2 h, and **f** 400 °C for 2 h

Figure 5.52 shows the SEM images of the surface and cross-section morphology of the Bi_2Te_3 film under different annealing conditions. The small picture in the upper right corner of each picture is a cross-section topography picture. It can be seen from the figure that the annealing temperature has little effect on the morphology of the Bi_2Te_3 film surface. When it is not annealed, it shows good crystallinity. The crystal grains grow in a direction perpendicular to the substrate, and each crystal grain grows in layers. From the cross-sectional morphology of Bi_2Te_3 thin films at different annealing temperatures, it can be seen that as the annealing temperature increases, the grains grown along the direction perpendicular to the substrate show a coarse trend, and the consistency of grain growth direction is destroyed. The higher

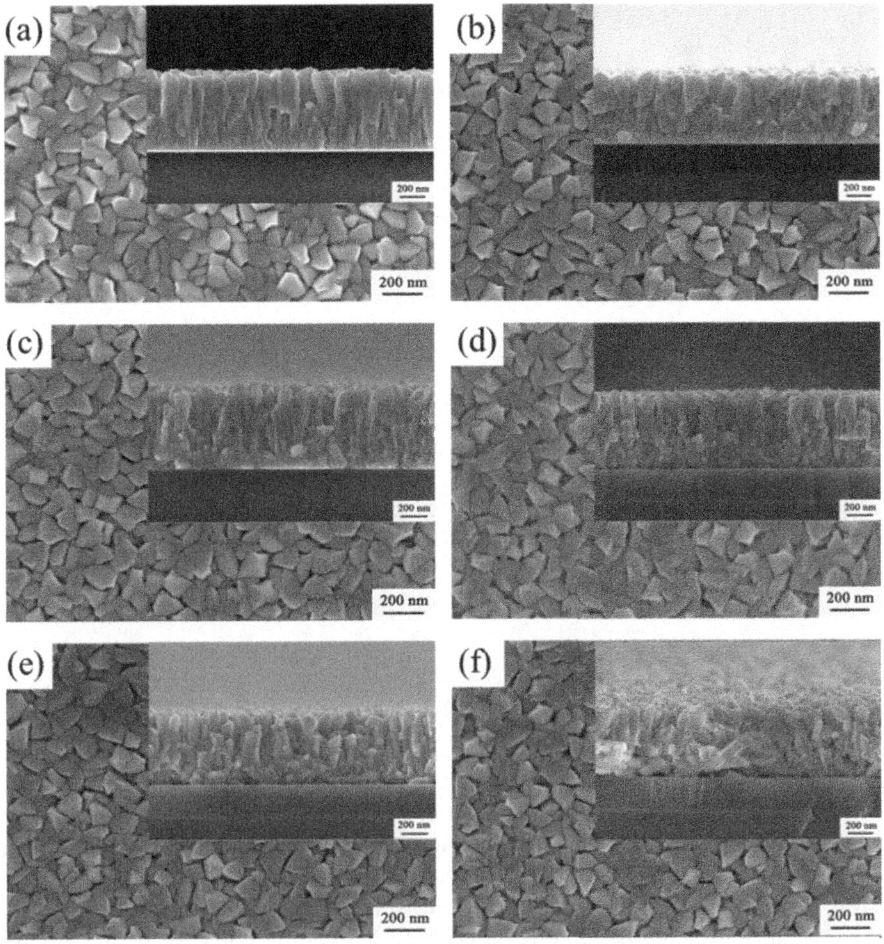

Fig. 5.52 The surface of the Sb_2Te_3 thin film at different annealing temperatures, that is, the cross-sectional topography structure. The small picture in the upper right corner of the figure is the corresponding section: **a** unannealed, **b** 200 °C annealed for 2 h, **c** 250 °C Anneal for 2 h, **d** 300 °C for 2 h, **e** 350 °C for 2 h, and **f** 400 °C for 2 h

the annealing temperature, the worse the consistency becomes. When the annealing temperature is increased to 400 °C, the orderly consistency of the grain growth of the Bi_2Te_3 thin film basically disappears, and the fusion between the grains appears, that is, the melting phase change recrystallization occurs. The morphology of the above Bi_2Te_3 film with annealing temperature is consistent with the XRD test results, as shown in Fig. 5.52b.

In summary, when the annealing temperature of Bi_2Te_3 film and Sb_2Te_3 film is increased to 400 °C, the structure of Bi_2Te_3 film exhibits high stability. This may be related to the stronger metallicity of the Bi element. In addition, the deposition

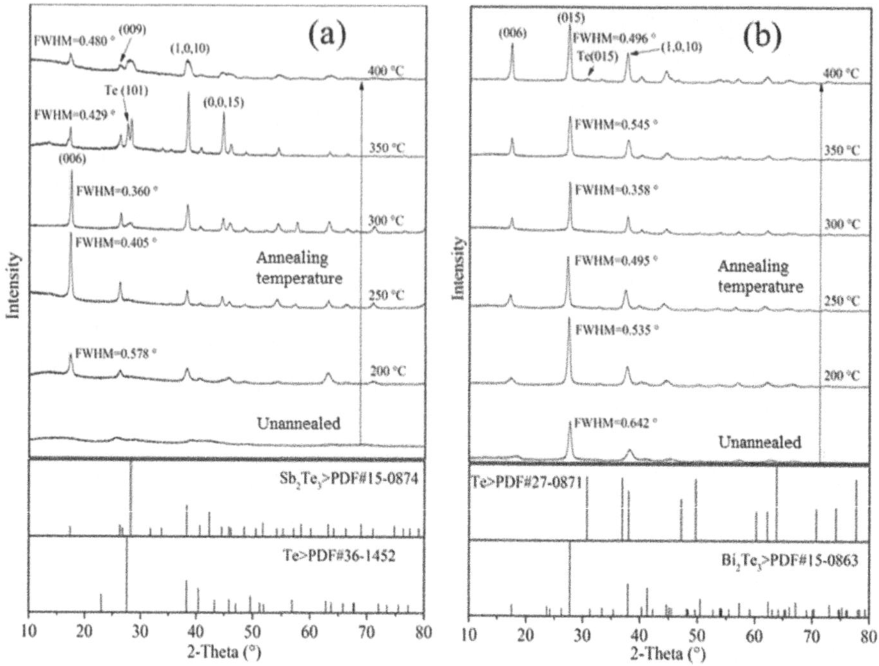

Fig. 5.53 XRD patterns of Sb_2Te_3 film and Bi_2Te_3 film at different annealing temperatures: **a** Sb_2Te_3 film, **b** Bi_2Te_3 film

pressure of Bi_2Te_3 film is higher than that of Sb_2Te_3 film, and the working pressure of sputtering has a certain effect on the morphology and structure of the film [25]. The deposition pressure of Sb_2Te_3 film is 1.1 Pa, and it has a good crystalline state before annealing. Since the melting point of Bi is only slightly higher than 270 °C, when the annealing temperature is higher than its melting point, Bi is in a molten state. When a large amount of Te volatilization is lost in the film, the molten Bi will diffuse in time to fill the gap caused by the volatilization loss of Te. Therefore, when the Bi_2Te_3 film is increased with the annealing temperature, holes are not formed due to the excessively high annealing temperature like the Sb_2Te_3 film.

Figure 5.53 shows the XRD patterns of Sb_2Te_3 film and Bi_2Te_3 film at different annealing temperatures. From the XRD pattern of the Sb_2Te_3 thin film shown in Fig. 5.53a, it can be known that according to the data of Sb_2Te_3 (JCPDS 15.0874), Sb_2Te_3 has a layered diamond structure with a space group of R-3 m. In the unannealed state, the XRD diffraction pattern has no obvious diffraction peaks and is amorphous. After annealing, the film changes from an amorphous state to a crystalline state. When the annealing temperature is increased from 200 to 300 °C, the main diffraction peaks are (006), (009), (1, 0, 10), and the crystals grow mainly in these three directions. When the annealing temperature is increased to 350 °C, the main direction of crystal growth changes, which indicates that the Sb_2Te_3 film has undergone recrystallization and phase transformation. At the same time, a new Te

(101) phase appeared, which is consistent with the results observed by XRD. When the annealing temperature is between 200 and 300 °C, the Sb_2Te_3 film is mainly along the (006) direction, and the full width at half maximum of the diffraction peak is reduced from 0.578° at 200 to 0.360° at 300 °C.

According to Debye-Scherrer formula, when the annealing temperature is increased from 200 to 300 °C, the grain size of Sb_2Te_3 film gradually increases. The grain size increases, the grain boundaries decrease, the scattering effect of carriers on the grain boundaries decreases, and the electrical conductivity of the film increases. When the annealing temperature is increased to 350 °C, although the full width at half maximum of the diffraction peak in the (006) direction becomes larger, that is, the grain size becomes smaller. However, due to phase recrystallization, the main growth directions of the grains are changed to (1, 0, 10) and (0, 0, 15), and the full width at half maximum of the diffraction peaks in these two directions is still 350. Decrease at °C, that is, the grain size of the Sb_2Te_3 film becomes larger, which causes the conductivity to increase. When the annealing temperature is increased to 400 °C, the width of all diffraction peaks increases, and the crystal grain size of the film decreases. This may be because the annealing temperature is too high, and the Te element is precipitated due to high vapor pressure volatilization, thereby forming grain nucleation centers. A large number of nucleation centers form more grains, resulting in a reduction in grain size. The grain size decreases, the grain boundaries increase, and the scattering effect of carriers on the grain boundaries is enhanced, which causes the conductivity of the film to decrease.

From the XRD pattern of the Bi_2Te_3 thin film shown in Fig. 5.53b, it can be known that according to the data of Bi_2Te_3 (JCPDS 15.0863), Bi_2Te_3 has a layered diamond structure with a space group of R-3 m. The diffraction pattern of Bi_2Te_3 thin film also has obvious diffraction peaks in the unannealed state, indicating that Bi_2Te_3 has good crystallinity. The main diffraction peaks of the diffraction pattern of Bi_2Te_3 thin film are (006), (015), (1, 0, 10), and the grains grow mainly in these three directions. The diffraction peak intensity in the (015) direction is the strongest. When the annealing temperature is 400 °C, a (015) diffraction peak appears as the Te phase. With the increase of the annealing temperature, the intensity of the (006) and (1, 0, 10) diffraction peaks gradually increases, and their full width at half maximum gradually decreases, which can indicate that the grain size in these two directions is increasing. For the (015) direction, when the annealing temperature is increased from 200 to 300 °C, the full width at half maximum of the diffraction peak gradually decreases. When the annealing temperature is 350 °C, the full width at half maximum of the diffraction peak increases. When the annealing temperature is 400 °C, the full width at half maximum of the diffraction peak decreases again. That is, when the annealing temperature is increased from 200 to 300 °C, the grain size in the (015) direction gradually increases. When the annealing temperature is 350 °C, the grain size decreases, and the annealing temperature is 400 °C. At this time, the grain size increases again. When the annealing temperature is increased from 200 to 300 °C, the grain size increases, grain boundaries decrease, carrier scattering decreases, carrier mobility increases, and electrical conductivity increases. At 350 °C, due to the high temperature, Te volatiles and precipitates to form a nucleation center, and Bi is in a molten state. When cooling, the precipitated Te is used to form a large number of new fine grains. When the annealing temperature is increased to 400 °C, the grain size formed by recrystallization increases. Due to the increase of phases in the (006) and (1, 0, 10) directions, the preferred orientation of the crystal grains in the film is destroyed due to the phase transition, and the disordered state of the crystal grains increases, as shown in Fig. 5.53f. As a result, the carrier scattering is enhanced, and the carrier mobility is reduced, which results in a decrease in the conductivity of the thin film.

5.4.3 Relationship Between Working Pressure and Sputtering Deposition Rate

During the sputter deposition of thin films, the changes in the sputter deposition rates of Sb_2Te_3 and Bi_2Te_3 thin films at different working pressures are shown in Fig. 5.54. It can be seen from the figure that as the working pressure increases, the sputtering deposition rate of the thin film increases first and then decreases. Regarding the influence of the sputtering working pressure on the sputtering rate of Sb_2Te_3 thin films, when the sputtering working pressure is increased from 0.2 Pa to 1.1 Pa, the sputtering rate is 5.2 Å/s at 0.2 Pa, and the deposition rate is increased to the experimental maximum of 5.4 Å/s when the working pressure is increased to 0.5 Pa. Then increase the sputtering working pressure again, the sputtering rate of the Sb_2Te_3 film decreases rapidly. When the sputtering working pressure is increased to 1.1 Pa, the deposition rate decreases to an experimental minimum of 3.0 Å/s. Regarding the effect of sputtering working pressure on the sputtering rate of Bi_2Te_3 film, when the sputtering working pressure is increased from 0.2 Pa to 1.1 Pa, the sputtering rate is 5.0 Å/s at 0.2 Pa, and the deposition rate is increased increase to the experimental maximum of 6.0 Å/s when the working pressure is increased to

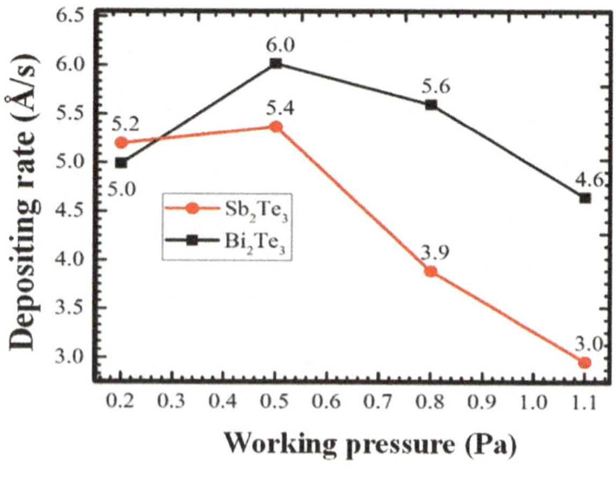

Fig. 5.54 Relationship between sputtering rates and working pressure of Sb_2Te_3 and Bi_2Te_3 films

0.5 Pa. Then increase the sputtering working pressure again, the sputtering rate of the Bi_2Te_3 film decreases rapidly. When the sputtering working pressure is increased to 1.1 Pa, the deposition rate decreases to the experimental minimum of 4.6 Å/s. For the reason that the film deposition rate increases and then decreases with the increase of the working pressure, the main reason is that during the film deposition process, the interaction between the sputtered target atoms and the working gas Ar atoms depends on the working gas concentration (that is working pressure). When the sputtering working pressure is low (such as 0.2 Pa to 0.5 Pa), the Ar atom concentration is low, the number of Ar^+ bombarded on the target surface is small, the number of target atoms bombarded is low, and the number of atoms that can be deposited on the surface of the substrate is small. The thickness of the film formed in a certain time is low, so the deposition rate is very low. However, with the increase of the working pressure, the concentration of Ar atoms in the working gas increases, the number of Ar^+ bombarded on the target surface increases, and the number of target atoms bombarded increases. The thickness of the film increases, so the deposition rate increases.

However, the increase in the concentration of the working gas particles increases the number of bombarded target atoms, and also increases the probability of collisions between the bombarded target atoms and the working gas atoms, resulting in a decrease in the average free path of the target atom motion, to a certain extent, reduces the number of target atoms reaching the substrate surface [26]. However, in the case of low working pressure, this weakening effect is less than the increasing effect of increasing working pressure on the number of deposited atoms. When the air pressure is gradually increased, the weakening effect of the working gas atoms is strengthened. When the effect of increasing the working pressure on the deposition rate is enhanced and weakened, the sputtering deposition rate of the thin film reaches the maximum, the working pressure is increased to 0.5 Pa. At the same time, the above two effects are mutually flat, and the film deposition rate reaches a maximum. When the working pressure continues to increase, the collision effect of the working gas particles on the deposited atoms continues to increase, and the weakening effect of the working gas particles on the film deposition rate exceeds its enhancing effect, causing the film deposition rate to decrease. The higher the working pressure is the stronger the weakening effect and the lower the sputtering rate. Therefore, when the working pressure increases from 0.5 to 1.1 Pa, the film deposition rate decreases.

5.4.4 Relationship Between Working Pressure, Annealing Temperature and Atomic Percentage of Thin Film

Sb_2Te_3 and Bi_2Te_3 are currently one of the best performing materials in room temperature thermoelectric materials. Strict stoichiometric ratio is a key factor affecting the thermoelectric performance of thermoelectric materials. The relationship between the atomic number ratio in Sb_2Te_3 and Bi_2Te_3 thermoelectric thin films and the

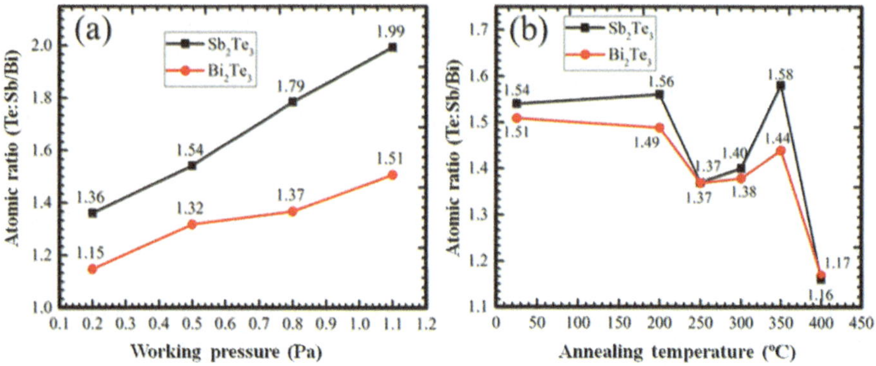

Fig. 5.55 The atomic ratio of Sb_2Te_3 and Bi_2Te_3 films varies with different experimental conditions: **a** the relationship between atomic ratio and working pressure, and **b** the relationship between atomic ratio and annealing temperature

sputtering working pressure and annealing temperature is shown in Fig. 5.55, where Fig. 5.55a is the atomic number of Sb_2Te_3 and Bi_2Te_3 thin films at the sputtering working pressure. Figure 5.55b shows the effect of annealing temperature on the atomic ratio of Sb_2Te_3 and Bi_2Te_3 films. It can be seen from Fig. 5.55a that the ratio of Te atoms to Sb/Bi atoms in Sb_2Te_3 and Bi_2Te_3 films increases with the increase of the sputtering working pressure. The ratio of Te atoms to Sb atoms in the Sb_2Te_3 film increased from 1.36 when the working pressure was 0.2 Pa to 1.99 when the working pressure was 1.1 Pa, and the ratio increased by 46.3%. The ratio of Te atoms to Bi atoms in the Bi_2Te_3 film increased from 1.15 at a working pressure of 0.2 Pa to 1.51 at a working pressure of 1.1 Pa, and the ratio increased by 31.3%. The increase ratio of the Te atom to Sb atom in the Sb_2Te_3 film is higher than the increase ratio of the Te atom to Bi atom in the Bi_2Te_3 film.

With the increase of the sputtering working pressure, the ratio of the number of Te atoms to Sb atoms and the ratio of the number of Te atoms to Bi atoms in the Sb_2Te_3 and Bi_2Te_3 films also gradually increased. There are two main reasons why the ratio of the two kinds of atoms in Sb_2Te_3 and Bi_2Te_3 films increases with the increase of sputtering working pressure: one is that the two atoms are bombarded from the target at different sputtering working pressure; another reason is that the two types of atoms reach the substrate under different sputtering working pressures to form a different number of films.

For the first reason, the Te target is located at the RF target, and the RF target is not affected by the conductivity of the target. Therefore, as the working pressure of the sputtering increases, the number of Te atoms generated by the Te target increases. However, for the Sb target, the Sb target is located on a DC target, and the sputtering of the DC target is affected by the conductivity of the target. Although Sb has a certain conductivity, it has poor conductivity. At the Sb target site, with the increase of the sputtering working pressure, the number of electrons ionized by the Ar gas in the working gas increases, and the accumulation of electrons on the Sb target surface has a certain weakening effect on the target electric field. Therefore, while the number of Sb atoms generated at the Sb target site increases with the increase in air pressure, it is also weakened by the electric field of the accumulated electrons on the target surface. As a result, the increase rate of the number of Te atoms with the increase of the sputtering working pressure is higher than the increase rate of the Sb atoms, which causes the ratio of the number of Te atoms to Sb atoms to increase with the increase of the sputtering working pressure. For the Bi_2Te_3 film under different sputtering working pressures, the increase in the ratio of the number of Te atoms and Bi atoms with the sputtering working pressure is similar to that of Sb_2Te_3 films, and it is also caused by the first reason. However, since the Bi target is more metallic and conductive than the Sb target, the electron aggregation effect on the Bi target is weaker than that of the Sb target. Therefore, the ratio of the Te atom to the Sb/Bi atom in the Sb_2Te_3 and Bi_2Te_3 films increases, and Bi_2Te_3 film is lower than the Sb_2Te_3 film.

For the second reason, the radius of the Te atom is smaller than the radius of the Sb atom and the Bi atom. During the movement of the sputtered atom toward the substrate, the atom with a larger radius has a higher probability of being collided in the working atmosphere. The average free path of the spheroid decreases and the number of atoms reaching the substrate decreases. Therefore, with the increase of the sputtering working pressure, the collision probability of Te atoms with a small radius is lower than that of Sb atoms and Bi atoms with a large radius. Therefore, the ratio of the number of Te atoms to Sb/Bi atoms increases with the sputtering working pressure becoming higher and increase.

It can be seen from Fig. 5.55b that before the annealing temperature is 200 °C, the ratio of the number of Te atoms and Sb/Bi atoms in the film remains basically stable. When the annealing temperature reaches 250 °C, the proportion of Te atoms decreases continually with increasing of the temperature. When the temperature rises to 350 °C, the measured atomic ratio of Te atoms in the film increases, and even in Sb_2Te_3, the atomic ratio of Te atoms in the film exceeds that of unannealed Te. When the annealing temperature is increased to 400 °C, the number of Te atoms in the film decreases sharply. The main reason for the above changes may be that the element Te has a high vapor pressure [27], especially in Sb_2Te_3 films. Due to the high vapor pressure of the Te element, it is easily volatile under vacuum conditions, especially at higher temperatures. However, in this experiment, annealing was performed under normal pressure in a nitrogen atmosphere. The Te element remained relatively stable at a lower annealing temperature (200 °C), and the proportion of Te atoms in the film was basically stable. However, with the increase of the annealing temperature, the Te element on the film surface began to volatilize, and some Te elements precipitated and crystallized on the film surface, causing the Te element in the film to decrease. As the annealing temperature continues to rise, the Te element inside the film precipitates on the surface and forms holes in the film. At the same time, Te rapidly grows and deposits on the film surface, and the Te precipitates grow up, as shown in Fig. 5.56. Te particles grow on the surface of the film, and the particle size is equivalent to the thickness of the film. The distribution of particles of this size on the surface of the film causes the Te element to be enriched at the near surface of the film. As a

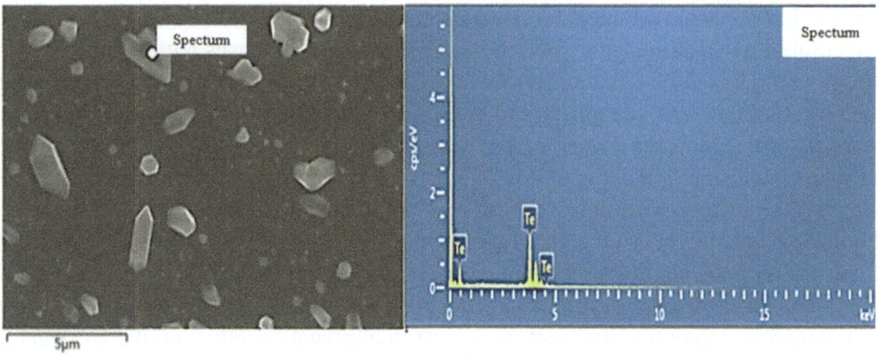

Fig. 5.56 Energy spectrum element analysis of precipitated particles on Sb_2Te_3 film annealed at 350 °C for 2 h

result, the Te atomic ratio obtained by surface scanning showed an increase. When the annealing temperature is increased to 400 °C, a large amount of Te element in the film is volatilized. Due to the high temperature, the Te element is volatilized seriously, and large holes are formed in the film. A large part of the volatile Te element is deposited on the inner wall of the annealing furnace cavity, and a part of the Te element precipitates and grows on the surface of the film. Due to the large film loss, the Te atomic ratio of the film is greatly reduced.

According to Fig. 5.55b, although the change of the atomic ratio of Sb_2Te_3 and Bi_2Te_3 films with annealing temperature is consistent, the change of the atomic ratio of Bi_2Te_3 film is smaller than that of Sb_2Te_3 film. The reason may be that Bi atoms are less electronegative than Sb atoms, and the strength of bonding with Te atoms is higher than that of Sb atoms. Therefore, the Te atom and Bi atom have a high binding force, which can suppress the volatilization of the Te element to a certain extent.

5.4.5 Relationship Between Sb_2Te_3 and Bi_2Te_3 Film Thickness and Annealing Temperature

Figure 5.57 shows the relationship between the thickness of Sb_2Te_3 and Bi2Te3 films and the annealing temperature. It can be seen from the figure that for the Sb_2Te_3 film, the thickness of the film after annealing increases with the increase of the annealing temperature. The main reason is that the Te element in the film precipitates and grows on the surface of the film. As the annealing temperature increases, more Te elements are precipitated from the inside of the Sb_2Te_3 film, and pores are formed in the film. The precipitated Te elements precipitate crystals on the surface of the film and grow into large crystal particles. The morphology of the film shown is changed. However, for the Bi_2Te_3 film, with the increase of annealing temperature, the thickness of the film after annealing decreases. When the annealing temperature

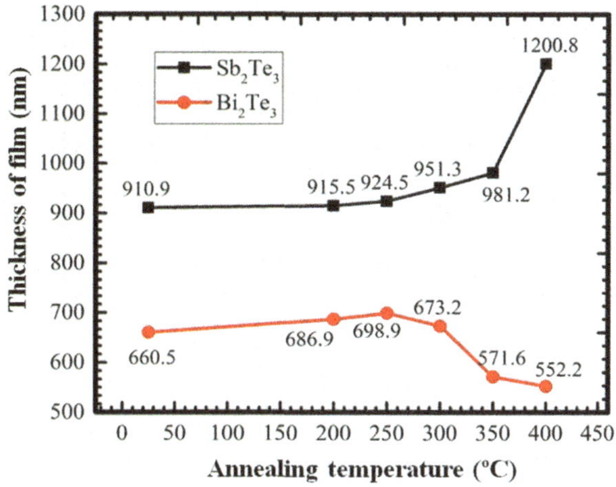

Fig. 5.57 Relationship between the thickness of Sb_2Te_3 and Bi2Te3 films and the annealing temperature

is increased, the Te element is volatilized due to the high vapor pressure. However, due to the low melting point of Bi, the diffusion mobility of Bi atoms is high at a higher annealing temperature. When annealing at high temperature, especially when the annealing temperature is higher than the melting point of Bi, Bi is in a molten state. Therefore, when the volatile element Te is lost, Bi and Te can be recrystallized quickly to restructure the thin film. The more Te loss, the thinner the Bi_2Te_3film will be after the film recombination.

5.4.6 Thermoelectric Properties of Sb_2Te_3 and Bi_2Te_3 Films

Figure 5.58 shows the relationship between the carrier concentration, the electrical conductivity, and the annealing temperature of the Sb_2Te_3 and Bi2Te3 films. According to Fig. 5.58a, the carrier concentration results obtained from the Hall test show that the carrier concentration value in the Sb_2Te_3 film is positive, indicating that the carrier type in the Sb_2Te_3 film is hole and the material is P-type. Thermoelectric materials. The negative value of the carrier concentration in the Bi_2Te_3 film indicates that the carrier type in the Bi_2Te_3 film is electron and the material is an N-type thermoelectric material. When the annealing temperature was increased from 200 to 350 °C, the absolute carrier concentration of Sb_2Te_3 and Bi_2Te_3 films decreased with the increase of annealing temperature. The carrier concentration of Sb_2Te_3 film decreased from 3.98×10^{20} to 0.77×10^{20} cm^{-3}, and the absolute value of carrier concentration of Bi_2Te_3 film decreased from 4.71×10^{20} to 1.10×10^{20} cm^{-3}. Continue to increase the annealing temperature to 400 °C, the carrier concentration of Sb_2Te_3 and Bi_2Te_3 films increased slightly, the carrier concentration in Sb_2Te_3 film increased to 0.91×10^{20} cm^{-3}, and the carrier concentration in Bi_2Te_3 film absolute value rose to 0.91×10^{20} cm^{-3}.

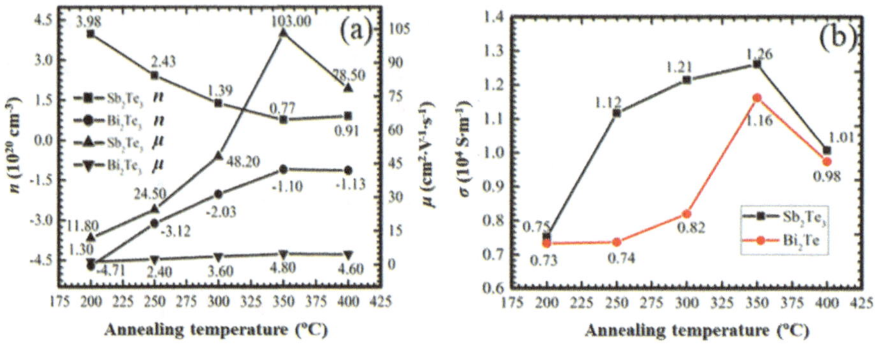

Fig. 5.58 Relations between carrier concentration, carrier mobility, conductivity, and annealing temperature of Sb_2Te_3 and Bi_2Te_3 films: **a** relationship between carrier concentration, mobility, and annealing temperature, **b** conductivity Relationship with annealing temperature

From Fig. 5.58b, it can be seen that the electrical conductivity of the Sb_2Te_3 and Bi_2Te_3 films increases with the increase of the annealing time, and reaches a maximum value at an annealing temperature of 350 °C. The electrical conductivity of the Sb_2Te_3 films from 200 °C The conductivity of Bi_2Te_3 thin film increased from 0.75×10^4 to 1.26×10^4 S m^{-1} from 0.73×10^4 S m^{-1} at 200 °C to 1.16×10^4 S m^{-1}. However, when the annealing temperature is increased to 400 °C, there is a significant decrease. The conductivity of the Sb_2Te_3 film decreases to 1.01×10^4 S m^{-1}, and the conductivity of the Bi_2Te_3 film decreases to 0.98×10^4 S m^{-1}.

The increase in conductivity may be due to the increase in annealing temperature, the crystal growth has a preferred orientation, and the growth of grains causes the reduction of grain boundaries. This reduces the effect of grain boundaries on carrier scattering and reduces the mobility of carriers. When the annealing temperature is too high, more crystal defects may be caused by the loss of Te atoms, which results in enhanced carrier scattering and reduced carrier mobility, resulting in a decrease in conductivity, as shown in Fig. 5.58a.

Figure 5.59 shows the curves of Seebeck coefficient and power factor of Sb_2Te_3 and Bi_2Te_3 thin film materials with annealing time. It can be seen from Fig. 5.65a that the absolute value of Seebeck coefficient increases with the increase of annealing temperature and reaches a maximum value at 350 °C. The Seebeck coefficient of Sb_2Te_3 film increases from 104.2 to 127.5 μV K^{-1}, the absolute value of Seebeck coefficient of Bi_2Te_3 thin film material increased from 108.8 to 134.4 μV K^{-1}. When the annealing temperature is increased to 400 °C, the absolute value of Seebeck coefficient decreases, and the Seebeck coefficient of Sb_2Te_3 thin film material decreases to 121.5 μV K^{-1}, and the absolute value of Seebeck coefficient of Bi_2Te_3 thin film material decreases to 127.9 μV K^{-1}.

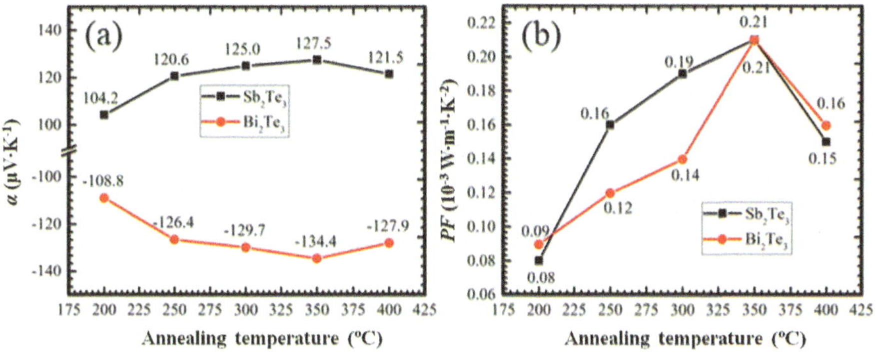

Fig. 5.59 Relationship between Seebeck coefficient, power factor and annealing temperature of Sb_2Te_3 and Bi_2Te_3 films

5.5 Chapter Summary

Using magnetron sputtering to prepare Sb_2Te_3, Bi_2Te_3 and Bi–Sb–Te based films, by adjusting the magnetron sputtering power, air pressure, substrate temperature, film thickness, annealing temperature, the structure and thermoelectric properties of the film were studied to adjust the film provides a reference for its structure and performance.

1. Sb_2Te_3 films were prepared by magnetron sputtering. The magnetron sputtering process parameters such as target sputtering power and deposition time were explored, and the effects of annealing on the crystallinity, morphology and thermoelectric properties of Sb_2Te_3 films were studied.

Using RF magnetron sputtering to deposit Sb_2Te_3 films at room temperature, the deposition rate of the films increases almost linearly with the increase of sputtering power. With the increase of the sputtering power, the resistivity of the Sb_2Te_3 film gradually decreases, while the Seebeck coefficient increases first and then decreases. When the RF power is 20 W, the Sb_2Te_3 film deposited at room temperature has a maximum power factor of 2.5 μW cm^{-1} K^{-2}. After annealing, the grain size and crystallinity of Sb_2Te_3 thin film increased, and the (1010) crystal plane preferentially grew. When the annealing temperature is 250 °C, Sb-rich nanocrystals are deposited on the surface of the film and a small amount of micro-nanopores appear. When the annealing temperature was further increased, part of the film on the surface of the film was peeled off and was spotted, and phase separation began to occur in the Sb_2Te_3 film. With the increase of annealing temperature, the carrier concentration of Sb_2Te_3 film decreases and the mobility increases. Further analysis shows that the average free path of the carriers in the film also gradually increases with the increase of the annealing temperature, but is much smaller than the grain size of the film. The resistivity of the Sb_2Te_3 film was significantly reduced after annealing, and the minimum value was 0.85 mΩ cm at an annealing temperature of 250 °C. When the temperature is further increased, the resistivity of the film increases sharply due to the deterioration of the film composition and microstructure. The Seebeck coefficient test shows that the Sb_2Te_3 film deposited at room temperature is a P-type semiconductor material. After optimized annealing conditions, the power factor of the Sb_2Te_3 film increased from 2.5 $\mu W\ cm^{-1}\ K^{-2}$ at room temperature to 18.09 $\mu W\ cm^{-1}\ K^{-2}$, and the thermoelectric performance was significantly improved.

When magnetron sputtering is used to deposit Sb_2Te_3 films with different thicknesses, as the sputtering time is extended, the relative temperature of the substrate is increased under the irradiation of the sputtering glow, the film is heated, and the atoms get enough energy and time to diffuse. The micro-crystal grains in the film are fused with each other and grow up, and the crystallinity of the film is gradually enhanced. As the thickness of the Sb_2Te_3 film increases, the film resistivity decreases gradually, and the Seebeck coefficient has no obvious trend of change. The maximum power factor of the Sb_2Te_3 film with a thickness of 270 nm after annealing at 250 °C is 19.55 $\mu W\ cm^{-1}\ K^{-2}$. Compared with annealing, the effect of film thickness on the thermoelectric properties of the material is weak.

2. Bi–Sb–Te alloy films with relatively uniform doping were prepared by magnetron co-sputtering. The magnetron sputtering process parameters such as target sputtering power, deposition time and substrate temperature were explored Bi–Sb–Te alloy film's morphological structure and thermoelectric properties; At the same time, the effects of annealing temperature on the crystallinity, morphology and thermoelectric properties of Bi–Sb–Te alloy films were analyzed.

When the Bi–Sb–Te alloy film is prepared by co-sputtering, the composition of Bi in the alloy film increases with the increase of the sputtering power of the Bi target. With the increase of Bi target sputtering power, the resistivity of the alloy thin film gradually increases while the Seebeck coefficient decreases. The XRD pattern of the Bi–Sb–Te alloy film deposited at room temperature has only a weaker Bi (220) diffraction peak. After annealing treatment, the alloy thin film is preferentially grown along the (001) direction to form a layered structure. When the fire temperature is 350 °C, the surface of the film becomes mottled and the composition deteriorates. Compared with Sb_2Te_3 film, the Bi–Sb–Te alloy film prepared by co-sputtering at room temperature has poor thermoelectric properties, but has better thermal stability.

With the increase of the annealing temperature, the carrier concentration of the Bi–Sb–Te alloy film gradually decreases and the mobility gradually increases, the resistivity of the film decreases and the Seebeck coefficient increases. Compared with the Sb_2Te_3 film, the average free path of the carrier of the Bi–Sb–Te alloy film is reduced, and the resistivity of the film is increased. When the annealing temperature is 300 °C, the power factor of the Bi–Sb–Te alloy film is 22.54 $\mu W\ cm^{-1}\ K^{-2}$. It can be seen that the alloy film doped with appropriate amount of Bi nano-grains has better thermoelectric properties and thermal stability than Sb_2Te_3 film after annealing optimization.

3. When Bi–Sb–Te alloy films with different thicknesses are prepared by co-sputtering, the resistivity and Seebeck coefficient of the alloy films gradually decrease with increasing thickness. After annealing, the thermoelectric properties of Bi–Sb–Te alloy films with different thicknesses are significantly improved. The maximum power factor of the film with a thickness of 280 nm after annealing (300 °C) is 26.41 $\mu W\ cm^{-1}\ K^{-2}$.
4. The substrate temperature has a significant effect on the morphology and microstructure of Bi–Sb–Te alloy films. When the substrate temperature is 150 °C, the film forms an (1010) phase-based alloy thin film; as the substrate temperature increases, the crystallinity of the alloy thin film increases and the film roughness increases. Compared with Bi–Sb–Te alloy film deposited at room temperature, the substrate-heated alloy film has lower resistivity and higher Seebeck coefficient. When the annealing temperature is 300 °C, the power factor of the film with a substrate temperature of 150 °C has a maximum value of 25.32 $\mu W\ cm^{-1}\ K^{-2}$. It can be seen that the proper substrate temperature can improve the thermoelectric properties of the film.
5. Sb_2Te_3/Te heterojunction thin film was prepared by magnetron co-sputtering. The crystallization and growth of Te in the heterojunction thin film were controlled by annealing, and a thin film with good thermoelectric properties was obtained. On the one hand, the crystallization of Te causes compressive stress on the main phase Sb_2Te_3, which may induce the ETT of Sb_2Te_3 and optimize the power factor. On the other hand, suitable Te nanoparticles introduce a large number of nano-interfaces, which strongly scatter phonons and reduce thermal conductivity. With the crystallization and growth of Te nanoparticles, the room temperature power factor increased from 0.6 to 11.2 $\mu W\ cm^{-1}\ K^{-2}$ and then decreased to 7.1 $\mu W\ cm^{-1}\ K^{-2}$. The corresponding thermal conductivity is 0.41 $W\ m^{-1}\ K^{-1}$, 0.55 $W\ m^{-1}\ K^{-1}$, and 0.69 $W\ m^{-1}\ K^{-1}$, respectively. Finally, the corresponding ZT value increased from 0.05 to 0.61 and then decreased to 0.31. In summary, Te nanoparticles of appropriate size and induced lattice strain can strongly scatter phonons and filter low-energy carriers to optimize the thermoelectric properties of the material.
6. Sb_2Te_3 and Bi_2Te_3 films were prepared by magnetron co-sputtering. Under the conditions of DC Sb target 15 W, DC Bi target 13 W, and RF Te target 10 W, the deposition rates of Sb_2Te_3 and Bi_2Te_3 films increased with the increase of sputtering working pressure. The deposition rate is the largest at 0.5 Pa. When the working pressure is increased, the deposition rate decreases. For Sb_2Te_3 thin films, the ratio of the number of atoms of Sb and Te atoms is close to 2:3 when the working pressure is 0.5 Pa. For the Bi_2Te_3 thin film, the atomic ratio of Bi atoms to Te atoms is close to 2:3 when the working pressure is 1.1 Pa. The annealing temperature can cause changes in the thickness of the deposited film, the change in the atomic ratio, and the electrical properties of the thermoelectric film. When the annealing temperature is 350 °C, the power factors of Sb_2Te_3 and Bi_2Te_3 films reach the maximum, which are 2.1 $\mu W\ cm^{-1}\ K^{-2}$.

References

1. Scherrer P (1918) Determination of the size and internal structure of colloidal particles using X-rays. Nachr Ges Wiss Göttingen 2:98–100
2. Pal K, Anand S, Waghmare UV (2015) Thermoelectric properties of materials with nontrivial electronic topology. J Mater Chem C 3(46):12130–12139
3. Shekar NVC, Polvani DA, Meng JF, Badding JV (2005) Improved thermoelectric properties due to electronic topological transition under high pressure. Phys B 358(1–4):14–18
4. Zhang H, Liu CX, Qi XL, Dai X, Fang Z, Zhang SC (2009) Topological insulators in Bi_2Se_3, Bi_2Te_3 and Sb_2Te_3 with a single Dirac cone on the surface. Nat Phys 5(6):438–442
5. Zhao K (2017) First principle study of electronic topological transition in Bi_2Te_3 and Sb_2Te_3 under pressure Doctor, Harbin Institute of Technology
6. Gomis O, Vilaplana R, Manjón FJ, Rodríguez-Hernández P, Pérez-González E, Muñoz A, Kucek V, Drasar C (2011) Lattice dynamics of Sb_2Te_3 at high pressures. Phys Rev B 84(17):174305
7. Zhao K, Wang Y, Xin C, Sui Y, Wang X, Wang Y, Liu Z, Li B (2016) Pressure-induced anomalies in structure, charge density and transport properties of Bi_2Te_3: a first principles study. J Alloy Compd 661:428–434
8. Bera A, Pal K, Muthu DVS, Sen S, Guptasarma P, Waghmare UV, Sood AK (2013) Sharp Raman anomalies and broken adiabaticity at a pressure induced transition from band to topological insulator in Sb_2Se_3. Phys Rev Lett 110(10):107401
9. Yu Z, Wang L, Hu Q, Zhao J, Yan S, Yang K, Sinogeikin S, Gu G, Hk Mao (2015) Structural phase transitions in Bi_2Se_3 under high pressure. Sci Rep 5:15939
10. Zhang J, Hu T, Yan J, Ke F, Wang J, Cui X, Li X, Ma Y, Yang J, Gao C (2017) Pressure driven semi-metallic phase transition of Sb_2Te_3. Mater Lett 209:78–81
11. Meng JF, Shekar NVC, Badding JV, Chung DY, Kanatzidis MG (2001) Multifold enhancement of the thermoelectric figure of merit in p-type $BaBiTe_3$ by pressure tuning. J Appl Phys 90(6):2836–2839
12. Polvani DA, Meng JF, Chandra Shekar NV, Sharp J, Badding JV (2001) Large improvement in thermoelectric properties in pressure-tuned p-type $Sb_{1.5}Bi_{0.5}Te_3$. Chem Mater 13(6):2068–2071
13. Yang HQ, Chen YJ, Wang XY, Miao L, Li XY, Han XD, Lu X, Wang GY, Zhou XY (2018) Realizing high thermoelectric performance in Te nanocomposite through Sb_2Te_3 incorporation. Cryst Eng Comm 20(47):7729–7738
14. Richter W, Becker CR (1977) A Raman and far-infrared investigation of phonons in the rhombohedral V_2–VI_3 compounds Bi_2Te_3, Bi_2Se_3, Sb_2Te_3 and $Bi_2(Te_{1-x}Se_x)_3$ $(0 < x<1)$, $(Bi_{1-y}Sb_y)_2Te_3(0 < y<1)$. Phys Status Solidi (b) 84(2):619–628
15. Pine AS, Dresselhaus G (1971) Raman spectra and lattice dynamics of tellurium. Phys Rev B 4(2):356–371. https://doi.org/10.1103/PhysRevB.4.356
16. Secor J, Harris MA, Zhao L, Deng H, Raoux S, Krusin-Elbaum L (2014) Phonon renormalization and Raman spectral evolution through amorphous to crystalline transitions in Sb_2Te_3 thin films. Appl Phys Lett 104(22):221908
17. Richter W, Renucci JB, Cardona M (1973) Hydrostatic pressure dependence of first-order Raman frequencies in Se and Te. Phys Status Solidi (b) 56(1):223–229
18. Wu Y, Chen Z, Nan P, Xiong F, Lin S, Zhang X, Chen Y, Chen L, Ge B, Pei Y (2019) Lattice strain advances thermoelectrics. Joule 3:1276–1288
19. Klemens PG (1958) Thermal conductivity and lattice vibrational modes. In: Solid State Physics, vol 7. Elsevier, pp 1–98
20. Carruthers P (1959) Scattering of phonons by elastic strain fields and the thermal resistance of dislocations. Phys Rev 114(4):995–1001
21. Liu E, Zhu B, Luo J (2008) Semiconductor physics, 7th edn. Electron Ind Press, Beijing
22. Zhu T, Hu L, Zhao X, He J (2016) New insights into intrinsic point defects in V_2VI_3 thermoelectric materials. Adv Sci 3(7):1600004

23. Noguchi S, Sakata H (1980) Electrical properties of undoped In_2O_3 films prepared by reactive evaporation. J Phys D Appl Phys 13(6):1129–1133
24. Li YY, Qin XY, Li D, Zhang J, Li C, Liu YF, Song CJ, Xin HX, Guo HF (2016) Enhanced thermoelectric performance of $Cu_2Se/Bi_{0.4}Sb_{1.6}Te_3$ nanocomposites at elevated temperatures. Appl Phys Lett 108(6):062104
25. Sun Y, Zhang E, Johnsen S, Sillassen M, Sun P, Steglich F, Bøttiger J, Iversen BB (2011) Growth of FeSb2 thin films by magnetron sputtering. Thin Solid Films 519(16):5397–5402
26. Nuthongkum P, Sakdanuphab R, Horprathum M, Sakulkalavek A (2017) [Bi]:[Te] control, structural and thermoelectric properties of flexible BixTey thin films prepared by RF magnetron sputtering at different sputtering pressures. J Electron Mater 46(11):6444–6450
27. Kim D-H, Byon E, Lee G-H, Cho S (2006) Effect of deposition temperature on the structural and thermoelectric properties of bismuth telluride thin films grown by co-sputtering. Thin Solid Films 510(1–2):148–153
28. Choi H, Jeong K, Chae J, Park H, Baeck J, Kim TH, Song JY, Park J, Jeong KH, Cho MH (2018) Enhancement in thermoelectric properties of Te-embedded Bi_2Te_3 by preferential phonon scattering in heterostructure interface. Nano Energy 47:374–384
29. Yoo I-J, Song Y, Lim DC, Myung NV, Lee KH, Oh M, Lee D, Do Kim Y, Kim S, Choa Y-H (2013) Thermoelectric characteristics of Sb_2Te_3 thin films formed via surfactant-assisted electrodeposition. J Mater Chem A 1(17):5430–5435
30. Yang HQ, Miao L, Liu CY, Li C, Honda S, Iwamoto Y, Huang R, Tanemura S (2015) A facile surfactant-assisted reflux method for the synthesis of single-crystalline Sb_2Te_3 nanostructures with enhanced thermoelectric performance. ACS Appl Mater Inter 7(26):14263–14271
31. Zhang Y, Bahk JH, Lee J, Birkel CS, Snedaker ML, Liu D, Zeng H, Moskovits M, Shakouri A, Stucky GD (2014) Hot carrier filtering in solution processed heterostructures: a paradigm for improving thermoelectric efficiency. Adv Mater 26(17):2755–2761
32. Ko DK, Kang Y, Murray CB (2011) Enhanced thermopower via carrier energy filtering in solution-processable Pt–Sb_2Te_3 nanocomposites. Nano Lett 11(7):2841–2844
33. Shi D, Wang R, Wang G, Li C, Shen X, Nie Q (2017) Enhanced thermoelectric properties in Cu-doped Sb_2Te_3 films. Vacuum 145:347–350
34. Zhang Y, Snedaker ML, Birkel CS, Mubeen S, Ji X, Shi Y, Liu D, Liu X, Moskovits M, Stucky GD (2012) Silver-based intermetallic heterostructures in Sb_2Te_3 thick films with enhanced thermoelectric power factors. Nano Lett 12(2):1075–1080
35. Kim J, Lee KH, Kim SD, Lim JH, Myung NV (2018) Simple and effective fabrication of Sb_2Te_3 films embedded with Ag_2Te nanoprecipitates for enhanced thermoelectric performance. J Mater Chem A 6(2):349–356
36. Schaumann J, Loor M, Ünal D, Mudring A, Heimann S, Hagemann U, Schulz S, Maculewicz F, Schierning G (2017) Improving the ZT value of thermoelectrics by nanostructuring: tuning the nanoparticle morphology of Sb_2Te_3 by using ionic liquids. Dalton T 46(3):656–668
37. Kim J, Zhang M, Bosze W, Park SD, Lim JH, Myung NV (2015) Maximizing thermoelectric properties by nanoinclusion of γ-SbTe in Sb_2Te_3 film via solid-state phase transition from amorphous Sb–Te electrodeposits. Nano Energy 13:727–734
38. Thankamma G, Kunjomana AG (2015) Studies on sulfur doping and figure of merit in vapor grown Sb_2Te_3 platelet crystals. J Cryst Growth 415:65–71
39. Shen S, Zhu W, Deng Y, Zhao H, Peng Y, Wang C (2017) Enhancing thermoelectric properties of Sb_2Te_3 flexible thin film through microstructure control and crystal preferential orientation engineering. Appl Surf Sci 414:197–204
40. Mukherjee S, Femi OE, Chetty R, Chattopadhyay K, Suwas S, Mallik RC (2018) Microstructure and thermoelectric properties of Cu_2Te-Sb_2Te_3 pseudo-binary system. Appl Surf Sci 449:805–814
41. Goncalves LM, Alpuim P, Rolo AG, Correia JH (2011) Thermal co-evaporation of Sb_2Te_3 thin-films optimized for thermoelectric applications. Thin Solid Films 519(13):4152–4157
42. Saleemi M, Ruditskiy A, Toprak MS, Stingaciu M, Johnsson M, Kretzschmar I, Jacquot A, Jagle M, Muhammed M (2014) Evaluation of the structure and transport properties of nanostructured antimony telluride (Sb_2Te_3). J Electron Mater 43:1927–1932. https://doi.org/10.1007/s11664-013-2911-6

43. Mun H, Lee KH, Yoo SJ, Kim HS, Jeong J, Oh SH, Snyder GJ, Lee YH, Kim YM, Kim SW (2018) Highly fluidic liquid at homointerface generates grain-boundary dislocation arrays for high-performance bulk thermoelectrics. Acta Mater 159:266–275
44. Liu EY, Lin FH, Yang ZR, Liu CJ (2018) A facile energy-saving route of fabricating thermoelectric Sb_2Te_3-Te nanocomposites and nanosized Te. Roy Soc Open Sci 5(10):180698
45. Van Thiet D, Van Quang N, Hai NTM, Huong NT, Cho S, Tuan DA, Dung DD, Van Tam T (2018) Optimizing the carrier density and thermoelectric properties of Sb_2Te_3 films by using the growth temperature. J Korean Phys Soc 72(8):915–919
46. Vieira EMF, Figueira J, Pires AL, Grilo J, F. Silva M, M. Pereira A, M. Goncalves L (2019) Enhanced thermoelectric properties of Sb_2Te_3 and Bi_2Te_3 films for flexible thermal sensors. J Alloy Compd 774:1102–1116
47. Schumacher C, Reinsberg KG, Akinsinde L, Zastrow S, Heiderich S, Toellner W, Rampelberg G, Detavernier C, Broekaert JA, Nielsch K (2012) Optimization of electrodeposited p-doped Sb_2Te_3 thermoelectric films by millisecond potentiostatic pulses. Adv Energy Mater 2(3):345–352
48. Shi W, Yu S, Liu P, Fan W (2013) Hydrothermal synthesis and thermoelectric transport properties of Sb_2Te_3–Te heterogeneous nanostructures. Cryst Eng Comm 15(15):2978–2985
49. Rowe IToECToECM (1995) CRC Handbook of thermoelectrics. CRC Press

Chapter 6
Growth of Sb_2Te_3 Films by Molecular Beam Epitaxial Method

The schematic diagram of the atomic layered structure of antimony telluride is shown in Fig. 6.1. Every five monatomic layers form a "quintuple layers (QL)", and the atoms are arranged along the c-axis direction as –[Te(1)–Sb–Te(2)–Sb–Te(1)]–, where (1) and (2) represent two different lattice positions of the Te atom. The adjacent two "five layers" are linked by van der Waals forces between Te(1)–Te(1). This rhombohedral structure determines the anisotropic transmission characteristics of antimony telluride. The structure and performance of the antimony telluride film are inseparable from its preparation process and technology. The antimony telluride film prepared by different preparation methods and conditions differs greatly in morphology and performance. These differences mainly come from the microstructure of the film prepared by different methods, The difference between the type and quantity of impurities or defects. The molecular beam epitaxy method is a near-ideal method for preparing high-quality thin film and thermoelectric heterostructure materials, and is compatible with MEMS processing technology. This section uses molecular beam epitaxy to grow Sb_2Te_3 thin films.

6.1 Sb-Rich Sb_2Te_3 Thin Film

In this section, molecular beam epitaxy is used to deposit Sb_2Te_3 thin films. First, the preparation process is explored to obtain thin films with a set stoichiometric ratio. Subsequently, Sb-rich Sb_2Te_3 thin films are prepared, and the effect of Sb on the thermoelectric properties of Sb_2Te_3 thin films is discussed.

Z. Hu and Z. Wu, *Nanostructured Thermoelectric Films*,
https://doi.org/10.1007/978-981-15-6518-2_6

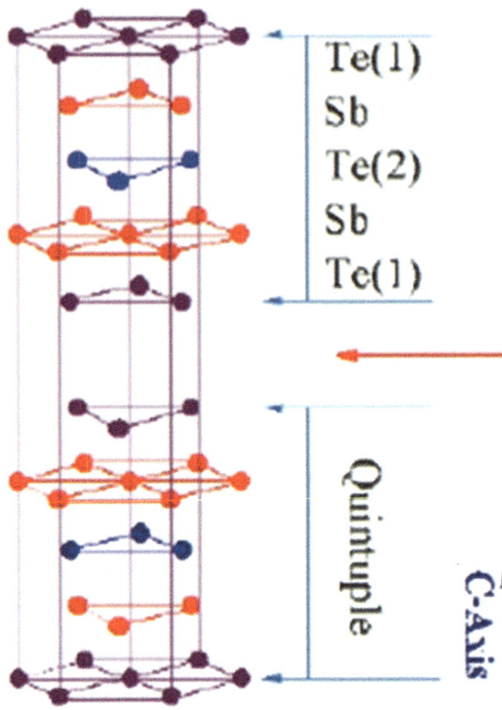

Fig. 6.1 Schematic diagram of antimony telluride lattice structure

6.1.1 Effect of Te and Sb Evaporation Temperature on the Stoichiometric Ratio of Sb_2Te_3-Based Films

In order to obtain Sb_2Te_3 films rich in Sb elements, the relationship between the respective deposition rates of Sb and Te as a function of temperature must be known first, so as to set experimental parameters, such as those used to prepare Sb_2Te_3 films with different contents of Sb elements. The evaporation temperature of Sb and Te can also be used to modify the parameters of the film thickness meter, which is convenient to directly monitor the Sb and Te deposition rate in real time.

The experimental method to explore the deposition rate of Sb and Te is as follows. In the experiment, high-resistance Si (111) and quartz glass wafers were used as substrates, and the size was about 2 cm × 2 cm. The substrate was subjected to a strict cleaning process. Sb and Te raw materials are selected from high-purity (99.999%) elemental granular raw materials, and put into different beam source furnaces according to a certain amount under high vacuum to explore the deposition rate of Sb and Te respectively. First, the deposition rate of Sb was explored. In the different experimental batches, the PID controller is used to give the beam source furnace different heating power, so that the beam source furnace maintains different temperatures. The batch experimental deposition time must be 30 min, and the substrate temperature should be kept at room temperature during the coating process. The thickness of the plated Sb film was measured with a step profiler, and the thickness was divided by the deposition time to obtain the deposition rate of the Sb element at different temperatures. In the same way, the deposition rate of Te at different temperatures can be obtained. The variation curve of the deposition rate of Sb and Te films with the temperature of the beam source furnace is shown in Fig. 6.2.

There is a certain relationship between the deposition rate of Sb and Te films and the stoichiometric ratio of the prepared Sb_2Te_3 thin film. From which the relationship between the stoichiometric ratio of the prepared Sb_2Te_3 thin film and the temperature of the beam source furnace can be obtained. During the experiment, Sb_2Te_3 films with different stoichiometric ratios can be obtained by adjusting the temperature of the Sb and Te beam source furnaces, providing theoretical guidance for the preparation of Sb-rich Sb_2Te_3 films. The quality of the film can be expressed in two different expressions, both of which are equal

$$\rho Sd = nM. \tag{6.1}$$

In the formula, ρ is the density of the material, S is the area of the material, d is the thickness of the material, n is the number of molecules or atoms in the material, and M is the mass of a single molecule or atom in the material. If Sb and Te are taken their block densities of the elements, the values are $\rho_{Sb} = 6.69$ g cm^{-3} and $\rho_{Te} = 6.25$ g cm^{-3}, which are substituted into the Eq. 6.1:

$$\frac{n_{Sb}}{n_{Te}} = \frac{M_{Te}\rho_{Sb}d_{Sb}}{M_{Sb}\rho_{Te}d_{Te}} \tag{6.2}$$

In order to prepare Sb_2Te_3 thin film that meets the stoichiometric ratio, $n_{Sb}/n_{Te}=$ 2:3 is required. From this, $d_{Sb}= 0.59d_{Te}$ is calculated, and the relationship between the deposition rate of Sb and Te elements is $v_{Sb}=0.59v_{Te}$. However, these conclusions are obtained under ideal conditions. In fact, the density of the film is not necessarily the same as the bulk value. The adsorption rate of different materials is not the same

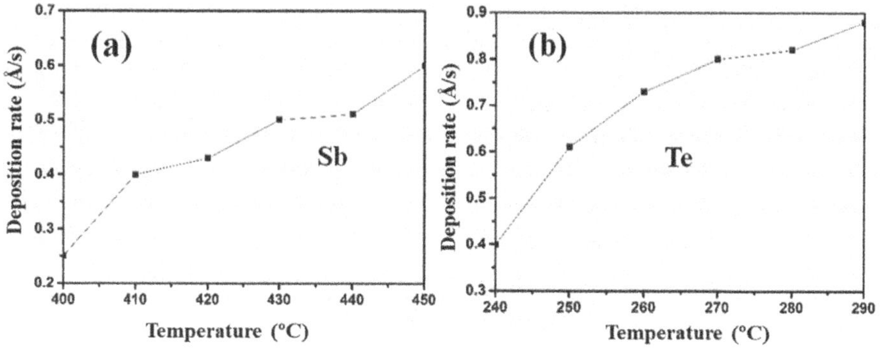

Fig. 6.2 **a** Sb and **b** Te film deposition rate as a function of beam source furnace temperature

under the influence of the substrate temperature. It is difficult for Te atoms to adsorb on the substrate at high temperature. In consideration of the influence of these factors, based on the theoretical derivation and combined with the actual test results, it was finally obtained that when the Sb and Te beam source furnace temperatures were 430 and 270 °C, and the substrate temperature was 150 °C, The Sb_2Te_3 thin film conforms to the stoichiometric ratio, and Sb_2Te_3 thin films with different stoichiometric ratios can be obtained by adjusting the temperature of the Sb or Te beam source furnace. Therefore, it laid a foundation for the research of preparing Sb-rich Sb_2Te_3 films.

6.1.2 Effect of Sb Rich on Microstructure and Thermoelectric Properties of Sb_2Te_3 Thin Films

It is found in the experimental results that Sb-rich Sb_2Te_3 films have higher Seebeck coefficients and better thermoelectric properties. Similar research results have also obtained numerical Seebeck coefficients and high thermoelectric properties in Pb-rich PbTe. The specific process of the experiment is as following that the experimental instrument is a molecular beam epitaxy was used to prepare Sb_2Te_3 films with different stoichiometric ratios by co-evaporation of Sb and Te elements. The substrate selected for the experiment was a high-resistance intrinsic Si (111) and quartz glass wafer, both of which undergo a fine cleaning process. After the substrate is processed, it is immediately transferred to a substrate holder in a vacuum chamber to be vacuumed for coating. The substrate hold was kept at room temperature without heating. When the vacuum reaches 10^{-9} Torr, turn on the PID controller and give independent heating to the Sb and Te beam source furnaces. The PID controller can accurately control the beam source furnace temperature and maintain the beam source furnace temperature at a fairly long time. After the beam source furnace is heated, high-purity Sb and Te are sprayed from the beam source furnace at a constant rate. During the real process, the temperature of the beam source furnace remains constant, and the deposition rate of Sb and Te also remains constant. In this way, Sb_2Te_3 films with a certain stoichiometric ratio can be prepared. For different batches of samples in the experiment, keeping the beam source furnace temperature of Te constant at 270 °C corresponds to a deposition rate of 0.8 Å s^{-1}, while the temperature of the Sb beam source furnace is different at 410 °C, 430 °C, and 450 °C, respectively. The corresponding deposition rates at are 0.4, 0.5, and 0.6 Å s^{-1}. The corresponding experiments obtained samples numbered S1, S2, and S3. After the sample deposition was completed, the prepared sample was characterized by XRD, TEM, EDS and other analytical methods for structure, morphology, and composition, and its Seebeck coefficient, electrical conductivity, and thermal conductivity were tested using homemade equipment. The Hall coefficient test has obtained properties such as carrier mobility and concentration of thin film materials.

By EDS analysis of the prepared Sb_2Te_3 thin film, the relationship between the Sb atomic content percentage and the Sb deposition rate in the thin film is shown in

Fig. 6.3. It can be seen from the figure that as the deposition rate of Sb increases, the composition of the Sb_2Te_3 thin film changes from Sb barren to stoichiometric ratio and finally to Sb surplus. In the S2 sample when the Sb deposition rate is 0.5 Å s^{-1}, Sb The ratio of the number of Te atoms to v is 2:3, that is, a stoichiometric sample was prepared experimentally, but when the deposition rate of Sb increased to 0.6 Å s^{-1}, the percentage of Sb atoms in the S3 sample increased to about 63%, this sample deviates from the stoichiometric ratio and is a sample containing the presence of excess Sb.

The X-ray diffractometer (D\max-2200X, Rigaku, Japan) was used to analyze the crystal structure of the prepared Sb_2Te_3 thin film sample. The parameters used in the test were mainly Cu target Kα rays as the X-ray diffraction source. The wavelength of this ray was $\lambda = 1.54056$ Å, the working voltage used in the measurement is 40 kV, the working current is 40 mA, the scanning range is 10°–90°, and the scanning is performed in steps of 0.02°. The XRD patterns of Sb_2Te_3 thin film samples with different Sb atom content are shown in Fig. 6.3. It can be seen from the figure that only a single phase exists in sample S2 and multiple phases exist in sample S3. In sample S2, diffraction peaks appeared at 18.42°, 26.34°, 38.25°, 44.59°, and 64.17°. These diffraction peaks correspond to (006), (009), (1, 0, 10) (0,0,15) and (0,0,21) of the Sb2Te3 crystal planes, all of these diffraction peaks correspond to the orthorhombic hexahedral phase (JCPDS 15-0874, R $\bar{3}$ m), thereby obtaining polymorphic crystals belonging to the space group R $\bar{3}$ m Sb2Te3 sample. Compared with the sample S2, the diffraction angle of the sample S3 is 18.42°, no Sb_2Te_3 phase diffraction peaks appear at 26.34, and there are two weaker diffraction peaks at 28.66° and 78.40° than the S2 sample, respectively. Corresponding to the (012) and (027) crystal planes of the metal Sb. This shows that, compared with S2 and S3, the metal Sb phase is more in the sample S3, that is, a sample with excess Sb is obtained when the Sb deposition rate is high (Fig. 6.4).

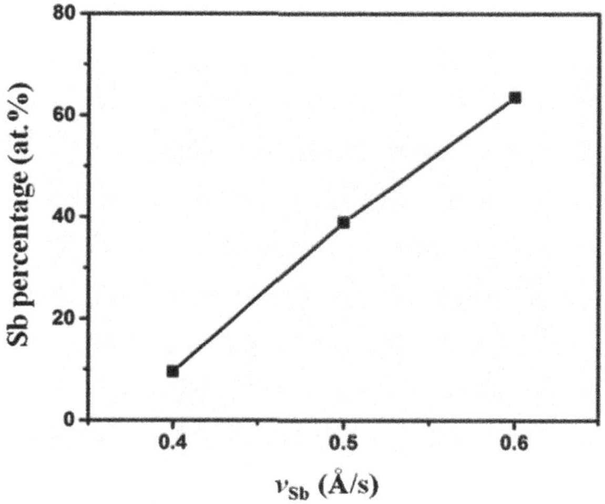

Fig. 6.3 Relationship between Sb atom content and Sb deposition rate v_{sb} in the prepared Sb_2Te_3 film

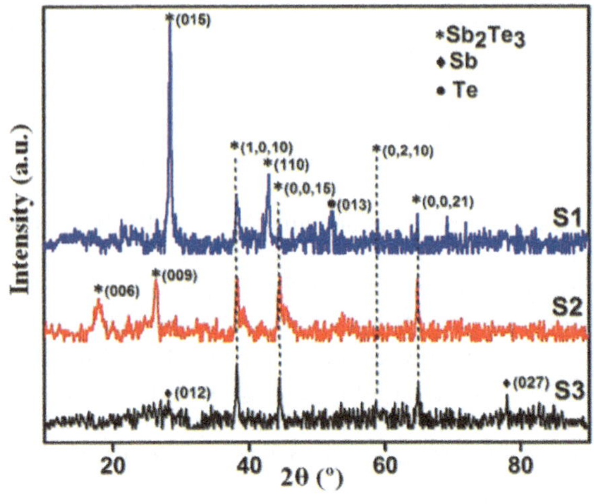

Fig. 6.4 XRD patterns of Sb_2Te_3 thin film samples with different Sb atom content

In order to further analyze the internal microstructure and the existence of excess Sb in the S3 sample, a deep analysis was performed on the sample S3 using a transmission electron microscope. The high-resolution transmission electron microscope (JEM-2010, JEOL, Japan) to observe the morphology of the material. TEM is similar to scanning electron microscope, but compared with SEM, the resolution of TEM is higher, especially for high-resolution HRTEM. The material's lattice fringes, etc., are even further analyzed into the morphology of the material. The TEM spectrum of the S3 sample is shown in the left image of Fig. 6.5. In order to more clearly observe the lattice arrangement of the S3 sample, a high magnification HRTEM analysis of the sample is performed as shown in the right image of Fig. 6.5. It can be seen from the left picture of Fig. 6.5 that the S3 sample exists in a polycrystalline state, and the

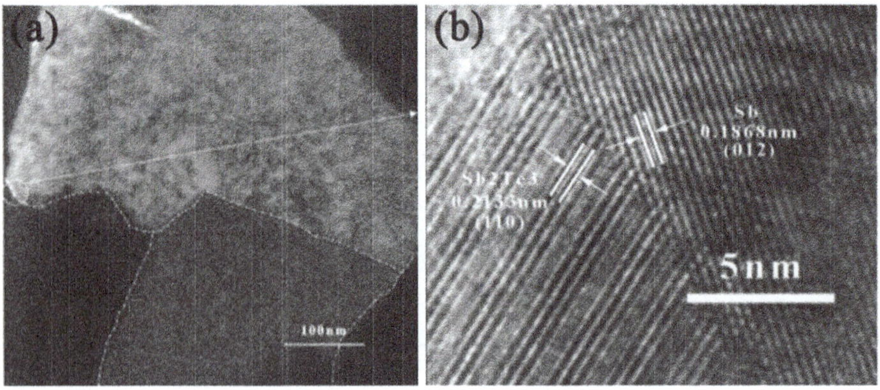

Fig. 6.5 TEM spectrum of the S3 sample (**a**) and local HRTEM image (**b**)

internal grain boundaries are very obvious. The white dotted lines in the figure are the boundaries of the grain boundaries. These grain boundaries may be Sb_2Te_3–Sb_2Te_3 or Sb_2Te_3–Sb demarcation. In order to further analyze the specific details of these micrograin boundaries, HRTEM was used to further analyze these grain boundaries. From the right figure of Fig. 6.5, it can be clearly seen that the grain boundaries of two different lattice constants clearly exist. Lattice constants found that on the left are lattice planes with a lattice constant of 0.2133 nm corresponding to the (110) crystal plane of Sb_2Te_3, and on the right are lattice planes with a lattice constant of 0.1868 nm corresponding to the (012) crystal plane of Sb. These crystal planes are also obtained in the XRD pattern Verification, consistent with the XRD test. The transmission electron microscope test showed that the Sb_2Te_3-Sb grain boundary exists in the sample. The existence of this grain boundary has a significant effect on the carrier and phonon transport in the sample, and thus has a certain positive effect on the thermoelectric properties of the Sb_2Te_3 film. This will be explained in the next analysis characterization.

After the microstructure and composition analysis, the next step is the test of electrical and thermal properties. First, the Sb_2Te_3 thin film samples with different Sb contents were tested by Seebeck coefficient and conductivity. Figure 6.6 shows the relationship between Seebeck voltage and temperature difference at room temperature. The slope of the straight line is the Seebeck coefficient of each sample. The Seebeck coefficient of sample S1 is 121 μV K^{-1}, the Seebeck coefficient of sample S2 is 90 μV K^{-1}, and the Seebeck coefficient of sample S3 is 536 μV K^{-1}, compared to samples S2 and S3 that meet the stoichiometric ratio and has a very high Seebeck coefficient value, almost 6 times the value of the S2 sample. Based on the Seebeck coefficient test, the Seebeck coefficient of Sb_2Te_3 thin film samples in which Sb is present is greatly improved compared to Sb_2Te_3 thin film samples without excess Sb.

For all Sb_2Te_3 thin film samples with different Sb contents, the Hall coefficient at room temperature (HL5500, Acceat Optical, UK) was tested. The Hall coefficient

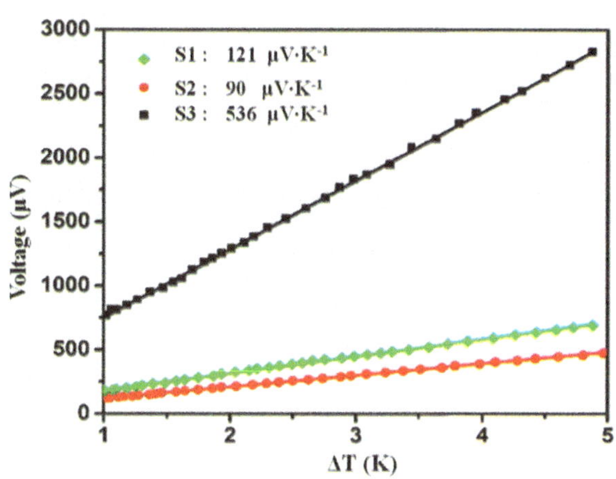

Fig. 6.6 Relationship between Seebeck voltage and temperature difference of Sb_2Te_3 thin film samples with different Sb contents. The slope of the straight line after fitting is the Seebeck coefficient of different film samples

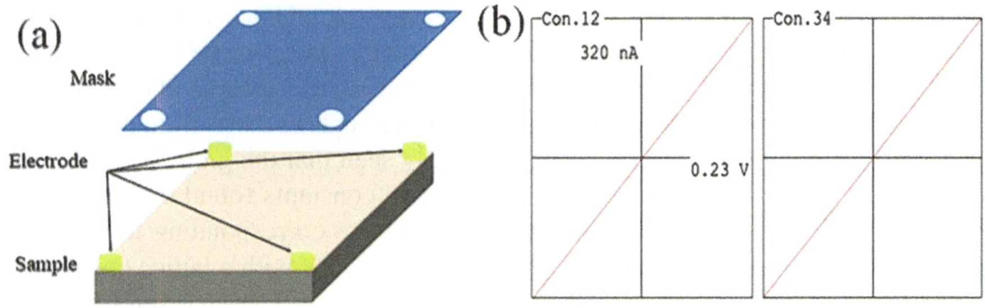

Fig. 6.7 **a** Sb_2Te_3 thin film sample and mask structure with electrodes, **b** Ohmic contact test results

test can obtain information such as carrier concentration and mobility. In order to test the Hall coefficient, electrodes with ohmic contact were prepared in advance. Semiconductor theory states that when a metal is in contact with a semiconductor, either an ohmic contact is formed or a Schottky contact is formed, which mainly depends on the work function of the metal and semiconductor. The work function of the Sb_2Te_3 bulk material is generally 4.45 eV, while the work function of the selected Ni metal is 5.15 eV, which is easy to form an ohmic contact. Electron beam evaporation deposition method used in electrode preparation is obtained in this paper. The sample size of the Hall coefficient test is a 1×1 cm^2 square. The four electrodes are on the four corners. To obtain the electrodes on the four corners, a specific mask is required. The sample with the electrode and the mask structure is shown in Fig. 6.7 of which the thickness of the plated Ni electrode is 500 nm. In order to verify whether the plated electrode meets the ohmic contact, an IV test is performed on the sample with the electrode. As shown in Fig. 6.7b, the forward and reverse voltages are applied to different electrodes, and the IV curve is a straight line. It is shown that the forward and reverse resistances are consistent, and thus an ideal ohmic contact electrode is obtained through experiments.

The Hall coefficient test results are shown in Table 6.1. The Sb_2Te_3 thin film samples with different Sb contents obtained from the Hall coefficient test and the values measured by the four-probe method set up in the laboratory have small deviations, indicating that the accuracy of the conductivity test results also verifies the

Table 6.1 Carrier mobility μ_H, carrier concentration n, conductivity σ_H and conductivity σ_F obtained by the four-probe method, obtained from the Hall coefficient test of different samples, obtained from the power factor $PF = \alpha^2\sigma_H$, 3ω method Numerical table of thermal conductivity κ and thermoelectric figure of merit ZT

Sample	μ_H (cm^2 V^{-1} s^{-1})	n (10^{19} cm^{-3})	σ_H (S cm^{-1})	σ_F (S cm^{-1})	PF (μW cm^{-1} K^{-2})	κ (W m^{-1} K^{-1})	ZT
S1	8.7	7.01	97.58	90.76	1.43	0.13	0.33
S2	11.7	7.73	144.71	126.79	1.17	0.17	0.21
S3	0.57	8.03	7.32	7.06	2.10	0.10	0.63

accuracy of the instruments built by the laboratory. The Hall coefficients of the Sb_2Te_3 thin film samples with different Sb contents are all positive values, which indicates that all the samples are semiconductors with holes as the majority carriers and holes conducting. From the results of the Hall coefficient test, the carrier concentration of Sb_2Te_3 thin film samples with different Sb contents is not much different, all on the order of 10^{19} cm^{-3}. However, the mobility of the samples with different Sb contents changes significantly. The carrier mobility was reduced by almost 20 times compared to the stoichiometric Sb_2Te_3 thin film sample.

The decrease in carrier mobility of Sb-rich Sb_2Te_3 thin film samples can be attributed to the interface grain boundaries formed between Sb and Sb_2Te_3. The existence of grain boundaries has been verified by HRTEM and XRD patterns. Studies have shown that if there is a difference in work function between materials in a nanocomposite, the energy level barrier of the carriers can be formed in the material to filter the carriers, which can be used to improve the thermoelectric properties of the materials. Figure 6.8 shows the energy band diagram of the contact between Sb–Sb_2Te_3 metal and semiconductor. Generally, Sb_2Te_3 is a degenerate semiconductor, and Fermi can be in the valence band. The band gap width (E_g), electron affinity (E_A) and work function (Φ) of Sb_2Te_3 in Fig. 6.8 are all taken from the bulk material. The band diagram of Sb–Sb_2Te_3 metal-semiconductor contact at equilibrium is shown in Fig. 6.8b. This figure shows that there is an energy level barrier at the Sb–Sb_2Te_3 interface. It can be seen from Fig. 6.8a that the work functions of Sb and Sb_2Te_3 are 4.55 eV and 4.45 eV, respectively. Therefore, there will be an equilibrium band diagram of band upturn in Fig. 6.8b after contact is about 0.1 eV. The theoretical barrier height of this value can effectively filter low energy carriers and hardly affect high energy carriers. Due to the existence of energy level barriers, at the interface of Sb and Sb_2Te_3, low-energy carriers (cold carriers) are strongly scattered, while high-energy carriers (hot carriers) are weakly scattered. High-energy carriers can cross the energy level barrier and low-energy carriers cannot, resulting in the separation

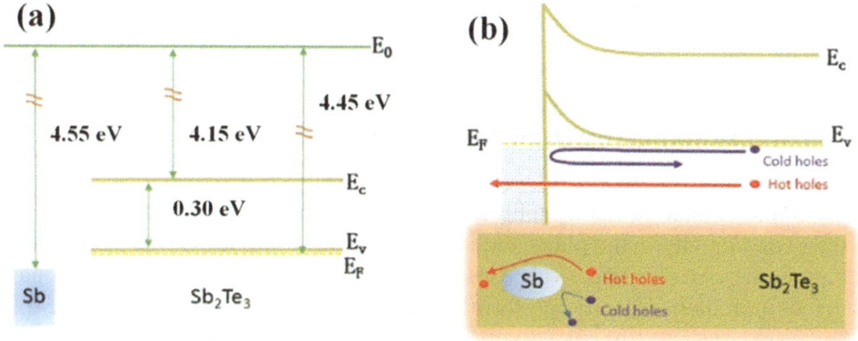

Fig. 6.8 Band diagram of Sb–Sb_2Te_3 metal-semiconductor contact: **a** is the band diagram before Sb and Sb_2Te_3 contact, and **b** is the band diagram after the two are in equilibrium, neglecting the influence of interface states

of hot and cold carriers. The strong scattering leads to a decrease in the mobility of the carriers, which explains why the mobility is significantly reduced in the Sb-rich samples. Cold low-energy carriers are detrimental to the Seebeck coefficient, and the separation of hot and cold carriers is conducive to increasing the Seebeck coefficient. On the other hand, from the Mott relationship, it can be seen that near the Fermi level, the Seebeck coefficient is a derivative of the energy-dependent conductivity versus energy, and is approximately inversely proportional to the carrier concentration and the carrier mobility. From the results of the Hall coefficient test in Table 6.1, it can be seen that the carrier concentration of Sb_2Te_3 thin film samples with different Sb content is not much different, all on the order of 10^{19} cm^{-3}, but the mobility of the samples with different Sb content changes significantly. The mobility of Sb_2Te_3 thin film samples is relatively small, which also explains the increase of Seebeck coefficient of Sb-rich Sb_2Te_3 thin film samples.

According to the theory of heat transfer, the existence of nanocrystalline interfaces will scatter medium and long wave phonons, and these phonons will transfer a large amount of heat. Nano inclusions have the effect of reducing the thermal conductivity of the parent material. In the experiment, the 3ω method was used to test the thermal conductivity of Sb_2Te_3 thin film samples with different Sb contents. Before the test, the sample should be specially treated. First, the sample and the blank substrate same as the sample substrate should be electron beam evaporated a layer of SiO_2 insulating layer after about 200 nm is applied, and then a silver metal wire is plated on the SiO_2 insulating layer using a semiconductor manufacturing process. The thermal conductivity test results are shown in Table 6.1. Among the Sb_2Te_3 film samples with different Sb contents, the Sb-rich Sb_2Te_3 film samples have the lowest thermal conductivity. Compared with the Sb_2Te_3 film samples that meet the stoichiometric ratio, the thermal conductivity is reduced to about 41%. The power factor P and thermoelectric figure of merit ZT of Sb_2Te_3 thin film samples with different Sb contents can be obtained by calculation. The calculation results are shown in Table 6.1. The results show that compared with the Sb_2Te_3 thin film samples that meet the stoichiometric ratio of the Sb-rich Sb_2Te_3 films. The sample power factor was increased by 80% and the thermoelectric figure of merit ZT was increased by a factor of three.

6.2 Directional Growth of Sb_2Te_3 Film

In the single crystal antimony telluride compound, the electrical conductivity and thermal conductivity in the plane perpendicular to the c-axis are 4 times and 2 times that along the c-axis, respectively [1]. Therefore, the antimony telluride film has a relatively high thermoelectric performance in the plane perpendicular to the c-axis. This section explores the successful preparation of stoichiometric (001) oriented antimony telluride films using molecular beam epitaxy, and investigates the effect of film thickness on their thermoelectric transmission characteristics, which is useful for understanding the role of the structure and performance is very important.

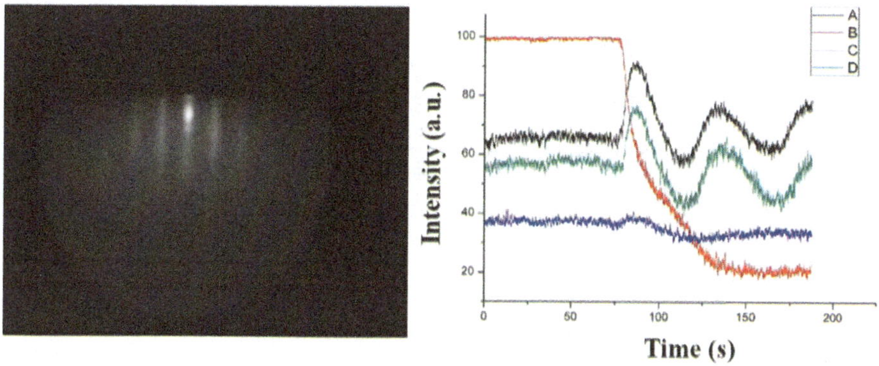

Fig. 6.9 RHEED diffraction pattern of Sb_2Te_3 thin film grown at an intrinsic Si (111) substrate temperature of 280 °C and the oscillation pattern of RHEED diffraction intensity during growth

Antimony telluride is a compound semiconductor with a band gap of 0.28 eV. Antimony telluride has a graphite-like orthorhombic hexahedral layered structure with five atoms in each primitive cell, and its lattice parameters are a = 0.4264 nm and c = 3.0458 nm. The schematic diagram of the atomic layer structure of antimony telluride is shown in Fig. 6.9. Every five monatomic layers form a "quintuple layers (QL)", and the atoms are arranged along the c-axis direction as –[Te(1)–Sb–Te(2)–Sb–Te(1)]–, where (1) and (2) represent two different lattice positions of the Te atom. Adjacent two "five-fold" are linked by van der Waals forces between Te(1)–Te(1). The structure of this orthorhombic hexahedron determines the anisotropic transmission characteristics of antimony telluride. In the single crystal antimony telluride compound, the electrical conductivity and thermal conductivity in the vertical c-axis plane are respectively 4 times and 2 times along the c-axis direction. Therefore, antimony telluride films have relatively high thermoelectric properties in a plane perpendicular to the c-axis. This section explores the successful preparation of stoichiometric (00l) oriented antimony telluride thin films using molecular beam epitaxy, and studies the effect of film thickness on its thermoelectric transmission characteristics. The role is quite important.

6.2.1 Film Preparation

In order to achieve epitaxial growth, a suitable substrate must first be selected. Molecular beam epitaxy requires that the deviation between the substrate's lattice constant and the lattice constant of the epitaxial layer should not be too large, generally maintained within 10% for epitaxial growth of Sb_2Te_3 films. The substrate is generally selected from BaF (111), Si (111), and the like. Since the lattice constant of the 1 × 1 reconstructed surface of the Si (111) surface is 0.384 nm, and the lattice mismatch degree of the lattice constant of 0.426 nm of the Sb_2Te_3 material is 9.86%, the intrinsic

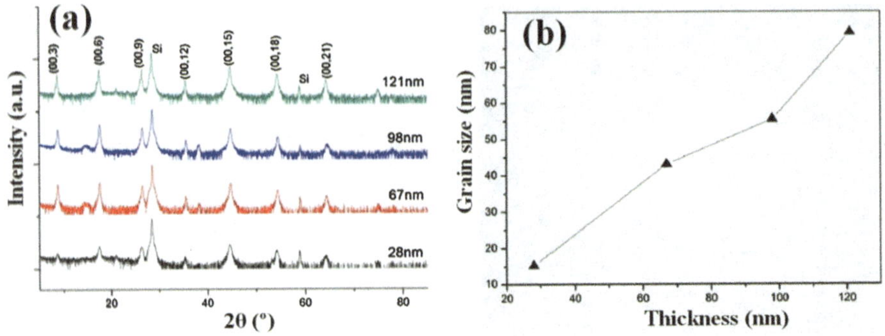

Fig. 6.10 **a** XRD diffraction patterns of Sb_2Te_3 films with different thicknesses; **b** variation of particle size with thickness

Si (111) 1 × 1 as a coated substrate, its resistivity is 1000 Ω cm. Before coating, the selected substrate must first be cleaned by RCA standard procedures. In this section, the co-evaporation method is used to grow epitaxial films. High-purity (99.999%) Sb and Te are placed in different beam source furnaces. In order to achieve the stoichiometric ratio of the prepared films, the evaporation rates of Sb and Te need to be made. Meet a certain proportion. The substrate was kept at room temperature during the coating process, and the prepared film thickness was tested with a step meter to obtain film formation rates at different temperatures. The curve of the film formation speed of antimony and tellurium films as a function of temperature is shown in Fig. 6.10.

After a series of exploration experiments, Sb, Te and Si substrates were selected under the conditions of $T_{Sb} = 430$ °C, $T_{Te} = 270$ °C, and $T_{Si} = 280$ °C. Figures 6.9 shows the intensity RHEED diffraction pattern of Sb_2Te_3 thin film deposited under substrate heating and the intensity oscillation pattern of diffraction fringes during film growth.

The clear RHEED diffraction fringes are repeated every 60°, showing that the prepared Sb_2Te_3 film has a plane six-fold symmetry. The combination of the lattice structure diagram of Sb_2Te_3 clearly reveals that the prepared film has a high (001) orientation, and its c axis is perpendicular to the substrate surface. The RHEED diffraction intensity oscillation chart can also reflect the atomic layered growth of the thin film during the growth process. Under the above growth conditions, a series of Sb_2Te_3 films with different thicknesses were prepared.

6.2.2 *Structural Characterization*

Figure 6.10a is a wide-angle XRD diffraction pattern of (001) oriented Sb_2Te_3 films with different thicknesses. Since the intrinsic peak of the Si (111) substrate is too strong, the logarithm of the diffraction intensity is taken on the ordinate in the XRD diffraction mode avoid masking the diffraction signal peaks of the film. Figure 6.10b shows the relationship between the particle size and the thickness of the thin film calculated according to the XRD diffraction mode.

For Sb_2Te_3 films of all thicknesses, all peaks correspond to the orthorhombic hexahedral phase (JCPDS 15-0874, R-3 m). Only the (00l) direction was observed, and the grain size is shown in Table 6.2.

Figure 6.11 is a SEM image of the 121 nm thick Sb_2Te_3 film surface. EDS analysis shows that all samples have a nominal stoichiometric ratio, in which the Te atom content is approximately equal to 60%. At the same time, the AFM test was performed on the prepared samples. Figures 6.12 is AFM 3D images of 67, 98, and 121 nm thick Sb_2Te_3 films. The test range of all images is 5×5 μm. It can be seen from the figure that as the thickness increases, the surface roughness of the film continuously decreases. This can be clearly shown by the AFM three-dimensional side view. The thicker the film, the smaller the thickness of the discontinuous layer on the surface. The corresponding surface roughness is 9.69, 8.58 and 2.85 nm. As the thickness of the film increases, the crystallites continue to grow and the interface between them shrinks, so the surface roughness decreases.

Table 6.2 Peak parameters of Sb_2Te_3 films with different thicknesses

Film thickness (nm)	Diffraction peak full width at half maximum					Particle size (nm)
	(003)	(006)	(009)	(0015)	(0018)	
28	0.379	0.386	0.410	0.506	0.503	15.3
67	0.195	0.276	0.300	0.418	0.450	43.1
98	0.184	0.190	0.180	0.265	0.286	55.5
121	0.148	0.153	0.142	0.194	0.239	79.3

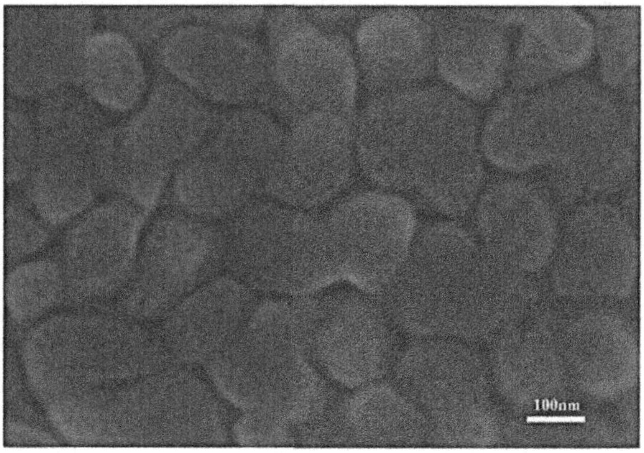

Fig. 6.11 SEM image of Sb_2Te_3 thin film (121 nm) (magnification × 100,000)

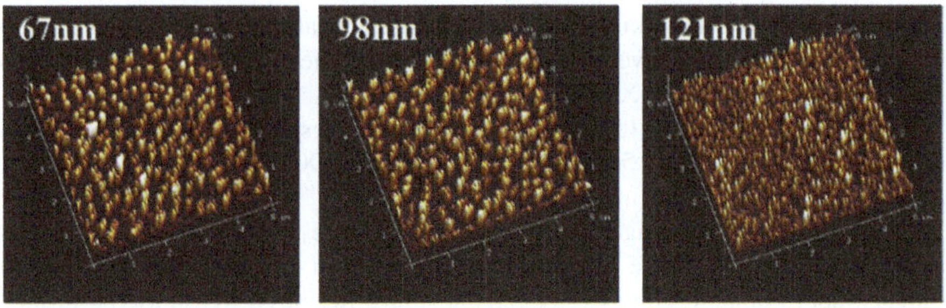

Fig. 6.12 3D AFM images of Sb_2Te_3 films with different thicknesses

6.2.3 Thin Film Electrical Transmission Characteristics

In order to have a deeper understanding of the electrical transmission characteristics of thin films, the Hall coefficient test (HL5500, Acceat Optical, UK) was used to obtain information on carrier concentration and mobility of the samples. Before carrying out the carrier concentration test, electrodes need to be plated on the tested thin film. In this paper, Al/Cu is deposited as the test electrode by electron beam evaporation. First, the film is cut into $1 \times 1\ cm^2$ squares according to requirements, and appropriate masks are used to deposit electrodes at the four corners of the squares. The specific structure is shown in Fig. 6.13.

The carrier charge of all films is positive (Tables 6.3 and 6.4), indicating that the films are mainly hole-conducting, and the carrier concentration is $2 \times 10^{19}\ cm^{-3}$. It is generally believed that the optimal carrier concentration for thermoelectric materials is in the range of $10^{19}\ cm^{-3}$, and the prepared thin film is just in this range.

Figure 6.14 shows the change of the electrical transmission characteristics of the Sb_2Te_3 film with the film thickness. It can be seen that the carrier concentration, mobility and conductivity of the film are sensitive to the film thickness. For the prepared Sb_2Te_3 film, the 28 nm-thick film has the highest carrier concentration

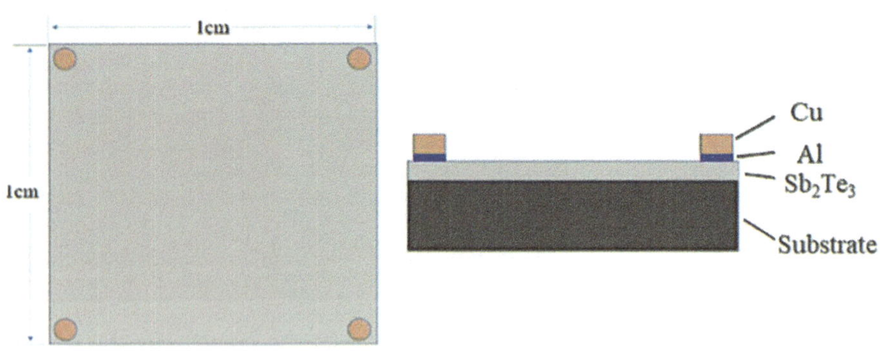

Fig. 6.13 Sb_2Te_3 thin film Hall coefficient test structure diagram

Table 6.3 Carrier concentration, mobility, electrical conductivity, and mean free path of carriers of Sb_2Te_3 films with different thicknesses

Film thickness (nm)	n ($\times 10^{19}$ cm^{-3})	μ (cm^2 V^{-1} s^{-1})	σ (S cm^{-1})
28	2.493	107	425.713
67	1.701	229	623.83
98	2.069	271	898.473
121	2.215	305	1036

Table 6.4 Mean free path of carriers, particle size and their ratios for Sb_2Te_3 films with different thicknesses

Film thickness (nm)	Particle size *Lp* (nm)	Mean free path *l* (nm)	*l/Lp*
28	15.3	6.37	0.42
67	43.1	12.01	0.28
98	55.5	15.20	0.27
121	79.3	17.22	0.22

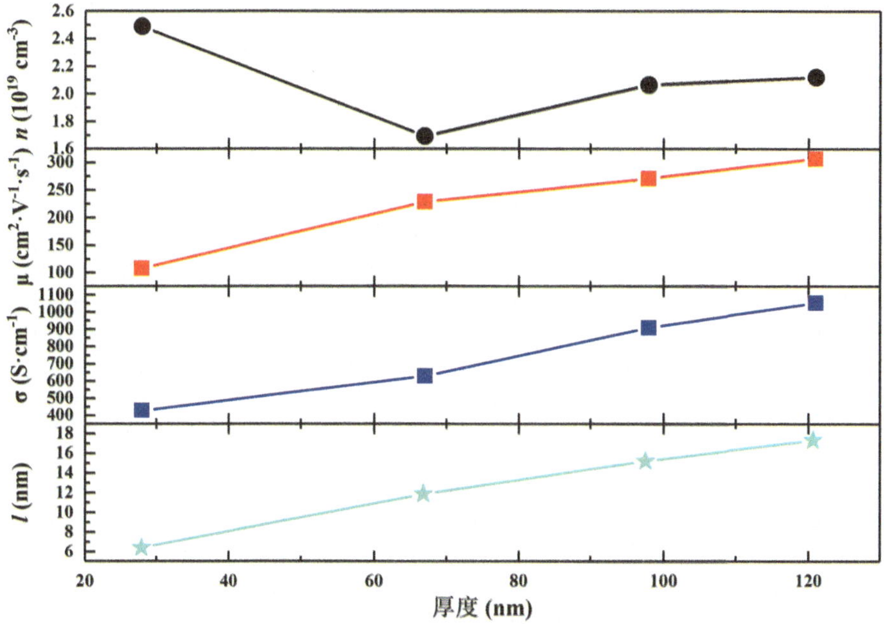

Fig. 6.14 Curves of carrier concentration, mobility, electrical conductivity, and mean free path of carriers with thickness of Sb_2Te_3 films (all data were obtained by testing at room temperature)

of 2.493×10^{19} cm^{-3}, which decreases sharply to 1.701×10^{19} cm^{-3} when the film thickness increases to 67 nm. This is because the existence of a large number of microcrystals in a 28 nm thin film results in a large grain boundary density. The atoms at the interface provide and scatter carriers, and therefore have the highest carrier concentration and low mobility in a 28 nm thick film. When the film continued to grow from 28 to 67 nm, a large number of microcrystals began to rapidly aggregate to form microcrystal growth, resulting in a reduction in interface density. This can be seen from the particle size and average free path of the film. Compared with the 28 nm film, the particle size in the 67 nm film has almost doubled, and the average free path has also doubled. This shows that the film has undergone a rapid growth process of microcrystals. The disappearance of some interfaces reduces the center of the provided carriers and reduces the scattering effect on the carriers. Therefore, the carrier concentration in the 67 nm film decreases rapidly and its mobility increases sharply.

When the film thickness continued to increase beyond 67 nm, the carrier concentration increased slightly but not as drastically as the initial decrease. As the thickness increases, the internal stress concentration of the thin film results in the formation of more impurities and defects in the microcrystals of the thicker film, which provides more carriers. This can be seen from the ratio of the mean free path of the carrier to the particle size. As the film thickness increases, this ratio decreases, indicating that carriers receive more scattering from defects in the crystallites of the thicker film. This is also evidenced by the RHEED diffraction pattern during growth. The decrease in carrier concentration at the beginning is dominated by the influence of particle boundaries in thinner films. When the film reaches a certain thickness, the influence of microcrystalline internal defects becomes more and more important. At the same time, as the thickness of the film increases, the influence of the substrate on the interior of the film becomes smaller and smaller, and the defect concentration in the film gradually stabilizes. Therefore, the carrier concentration becomes slower and slower as the thickness increases. For all films, the mobility and mean free path increase monotonically with thickness, indicating that carrier scattering from particle boundaries is more pronounced than defects. At room temperature, the 121 nm thick film has the largest carrier mobility of 305 $cm^2\ V^{-1}\ s^{-1}$ and the largest mean free path of 17.22 nm. Even for the 28 nm-thick film, its carrier mobility and the average free path also reached 107 $cm^2\ V^{-1}\ s^{-1}$ and 6.37 nm. It can be seen from Fig. 6.14c that the conductivity monotonically increases with the thickness of the film. When the film thickness increases from 28 to 121 nm, the conductivity increases from 425.7 to 1036 $S\ cm^{-1}$. The slow increase in conductivity at the beginning is mainly due to the increased carrier mobility which partially compensates for the decrease in carrier concentration.

Figure 6.15 shows the resistivity of the films at 67, 98, and 121 nm as a function of temperature. At room temperature to 150 °C, the resistivity of all films increases with increasing temperature. The relationship between resistivity and temperature is approximately linear, and the fitting slopes of all films are basically equal. The sheet resistance temperature coefficients corresponding to 67, 98, and 121 nm are 4.15 $\mu\Omega\ cm\ K^{-1}$, 3.85 $\mu\Omega\ cm\ K^{-1}$, and 3.35 $\mu\Omega\ cm\ K^{-1}$, respectively. For Sb_2Te_3 films with stoichiometric ratio, the main impurities are intrinsic point defects, which are generally shallow doped energy levels. The ratio of the average free path of the carrier of the film to the particle size is slightly less than 0.5, indicating that there are fewer impurities or defects in the microcrystals in the film.

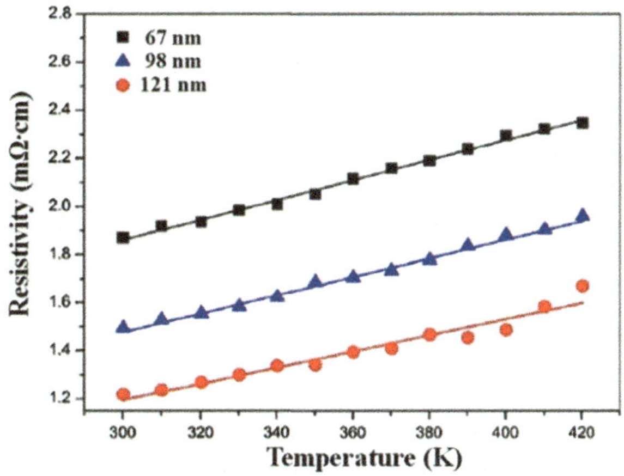

Fig. 6.15 Curve of resistivity of Sb2Te3 thin film as a function of temperature

6.3 Bi-Generation Doped Sb_2Te_3 Film

This section uses molecular beam epitaxy to introduce Bi atoms into the lattice sites of Sb_2Te_3. Because Bi_2Te_3 and Sb_2Te_3 have the same graphite-like orthorhombic hexahedral layered structure, every five monatomic layers form a "quintuple layer (QL)", and the atoms are arranged along the c-axis direction as –[Te(1)–Sb(Bi)–Te(2)–Sb(Bi)–Te(1)]–, where (1) and (2) represent two different lattice positions of the Te atom. An attempt was made to design the introduction of Bi atoms into the lattice sites of Sb_2Te_3 by molecular beam epitaxy under the condition of substrate heating, and partially replace the Sb sites to form the structure shown in Fig. 6.16.

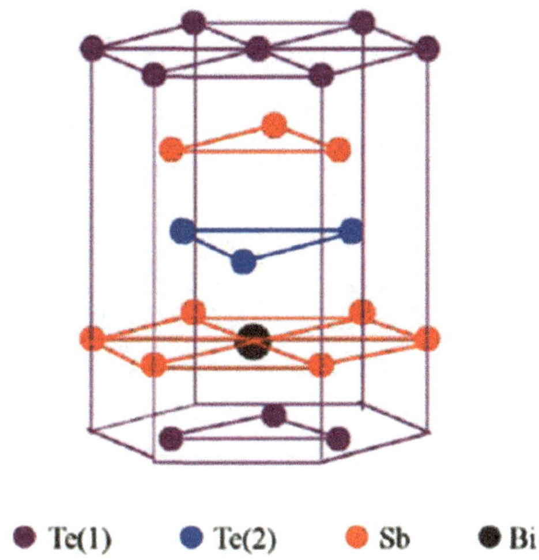

Fig. 6.16 Lattice structure of Bi atoms instead of doped Sb_2Te_3

6.3.1 Film Preparation

In this section, an intrinsic Si (111) substrate (resistivity 1000 Ω cm) and a glass substrate are used to prepare Sb_2Te_3 films and Bi atoms instead of doped $Bi_xSb_{2-x}Te_3$ films by molecular beam epitaxy. Before coating, the substrate is first cleaned by RCA standard procedures. In this section, epitaxial films are still grown by co-evaporation. Bi, Sb, and Te with high purity (99.999%) are placed in different beam source furnaces. The evaporation rate ratio is appropriately adjusted to prepare films that meet the needs. The relationship between the deposition rate of Sb and Te elements and the temperature has been obtained previously. Here we design experiments to study the relationship between the deposition rate of Bi elements and the temperature of the beam source furnace. The substrate was kept at room temperature during the coating process, and the prepared film thickness was measured with a step meter to obtain film formation rates at different temperatures. The curve of the film formation speed with temperature for bismuth films is shown in Fig. 6.17.

In this section, the epitaxially grown thin film is selected under the substrate heating condition, and the substrate temperature is set to 280 °C. After a series of exploration experiments, the Bi, Sb and Te temperatures were selected to be $T_{Bi} = 510$ °C, $T_{Sb} = 430$ °C, and $T_{Te} = 310$ °C. Bi-doped Bi–Sb–Te films were deposited under the conditions of EDS. Tests showed that the atomic composition of the prepared film was $Bi_{0.4}Sb_{1.6}Te_3$; the Sb_2Te_3 film was prepared under the conditions of $T_{Sb} = 430$ °C and $T_{Te} = 270$ °C. By controlling the coating time, Sb_2Te_3 thin film and $Bi_{0.4}Sb_{1.6}Te_3$ thin film with a thickness of about 200 nm were obtained. At the same time, the error was less than 5% by using a step profiler.

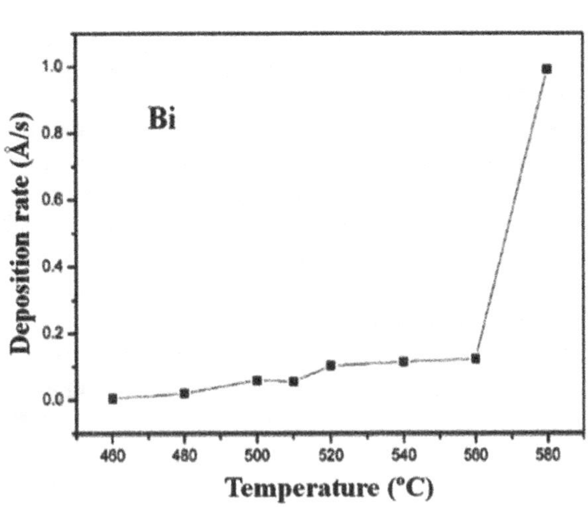

Fig. 6.17 Changes in film formation rate of bismuth films with temperature

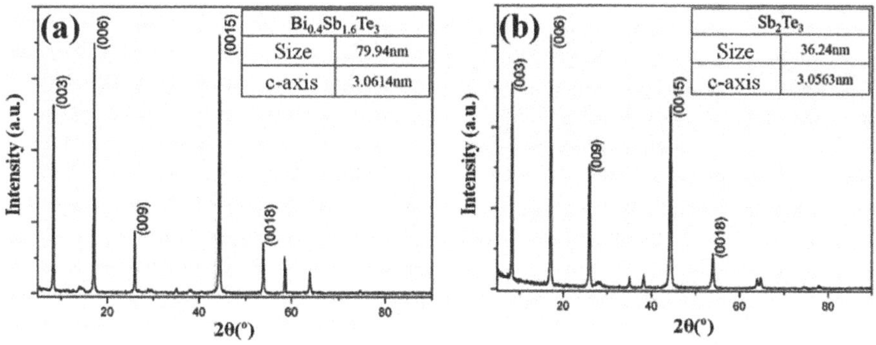

Fig. 6.18 Wide-angle XRD diffraction patterns of **a** $Bi_{0.4}Sb_{1.6}Te_3$ and **b** Sb_2Te_3 films

6.3.2 Structural Characterization

The wide-angle XRD patterns of $Bi_{0.4}Sb_{1.6}Te_3$ film and Sb_2Te_3 film are shown in Fig. 6.18, respectively. For the Sb_2Te_3 film, all obvious diffraction peaks correspond to the orthorhombic hexahedral phase (JCPDS 15-0874, R-3 m), and the film has a (00l) orientation. The particle size of $Bi_{0.4}Sb_{1.6}Te_3$ thin film reached 79.94 nm, which is about 2.2 times the particle size of Sb_2Te_3 thin film 36.24 nm. This may be due to the introduction of larger Bi atoms into the lattice sites of Sb_2Te_3 to offset the shrinkage stress in the inner part of the film, making the lattice structure more stable and promoting the growth of microcrystals. The enhanced crystallinity can be reflected by the sharper and stronger diffraction peak signal of $Bi_{0.4}Sb_{1.6}Te_3$ film.

The prepared $Bi_{0.4}Sb_{1.6}Te_3$ thin film was examined by transmission electron microscope (TEM). The film deposited on the Si substrate was scraped off, washed with anhydrous ethanol onto the copper mesh for testing, and then allowed to dry naturally to prepare a test sample. Figure 6.19 is the low and high magnification TEM images of $Bi_{0.4}Sb_{1.6}Te_3$ film. The inset of Fig. 6.19a shows the selected area electron diffraction pattern. The detection area is marked with a white circle in the figure. The bright hexagonal symmetrical selective electron diffraction mode shows that the crystallites have a good single crystal structure, which clearly indicates that Bi atoms were successfully introduced into the Sb lattice sites of Sb_2Te_3 to form a meta-doped thin film. From Fig. 6.19b, we can clearly see the obvious lattice fringes. The spacing between adjacent fringes is about 0.332 nm, which is very close to the (2011) plane spacing in the Sb_2Te_3 lattice structure.

6.3.3 Thermoelectric Performance Test of Thin Film

A Seebeck coefficient test was performed on a thin-film sample prepared on a glass substrate using a home-made Seebeck coefficient test device. The test results are

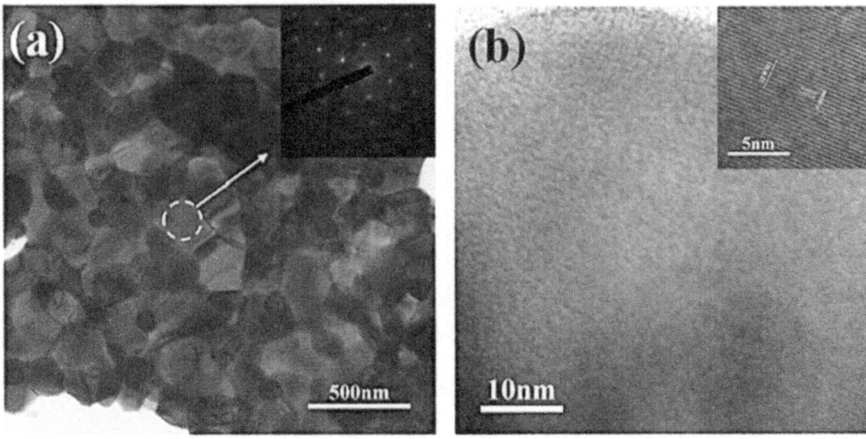

Fig. 6.19 **a** Low magnification TEM image (12,000×) of $Bi_{0.4}Sb_{1.6}Te_3$ thin film; the illustration is the selected electron diffraction mode (the detection area is marked with a white circle in the figure). **b** High magnification TEM image of $Bi_{0.4}Sb_{1.6}Te_3$ film (500,000 times)

shown in Fig. 6.20. The Seebeck coefficients of Sb_2Te_3 film and $Bi_{0.4}Sb_{1.6}Te_3$ film are 159.86 μV K^{-1} and 134.71 μV K^{-1}, respectively. Substitution of Bi atoms in the sub-site makes the Seebeck coefficient of $Bi_{0.4}Sb_{1.6}Te_3$ film slightly lower than that of Sb_2Te_3 film, but the effect is not obvious (Fig. 6.20).

Hall coefficient test was performed on the prepared $Bi_{0.4}Sb_{1.6}Te_3$ film and Sb_2Te_3 film to obtain information on carrier concentration and mobility. The results of the carrier concentration and mobility tests, as well as the calculated film particle size and carrier mean free path, are shown in Table 6.5. The measured carriers for the $Bi_{0.4}Sb_{1.6}Te_3$ film and the Sb_2Te_3 film are all positive, indicating that hole conduction is mainly in the two samples. The $Bi_{0.4}Sb_{1.6}Te_3$ film has a low carrier concentration of 7.9×10^{18} cm^{-3} and a small mobility of 50.6 cm^2 V^{-1} s^{-1}. The corresponding values of the Sb_2Te_3 film are 1.1×10^{19} cm^{-3} and 457 cm^2 V^{-1} s^{-1}, respectively.

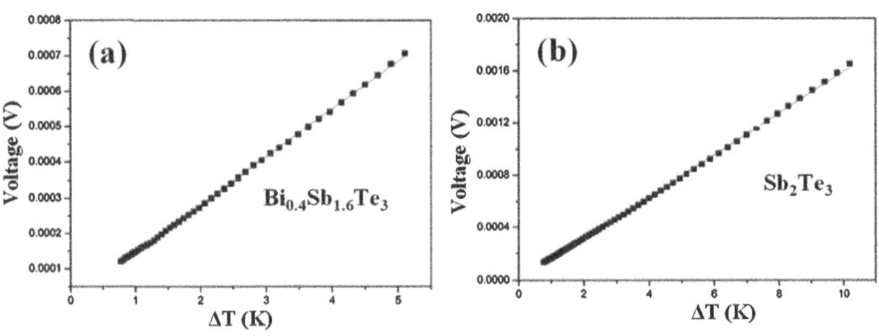

Fig. 6.20 Seebeck coefficient test results of **a** $Bi_{0.4}Sb_{1.6}Te_3$ and **b** Sb_2Te_3 films

Table 6.5 Electrical transmission characteristics of $Bi_{0.4}Sb_{1.6}Te_3$ and Sb_2Te_3 thin films (*Lp*, *n*, μ, *l*, and σ represent the particle size, carrier concentration, mobility, mean free path, and conductivity of the film, respectively)

Sample	*Lp* (nm)	n ($\times 10^{19}$ cm^{-3})	μ (cm^2 V^{-1} s^{-1})	l (nm)	σ (S cm^{-1})
$Bi_{0.4}Bi_{1.6}Te_3$	79.94	0.79	50.6	2.06	64.3
Sb_2Te_3	36.24	1.1	457	9.62	801

It can be seen that the introduction of Bi atoms simultaneously reduces the carrier concentration and mobility. Compared with Sb_2Te_3film, $Bi_{0.4}Sb_{1.6}Te_3$ film has a smaller mean free path of carriers, although its particle size is larger than the former. The large particle size means that the scattering of carriers at the interface of the crystallites will be weak, so the low carrier concentration and small mean free path indicate that there are more scattering centers in the crystallites. These scattering centers are probably derived from meta-doped Bi atoms, because $Bi_{0.4}Sb_{1.6}Te_3$ films have higher crystallinity. The work functions of Bi and Sb are 4.22 eV and 4.55 eV, respectively. Therefore, the positions of the conduction band and valence band are different in Bi_2Te_3 and Sb_2Te_3 materials. The conduction band of Bi_2Te_3 material is slightly lower than that of Sb_2Te_3 film, and its valence band is slightly higher than that of Sb_2Te_3 material, that is, the band gap of Bi_2Te_3 material is in the band gap of Sb_2Te_3 material. A schematic diagram of the energy levels for these two materials is shown in the lower left corner of Fig. 6.21. When a small number of Bi atoms is introduced into the Sb_2Te_3 lattice structure to occupy the Sb atomic position, it will lead to the formation of new energy levels at the bottom of the conduction band

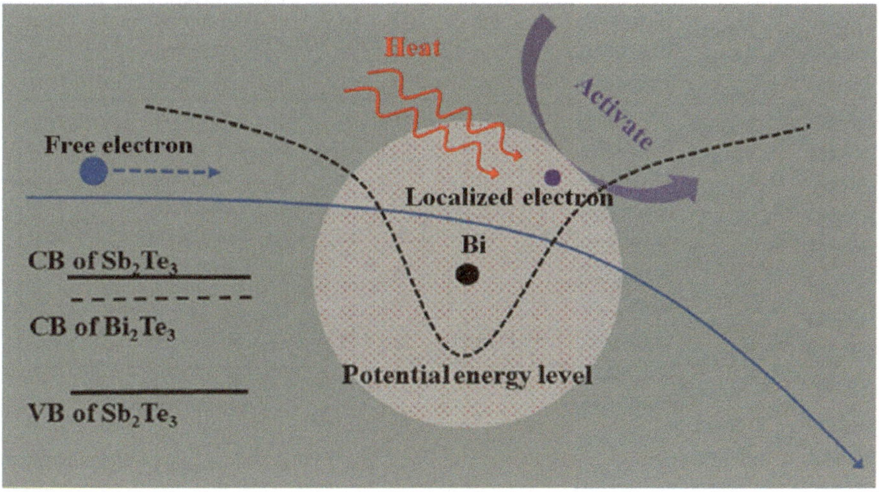

Fig. 6.21 Schematic diagram of the structure of the Atomic Bi atom doped into the Sb_2Te_3 lattice site, and the influence of the Atomic Bi atom on the transport of electrons (the black line represents the potential energy of the free point when it is around the Bi atom)

and the top of the valence band. Electrons (transport in the conduction band) and holes (transport in the valence band) will be somewhat attractive by these newly generated energy levels. Carriers will therefore be subject to increased scattering, resulting in reduced relaxation time. The schematic diagram of the electron transfer in the conduction band of the meta-atom Bi atom is shown symbolically in Fig. 6.21.

In order to further understand the thin film electrical transmission characteristics, the $Bi_{0.4}Sb_{1.6}Te_3$ thin film and the Sb_2Te_3 thin film resistance were tested in the range of −50 to 150 °C, and then the conductivity of the prepared thin film samples was measured as a function of temperature as shown in Fig. 6.22. The conductivity of Sb_2Te_3 thin films decreases monotonically with increasing temperature. Similar phenomena have been reported in previous literatures. Pure Sb_2Te_3 films have shallow-level impurities, most of which have been ionized at lower temperatures below room temperature. As the temperature rises further, the increased lattice vibration increases the resistance to electron transport, leading to a decrease in electrical conductivity. However, for the $Bi_{0.4}Sb_{1.6}Te_3$ film, its electrical conductivity increases approximately as the temperature increases with temperature. In $Bi_{0.4}Sb_{1.6}Te_3$ thin films, the substitution of doped Bi atoms will have a certain inhibitory effect on the ionization of local electrons that are impurities or defects, making the conductivity activation energy increase. Therefore, as the temperature increases, more carriers are excited, which increases the conductivity of the $Bi_{0.4}Sb_{1.6}Te_3$ film.

The relationship between the logarithm of the conductivity of $Bi_{0.4}Sb_{1.6}Te_3$ thin film and the inverse temperature is shown in Fig. 6.23. The calculated conductivity activation energy of $Bi_{0.4}Sb_{1.6}Te_3$ film is 43.2 meV, which is slightly higher than the corresponding energy of room temperature of 38.8 meV ($E_0 = 32\,k_BT$). Around room temperature, the ionization degree of impurities is relatively low, so the $Bi_{0.4}Sb_{1.6}Te_3$ thin film shows obvious semiconductor conductivity characteristics in the tested temperature range.

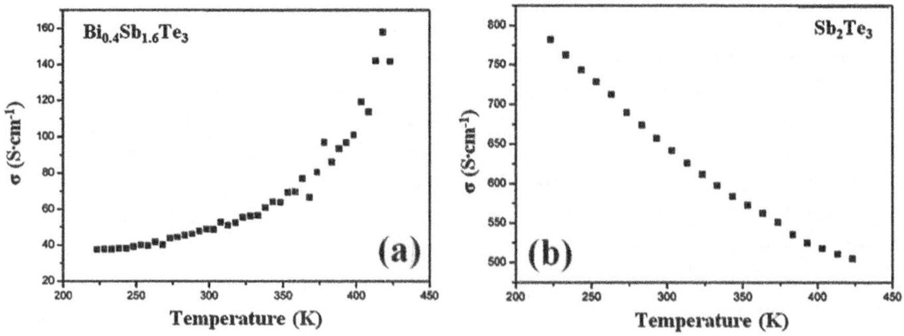

Fig. 6.22 **a** Variation of electrical conductivity of $Bi_{0.4}Sb_{1.6}Te_3$ and Sb_2Te_3 films with temperature

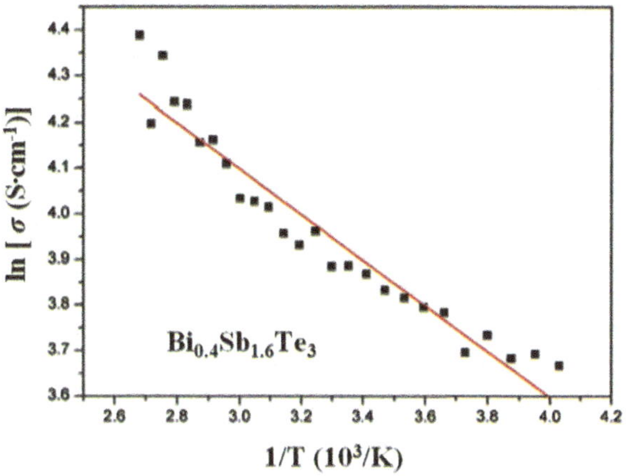

Fig. 6.23 The relationship between the logarithm of the conductivity of the $Bi_{0.4}Sb_{1.6}Te_3$ film and the inverse temperature

6.4 Te-Doped Sb_2Te_3 Film

Sb_2Te_3 based thermoelectric materials and their derivatives are one of the materials with the best thermoelectric properties at room temperature. The low dimensionality is beneficial to further improve the thermoelectric properties of Sb_2Te_3 based thermoelectric materials. The introduction of nanoparticles into thermoelectric materials forms the energy level barrier of carriers. Based on the effect of carrier filtering, improving the thermoelectric properties of thermoelectric materials is a very effective method, which has been extensively studied. The molecular beam epitaxy method is a nearly ideal method for preparing high-quality thin films and thermoelectric heterostructure materials, and is compatible with MEMS processing technology. Using molecular beam epitaxy equipment, nanoparticles of different diameters and numbers of Te were successfully introduced into Sb_2Te_3 thin film thermoelectric materials. The existence of Te nanoparticles can significantly affect the microstructure and thermoelectric properties of Sb_2Te_3 films. XRD, SEM, AFM, TEM were used to characterize the microstructure of Te nanoparticles with different contents, and the effect of Te nanoparticle content on the microstructure of Sb_2Te_3 films was studied. At the same time, samples of Te nanoparticles with different contents were also subjected to thermoelectricity characterization of performance. The effects of Te nanoparticle content on the thermoelectric properties of Sb_2Te_3 films were analyzed by the Seebeck coefficient test, Hall coefficient test, and thermal conductivity test results.

6.4.1 Film Preparation

The thermoelectric preparation of Sb_2Te_3 thin films containing different Te nanoparticles is performed in a molecular beam epitaxy device (DZS-700, SKY Tech, China).

The Sb_2Te_3 thin film layer and the Te layer are grown alternately. When the Te layer is thin, discontinuous nanometers are formed Particle as shown in Fig. 6.24.

The Sb_2Te_3 thin film was obtained by co-evaporation of high purity (99.999%) Sb and Te elemental materials. The introduction of Te nanometers was obtained by separately evaporating the Te elementary material. When preparing samples of different content of Te nanoparticles, keep the thickness of the Sb_2Te_3 layer at 5 nm, the thickness of the Te layer varies from 0, 1, 2 to 4 nm for different samples, and alternately grows for 50 cycles. The substrate selection includes high-resistance Si (111) and quartz glass, which have undergone strict cleaning processes. During the film preparation process, the substrate temperature is maintained at 150 °C, and the cavity pressure is 10^{-9} Torr.

First, the processed substrate was placed into the chamber of the molecular beam epitaxy equipment then started to evacuate. When the vacuum reached 10^{-9} Torr or more, the PID controller was turned on to independently heat the Sb and Te beam source furnaces and substrate stages. When the temperature of the Sb and Te beam source furnaces rose to 430 and 270 °C, and the substrate temperature was 150 °C, samples were deposited. The beam source furnaces each had a baffle to control the deposition beam on or off. When the Sb_2Te_3 layer is prepared, the two baffles of the Sb and Te beam source furnaces were both opened at the same time. While the thickness was monitored by a film thickness meter to 5 nm, the furnace baffle of Sb was closed and the furnace ballle of Te was opened that only allowed the ejection of Te beam. According to different batches the thickness of the sample Te deposited varies from 0, 1, 2 to 4 nm. Sb_2Te_3 layers and Te layers were alternately grown for 50 cycles like this, and Sb_2Te_3 films with different Te contents were prepared, and the thicknesses were 250, 300, 350, and 450 nm, respectively, and the corresponding Te volume contents were 0,17, 29 and 45% of the samples, respectively.

After the sample deposition is completed, the prepared sample is characterized by XRD, SEM, AFM, TEM and other structural methods, and its Seebeck coefficient, electrical conductivity, and thermal conductivity are tested by self-made equipment. At the same time, the properties of carrier mobility and concentration of the thin film material were obtained by Hall coefficient test.

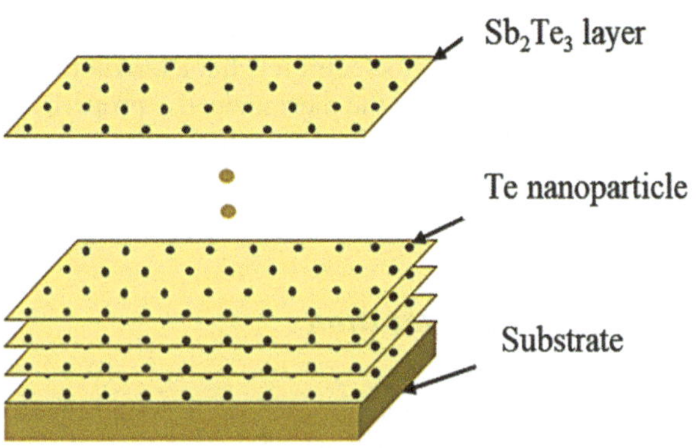

Fig. 6.24 Schematic diagram of the introduction of Te nanoparticles by alternate Sb_2Te_3 and Te layers

6.4.2 *Effect of Te Nanoparticle Content on Microstructure of Sb_2Te_3-Based Films*

6.4.2.1 X-Ray Structural Analysis (XRD)

The XRD patterns of Sb_2Te_3 films with different Te contents are shown in Fig. 6.25. The volume content of Te increased from 0% (Te0), 17% (Te1), 29% (Te2) to 45% (Te4). It can be seen from the figure that the diffraction peaks of all the samples on the whole correspond to the orthorhombic hexahedral phase (JCPDS 15-0874, R-3 m), thus obtaining the Sb_2Te_3 sample belonging to the R $\bar{3}$ m space group. When Te nanoparticles are introduced, as the Te content increases, the effects of Te nanoparticles on the sample are mainly concentrated in two aspects. When Te nanoparticles are introduced, as the Te content increases, the effects of Te nanoparticles on the sample are mainly concentrated in two aspects. On the one hand, the peak intensity of the weaker initial intensity is enhanced, and some of the diffraction peaks of the strong initial intensity are enhanced. At mean time some initially strong peaks became weaker, and this change was attributed to the effect of Te on the growth orientation of the Sb_2Te_3 film. With the increase of Te content, the Sb_2Te_3 film gradually grows in the (00l) direction. On the other hand, with the increase of Te content, new diffraction peaks appear at the diffraction angles of 45.89° and 64.37°, corresponding to (003) (JCPDS 65-3370) and (116) (JCPDS 23-1000) crystal plane. This proves that Te nano-grains were introduced into the Sb_2Te_3 film, and the existence of the (003) crystal plane of Te may help the Sb_2Te_3 film grow along the (00l) direction. The grain size of the Sb_2Te_3 film can be calculated by the Scherrer formula. As the Te content increases, the grain size of Sb_2Te_3 gradually increases, from 13.5 nm (Te0), 14.7 nm (Te1), 16.3 nm (Te2) up to 21.9 nm (Te4). The increase in grain size may be due to the increase in the thickness of the sample. However, even if the sample

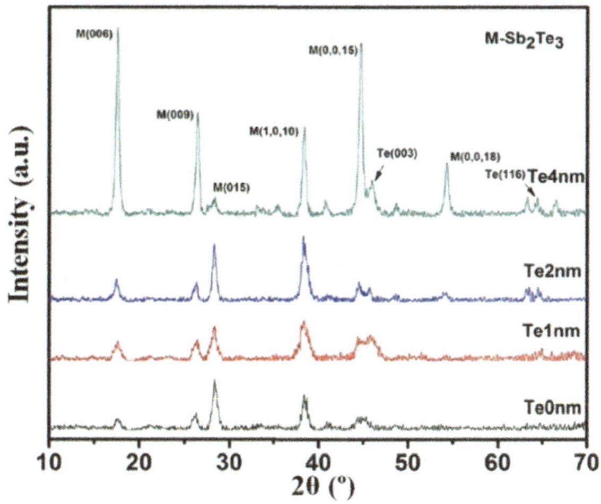

Fig. 6.25 XRD patterns of Sb_2Te_3 films with different Te contents. Te content increased from 0% (Te0), 17% (Te1), 29% (Te2) to 45% (Te4)

thickness reaches hundreds of nanometers, the grain size does not reach hundreds of nanometers. This is due to the insertion of Te, which blocks the growth of grains (Fig. 6.25).

6.4.2.2 Field Emission Scanning Electron Microscope (FESEM)

The surface morphology of the Sb_2Te_3 thin film sample with the Te content can be clearly seen from the SEM test data in Fig. 6.26. In the sample Te0 without Te nanoparticles, the surface of the sample was composed of uniform Sb_2Te_3 grains, and no obvious nanoparticles appeared. When the Te layer was alternately introduced into the Sb_2Te_3 sample, nanoparticles appeared in the sample, and the Te nanoparticles grew in the form of islands. The size of the islands was directly proportional to the Te content. It can also be clearly seen from Fig. 6.26 that with the Te content increasing the diameter of the nanoparticles increases gradually.

As shown in Fig. 6.27, the surface and cross-section SEM images of the Sb_2Te_3 film at a Te content of 17%. From the surface SEM, it can be seen that the average diameter of Te nanoparticles is about 5 nm and the average areal density is about 160 μm^{-2}. The Te nanoparticles are evenly distributed on the surface of the Sb_2Te_3 film sample. Te nanoparticles are also uniformly distributed. From the cross-sectional view of the Sb_2Te_3 thin film sample, it is seen that Te nanoparticles are uniformly embedded in the Sb_2Te_3 thin film.

6.4.2.3 Atomic Force Microscope (AFM)

The AFM images of Sb_2Te_3 films with different Te contents are shown in Fig. 6.28. The scanning size of all the images is $1 \times 1\ \mu m^2$. The surface roughness of the sample can be obtained by AFM testing, and the direct change in the size of Te nanoparticles at the surface can be inferred. When the Te content is 0% (Te0), 17% (Te1), 29% (Te2), and 45% (Te4), the surface roughness of the corresponding Sb_2Te_3 thin film samples is 2.08, 2.32, 2.67, and 4.78 nm. The change in surface roughness with Te content is due to the influence of Te nanoparticles.

6.4.2.4 Transmission Electron Microscope (TEM)

When the Te content is 17%, the TEM images of the Sb_2Te_3 thin film samples are shown in Fig. 6.29 from which the Te nanoparticles are uniformly distributed in the Sb_2Te_3 thin film samples. At 5 nm this is consistent with the data obtained by SEM.

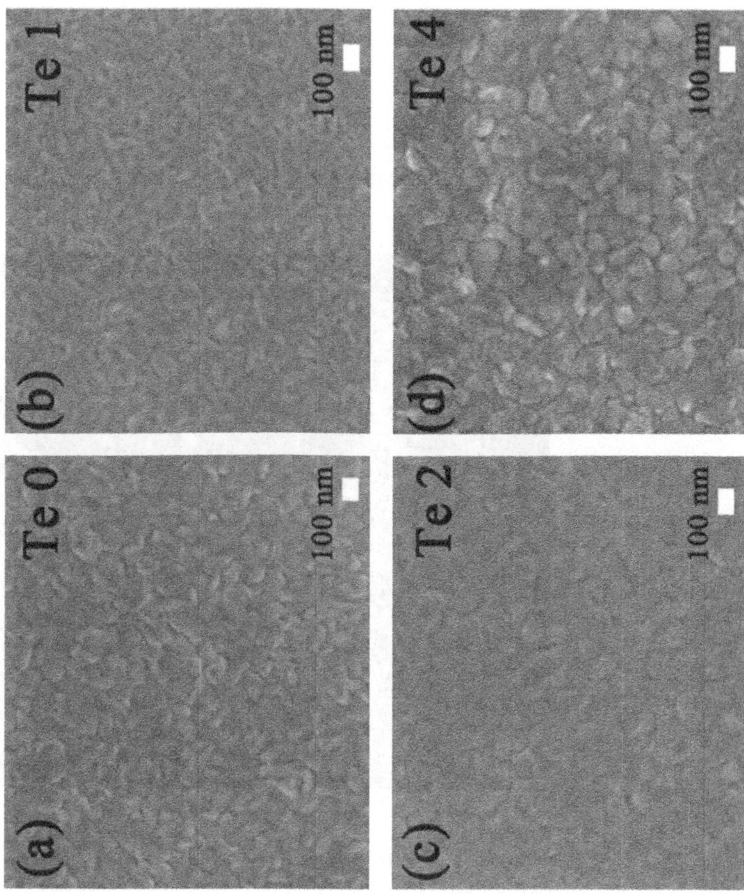

Fig. 6.26 SEM images of Sb_2Te_3 films with different Te contents. Te content increased from 0% (Te0), 17% (Te1), 29% (Te2) to 45% (Te4)

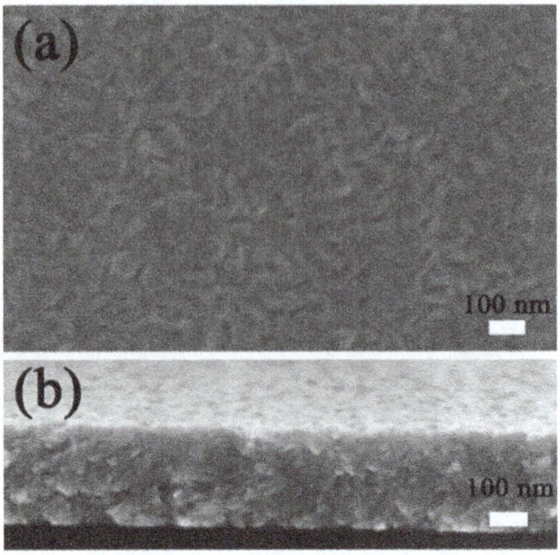

Fig. 6.27 SEM image of the surface and cross section of the Sb_2Te_3 film when 29 Te content is 17%

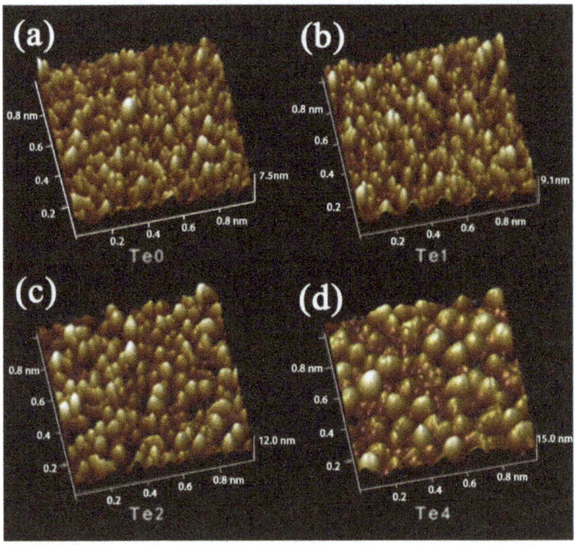

Fig. 6.28 AFM images of Sb_2Te_3 films with different Te contents. Te content increased from 0% (Te0), 17% (Te1), 29% (Te2) to 45% (Te4)

6.4.3 Effect of Te Nanoparticle Content on the Thermoelectric Properties of Sb_2Te_3 Based Films

6.4.3.1 Hall Coefficient Test

It can be seen from Table 6.6 that the Hall coefficients of Sb_2Te_3 films with different Te contents are all positive values, indicating that all samples are hole-conducting, and the samples obtained by the experiments are all P-type semiconductors. The relationship between the carrier mobility and carrier concentration of Sb_2Te_3 films

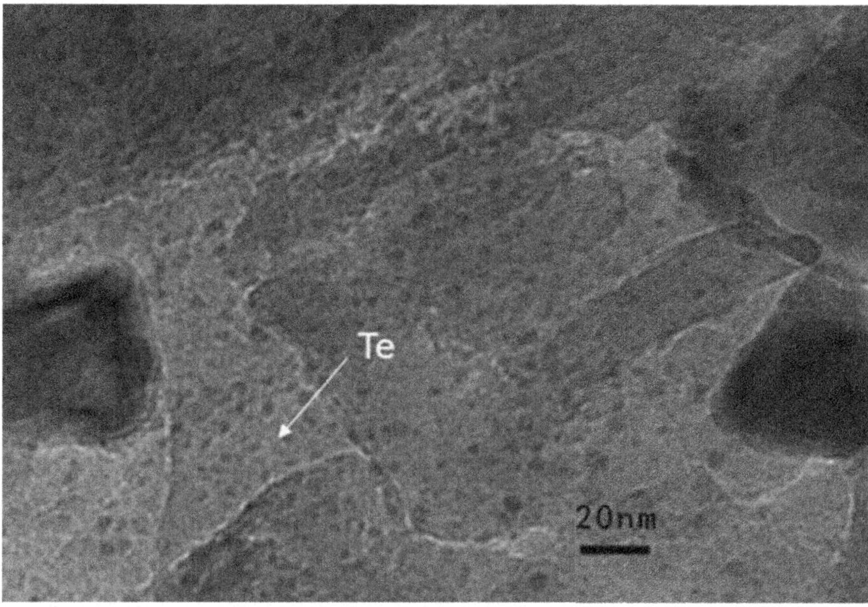

Fig. 6.29 TEM image of Sb2Te3 thin film when Te content is 17%

Table 6.6 Hall coefficient RH, carrier mobility μ, carrier concentration n, conductivity σ_H, and conductivity σ_F obtained by the four-probe method for samples with different Te contents

Te volume (%)	R_H (10^{-3} cm^3 C^{-1})	μ (cm^2 V^{-1} s^{-1})	N (10^{20} cm^{-3})	σ_H (S cm^{-1})	σ_F (S cm^{-1})
0	73.5	33.6	0.85	456.96	512.07
17	25.6	10.9	2.44	425.54	486.78
29	21.4	5.61	2.92	262.09	261.68
45	19.3	1.93	3.23	99.74	100.97

with different Te contents as a function of Te content is shown in Fig. 6.30. As the Te content increases, the sample mobility decreases rapidly, and the carrier concentration slowly increases. The rate of decrease in mobility is greater than the rate of increase in carrier concentration, which ultimately leads to a decrease in the conductivity of the sample as the Te content increases.

The decrease in mobility with increasing Te content can be attributed to the influence of the interface grain boundary formed between Te and Sb_2Te_3. If there is a work function difference between the materials in the nanocomposite, the energy level potential of the carriers can be formed in the material. The barrier acts as a filter for carriers can be used to improve the thermoelectric properties of the material. Figure 6.31 shows the band diagram of the Te–Sb_2Te_3 contact. Figure 6.31b shows the band diagram of the Te–Sb_2Te_3 contact at equilibrium. This figure shows

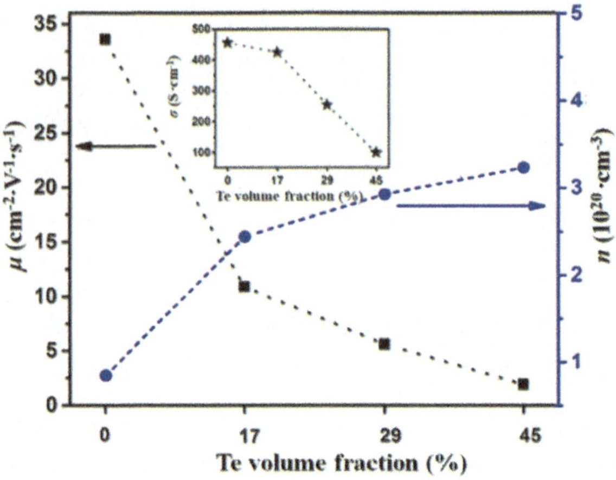

Fig. 6.30 Relationship between carrier mobility and carrier concentration of Sb_2Te_3 films with different Te contents as a function of Te content. The upper right illustration shows the relationship between the conductivity and the Te content

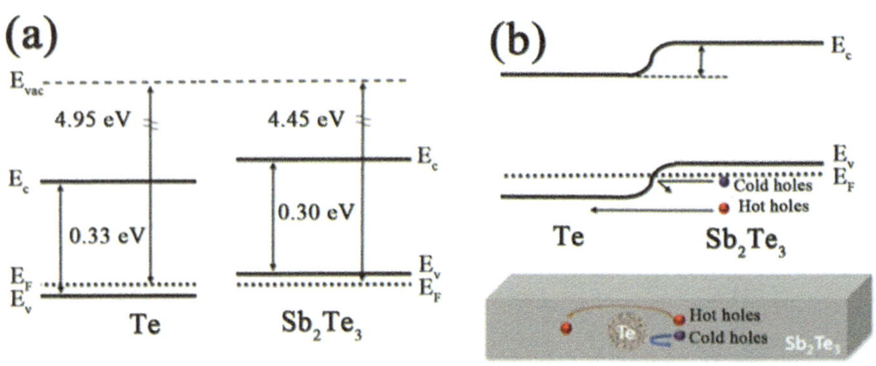

Fig. 6.31 Energy band diagram of Te in contact with Sb_2Te_3. **a** is the band diagram before Te and Sb_2Te_3 are in contact, and **b** is the band diagram after equilibrium between the two. The influence of the interface state is ignored

that there is an energy level barrier at the Te–Sb_2Te_3 interface. It can be seen from Fig. 6.31a that the work functions of Te and Sb_2Te_3 are 4.95 eV and 4.45 eV, respectively. Therefore, there will be an equilibrium band diagram of band bending in Fig. 6.31b after contact. Due to the existence of energy level barriers, at the interface of Te and Sb_2Te_3, low-energy carriers (cold carriers) are strongly scattered, while high-energy carriers (hot carriers) are weakly scattered. High-energy carriers can cross the energy level barrier and low-energy carriers cannot, resulting in the separation of hot and cold carriers. The strong scattering leads to a decrease in the mobility of the carriers, which explains the reason for the decrease in the mobility of the samples containing Te nanoparticles. As the Te content increases, the mass of the Te nanoparticles becomes larger and larger. The interface of Te–Sb_2Te_3 is more and more obvious, and the scattering of cold carriers is also more and more intense. Therefore, the mobility of the carriers decreases as the Te content increases.

6.4.3.2 Seebeck Coefficient Test

The Seebeck coefficient and conductivity testing device built in the laboratory were used to test Sb_2Te_3 thin film samples with different Te contents as shown in Fig. 6.32. A positive Seebeck coefficient value indicates that the samples are all P-type semiconductors, which is consistent with the Hall coefficient test results. As the Te content increases, the Seebeck coefficient increases rapidly from 121 $\mu V\ K^{-1}$ of the Sb_2Te_3 thin film without Te nanoparticles and in a stoichiometric ratio to 195 $\mu V\ K^{-1}$ of the Sb_2Te_3 thin film with 45% Te content. The power factor can be calculated by Seebeck coefficient and conductivity. With the increase of Te content, the power factor first increases and then decreases. When the Te content is 17%, the power factor value reaches a maximum of 9.3 $\mu W\ cm^{-1}\ K^{-2}$ by comparing with the Sb_2Te_3 film (6.4 $\mu W\ cm^{-1}\ K^{-2}$) without Te nanoparticles meeting the stoichiometric ratio, it has an improvement of ~50%. When the Te content continues to increase, the power factor decreases, because the conductivity decreases rapidly with the increase of the Te content, and the Seebeck coefficient increases at a faster rate.

From the analysis of Fig. 6.31, it is found that the presence of Te nanoparticles will form a Te–Sb_2Te_3 interface barrier in the sample. Due to the existence of this interface barrier, low-energy carriers (cold carriers) at the interface of Te and Sb_2Te_3 is strongly scattered, while high-energy carriers (hot carriers) are weakly scattered, so that high-energy carriers can cross the energy level barrier and low-energy carriers cannot resulting in hot and cold loads. Separation of flow carriers cold low-energy carriers are not good for Seebeck coefficients, and the separation of hot and cold carriers is helpful to increase the Seebeck coefficients. This explains why the Seebeck coefficients increase with the increase of Te content in Fig. 6.32. On the other hand, from the Mott relationship, it can be seen that near the Fermi level, the Seebeck coefficient is a derivative of the energy-dependent conductivity versus energy, and is approximately inversely proportional to the carrier concentration and the carrier mobility. From the results of the Hall coefficient test, the carrier concentration of Sb_2Te_3 thin film samples with different Te content is not much different, all on the order of 10^{20} cm^{-3}, but the mobility of the sample changes significantly with different Te content. The higher the content, the smaller the mobility in the Sb_2Te_3 film samples, which also explains the change in the Seebeck coefficient of Sb_2Te_3 film samples with different Te contents.

At room temperature, the relationship between Seebeck coefficient and carrier concentration of Sb_2Te_3 film samples with different Te contents (Pisarenko plot) is shown in Fig. 6.33 as the circled values. Assuming that the Sb_2Te_3 film is a degenerate semiconductor with a single-cycle energy band, the following formula is obtained under this model:

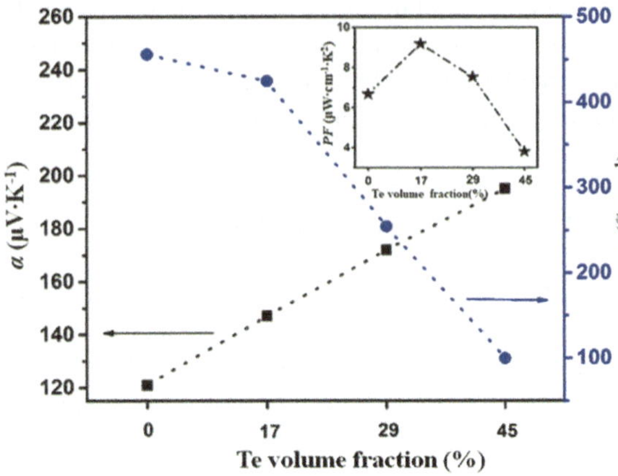

Fig. 6.32 Relationship between the Seebeck coefficient and conductivity of Sb_2Te_3 films with different Te contents as a function of Te content. In the upper right illustration is the relationship between the power factor and the Te content

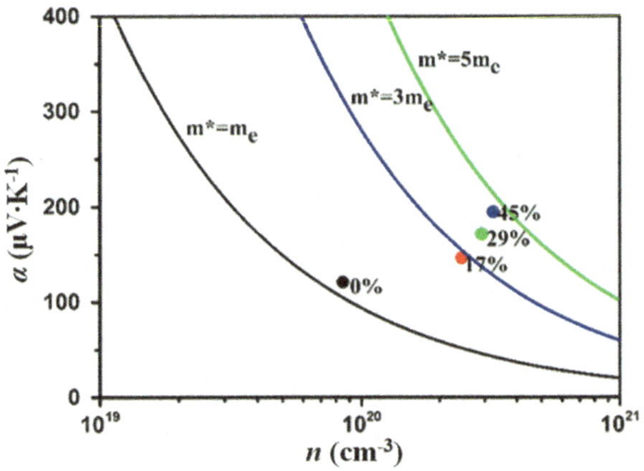

Fig. 6.33 Seebeck coefficient and carrier concentration (Pisarenko plot). The curves represent the theoretical relationship diagrams under different effective masses. The solid circles in the figure represent the numerical relationships at different Te contents in the experiment

$$\alpha = \frac{8\pi^2 k_B^2 T}{3eh^2}\left(\frac{\pi}{3n}\right)^{\frac{2}{3}} m^*. \tag{6.3}$$

In the formula, k_B is Boltzmann constant, T is absolute temperature, e is electron quantity, h is Planck constant, n carrier concentration, m^* effective carrier mass. When the effective carrier masses are taken as m_e, $3m_e$, and $5m_e$, three different Seebeck coefficients and carrier concentration curves are obtained as shown in Fig. 6.33. It can be seen from the curve that the Seebeck coefficient is inversely related to the carrier concentration. Given the carrier concentration, the Seebeck coefficient increases with the effective carrier mass. The circle values in Fig. 6.33 are the experimental values of the samples with different Te contents. The increase of the Seebeck coefficient with the increase of Te content is due to the increase of the effective carrier mass. The effective mass of the carrier is a reflection of the internal force of the material. This also shows that in the film containing Te nanoparticles, the low-energy carriers are strongly scattered due to the existence of the Te–Sb_2Te_3

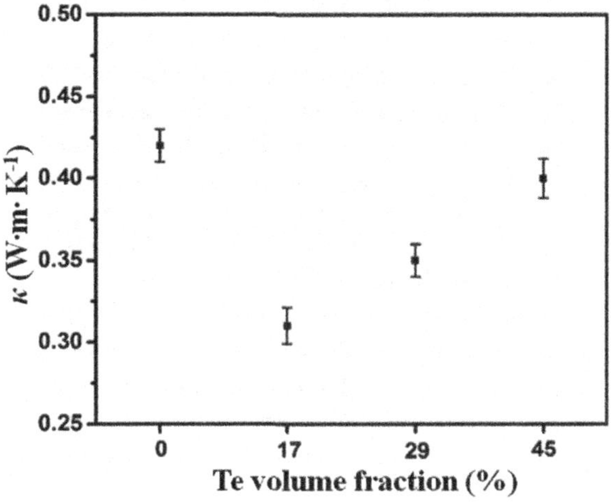

Fig. 6.34 Relationship between the thermal conductivity of Sb_2Te_3 films with different Te contents as a function of Te content

interface, low-energy carriers are strongly scattered, which increases the effective mass of the carriers.

6.4.3.3 Thermal Conductivity Test

According to the theory of heat transfer, the existence of nanocrystalline interfaces will scatter medium and long wave phonons, and these phonons will transfer a large amount of heat. Nano inclusions have the effect of reducing the thermal conductivity of the parent material. In the experiment, the Sb_2Te_3 thin film samples with different Te content were treated with 3ω method at room temperature, and the thermal conductivity was tested. The thermal conductivity test results are shown in Fig. 6.34. The error bars indicate the repeated measurement results. The thermal conductivity of the Sb_2Te_3 thin film sample without Te nanoparticles is 0.42 W m^{-1} K^{-1}. When Te nanoparticles are introduced into the material system, the thermal conductivity is 0.31–0.40 W m^{-1} K^{-1}. When the Te content is 17%, compared with the Sb_2Te_3 thin film sample without Te nanoparticles, the maximum thermal conductivity reduction is ~26%, and the thermoelectric figure of merit is also 0.90 at the maximum, while the Sb_2Te_3 film without Te nanoparticles the pyroelectric figure of merit of the sample is 0.46.

6.5 Chapter Summary

1. Using the molecular beam epitaxy (MBE) system, the relationship between the deposition rate of Sb and Te and the temperature of the beam source furnace was first explored. Then, the deposition rate of Te was fixed, and the Sb deposition

rate was changed to obtain Sb–Te films with different compositions. The thin film structure and thermoelectric properties were then characterized. Compared with the Sb_2Te_3 thin film sample that meets the stoichiometric ratio, the Sb_2Te_3 thin film with a higher Sb deposition rate has a metallic Sb phase, and the sample exists in a polycrystalline state. Sb-rich Sb_2Te_3 thin film samples have Sb nanoparticles and Sb_2Te_3–Sb grain boundaries. The existence of this grain boundary has a significant impact on the carrier and phonon transport in the sample. The Sb_2Te_3 thin film sample obtained from Sb–rich Sb_2Te_3 thin film has a larger Seebeck coefficient (Seebeck coefficient increased by nearly 6 times) and a lower thermal conductivity (the thermal conductivity is reduced by 41%)), The larger the power factor (the power factor is increased by 80%), and finally has a thermoelectric figure of merit ZT that is almost three times higher.

2. Sb_2Te_3 film with high (00l) orientation and satisfying stoichiometric ratio was prepared on the intrinsic Si (111) substrate by MBE. The film is highly crystalline and grows along the (00l) crystal direction perpendicular to the surface of the substrate. As the film thickness increases, the particle size increases significantly. Hall tests show that the carrier concentration and mobility of the film are sensitive to the film thickness. By analyzing the particle size of the film and the mean free path of the carriers, it is found that defects at the grain boundaries and inside the crystallites have a strong influence on the carrier concentration and mobility. In thicker films, the influence of microcrystalline internal defects on carrier transport performance is becoming increasingly important. At room temperature, the 121 nm film has a maximum carrier mobility of 305 $cm^2\ V^{-1}\ s^{-1}$ and a moderate carrier concentration of $2.125 \times 10^{19}\ cm^{-3}$. In the vicinity of room temperature, an approximately linear increase in resistivity with temperature was observed, which can be explained by the low doping concentration and small ionization energy.
3. Using molecular beam epitaxy technique to prepare $Bi_{0.4}Sb_{1.6}Te_3$ film and Sb_2Te_3 film with Bi atomic substitution on intrinsic Si (111) and glass substrate. $Bi_{0.4}Sb_{1.6}Te_3$ film has higher crystallinity and lower carrier concentration and mobility than Sb_2Te_3 film. A small number of Bi atoms are introduced into the Sb atomic sites of the Sb_2Te_3 film to form a substitutional doping, which offsets the compressive stress in the film, which promotes the growth of crystallites and increases the crystallinity. Analysis of the particle size and the mean free path of the carriers shows that the carrier mobility is greatly affected by the Bi atoms doped in place. The reason for the enhanced carrier scattering effect is probably due to the Coulomb force of the meta-doped Bi atom. In the range of −50 to 150 °C, $Bi_{0.4}Sb_{1.6}Te_3$ film and Sb_2Te_3 film show different conductivity temperature changes. The introduced Bi atom reduces the energy level of the impurity, resulting in a higher conductivity activation energy of 43.2 meV.
4. The effects of Te nanoparticles on the thermoelectric properties of Sb_2Te_3 thermoelectric films were studied. Using a molecular beam epitaxy system, first, under ultra-high vacuum, Sb_2Te_3 thin film layers and Te layers are alternately grown, and discontinuous nanoparticles are formed when the Te layer is thin, and Sb_2Te_3 films with different contents of Te nanoparticles are obtained. Results

The samples were prepared with thicknesses of 250, 300, 350, and 450 nm, and the corresponding Te volume contents were 0%, 17%, 29%, and 45%, respectively. The surface roughness of the samples was 2.08 nm, 2.32 nm, 2.67 nm and 4.78 nm, respectively. The change in surface roughness with Te content is due to the influence of Te nanoparticles. When the Te content is 17%, the Te nanoparticles are uniformly distributed in the Sb_2Te_3 thin film sample with a diameter of about 5 nm. The Seebeck coefficient test shows that as the Te content increases, the Seebeck coefficient increases rapidly, from 121 $\mu V\ K^{-1}$ of the Sb_2Te_3 film that does not contain Te nanoparticles and meets the stoichiometric ratio, to 195 $\mu V\ K^{-1}$ with Te content of 45% in the Sb_2Te_3 film. The power factor increases first and then decreases with increasing Te content. When the Te content is 17%, the maximum power factor value is 9.3 $\mu W\ cm^{-1}\ K^{-2}$ compared to the Sb_2Te_3 film (6.4 $\mu W\ cm^{-1}\ K^{-2}$) There is about ~50% improvement. The thermal conductivity of Sb_2Te_3 thin film samples without Te nanoparticles is 0.42 $W\ m^{-1}\ K^{-1}$. When Te nanoparticles are introduced into the material system, the thermal conductivity decreases to 0.31–0.40 $W\ m^{-1}\ K^{-1}$ depending on the Te content. When the Te content is 17%, compared with the Sb_2Te_3 thin film sample without Te nanoparticles, the maximum thermal conductivity reduction is ~26%, and the thermoelectric figure of merit is 0.90 at this time.

Reference

1. Madelung O (2003) Semiconductor: date handbook. Springer

Chapter 7
Molecular Beam Epitaxial Growth of Bi_2Te_3 Thin Films

Bismuth telluride, as the material with the best thermoelectric performance at room temperature [1], has attracted more and more attention. At the same time, theoretical calculations have shown that the quantum well structure of bismuth telluride material can reach 13 times the ZT value of the bulk material [2], so the low dimensionality of bismuth telluride material is the focus of scientists. Plucinski et al. Systematically studied the surface electrical properties of Bi_2Te_3 nanofilms prepared by molecular beam epitaxy [3], Wang used molecular beam epitaxy to prepare one-dimensional nanorod materials of bismuth telluride [4], and Bi_2Te_3 is still a very good Topological insulator materials, there have been a lot of reports in the literature to prepare Bi_2Te_3 topological insulator materials by molecular beam epitaxy [5–7].

7.1 Effect of Cross-Type Bi_2Te_3 Nanosheet Film and EG Heat Treatment on Thermoelectric Properties

Nanopores can effectively reduce the thermal conductivity of materials. For example, by making holes in bulk Bi_2Te_3 [8], Cu_2Se [9], and SnSe [10], the thermoelectric performance can be improved. Dun et al. [11] synthesized Bi_2Te_3 nanosheets by solution method and placed the nanosheets in ethylene glycol (EG) in a hydrothermal reactor to control different temperatures and different times. A hole of 5–100 nm is generated in the center, and the atomic ratio of Bi and Te in Bi_2Te_3 is dissolved at 2:3. In this section, Bi_2Te_3 nanosheets were prepared, combined with ethylene glycol heat treatment, to study the behavior and thermoelectric properties of the nanosheets.

Z. Hu and Z. Wu, *Nanostructured Thermoelectric Films*,
https://doi.org/10.1007/978-981-15-6518-2_7

7.1.1 Preparation of Nanosheet Films and Ethylene Glycol Heat Treatment

The MBE equipment was used to deposit the Bi_2Te_3 nanosheet film. The raw materials were Bi (99.999%) and Te (99.999%) with high purity. Si substrates with 600 nm SiO_2 on the surface were selected. The substrate after the silicon oxide wafer cleaning process is placed in the cavity, and the background vacuum is 6×10^{-9} Torr. Prior to film deposition, the substrate was heated to 473 K and held for 1 h. The deposition rate of Bi is about 0.14 Å/s, the deposition rate of Te is about 0.24 Å/s, and the working vacuum during coating is 8×10^{-8} Torr. In the first step, 8 nm of Bi_2Te_3 was deposited. In the second step, the temperature of the substrate table was raised to 623 K at a rate of 5 K/min and held for 1 h, and then the temperature was decreased to 523 K at a rate of 5 K/min. In the third step, Bi_2Te_3 was deposited at 50 nm, and the substrate temperature was lowered to room temperature at a rate of 5 K/min in a high-vacuum cavity (this sample was named B). In order to explore the feasibility of physical method for preparing nanosheets and combining solutions to make holes, sample B was placed in a hydrothermal reactor (capacity 100 mL) filled with ethylene glycol (EG, purity 99.99%), and then hydrothermally The kettle was placed in an oven and heated to 343 K, maintained for 30 min, 60 min, and 90 min, and the samples were named B30, B60, and B90, respectively. Subsequently, the corresponding characterization and thermoelectric properties of Bi_2Te_3 nanosheet films were tested.

7.1.2 Morphology and Structure Characterization of Cross-Type Nanosheet Films

First, the surface of the film was inspected, as shown in Fig. 7.1. The morphology of the initial Bi_2Te_3 thin film appears as a cross-shaped nanosheet, and the size of the nanosheet is small. After the solution treatment, the edges of the nanosheets became blurred, and some dots appeared, as shown in Fig. 7.1b. With the increase of the heat treatment time of ethylene glycol, the morphology of the nanosheets becomes more and more blurred, which is due to the decomposition of Bi_2Te_3 during the heat treatment of ethylene glycol. The irregular and cross-shaped morphology of the initial nanosheets results in different energies for dissolving different sites on the Bi_2Te_3 nanosheets, so there is no obvious regularity of the nanosheet morphology after ethylene glycol heat treatment.

In order to test whether the heat treatment of ethylene glycol has an effect on the composition ratio of the Bi_2Te_3 nanosheet film, the elemental distribution of the composition of the film was measured using a scanning electron microscope. The Te/Bi atomic percentages in samples B, B30, B60, and B90 were 63.8/36.2, 63.23/36.77, 63.56/36.44, and 64.30/35.70, respectively. The atomic ratio of Te/Bi is about 1.75, the Te component in the film is excessive, and the composition fluctuation

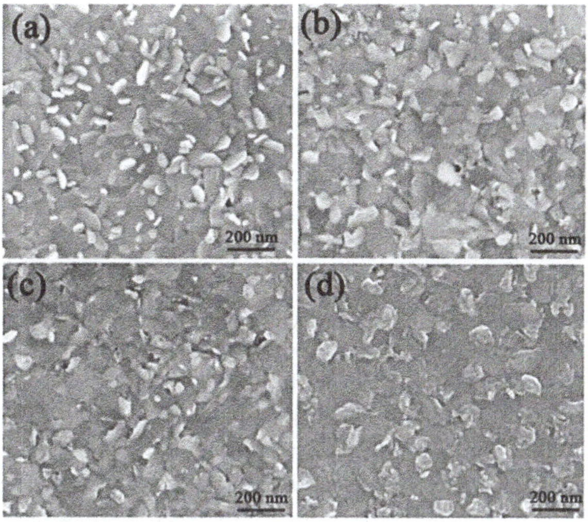

Fig. 7.1 Surface morphology of the sample: **a** Sample B; **b** Sample B30; **c** Sample B60; **d** Sample B90

of each film is within the error of energy spectrum detection. The results show that is a small effect of the heat treatment of ethylene glycol on the composition of the Bi_2Te_3 nanosheet film. The change in the surface morphology of the Bi_2Te_3 nanosheet film is mainly caused by the dissolution of Bi_2Te_3 in a certain ratio (Bi/Te = 2:3) during the heat treatment of ethylene glycol, rather than a certain ratio. A change in composition.

The surface morphology of the nanosheet film was observed by scanning electron microscopy. In order to further explore its microstructure, the surface morphology of the film was characterized by AFM. The observation area was 5 μm × 5 μm, as shown in Fig. 7.2. The surface root-mean-square roughness of samples B, B30, B60, and B90 were 7.2 nm, 7.0 nm, 6.3 nm, and 9.6 nm, respectively. The roughness of

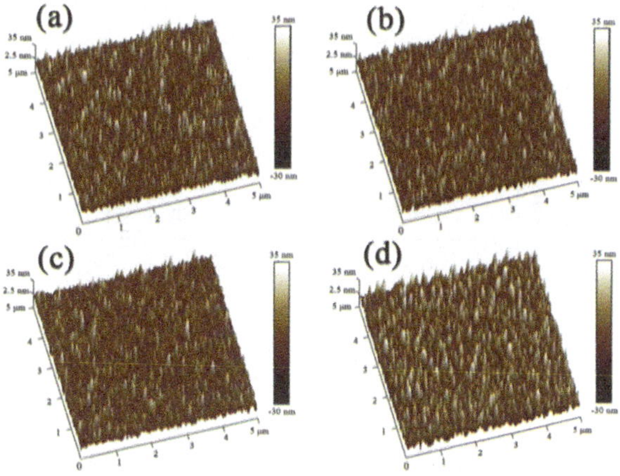

Fig. 7.2 AFM images of sample surface morphology: **a** B, **b** B30, **c** B60, and **d** B90, respectively

Table 7.1 Basic parameters of Bi_2Te_3 film

Sample	Atomic percentage (Te/Bi)	EG heat treatment time (min)	Thickness (nm)	RMS (nm)
B	63.8/36.2	0	~60	7.2
B30	63.23/36.77	30		7.0
B60	63.56/36.44	60		6.3
B90	64.30/35.70	90		9.6

the sample surface decreases first and then increases, which indicates that during the heat treatment of ethylene glycol, the raised areas of the nanosheets are preferentially dissolved first, so that the film gradually becomes flat. With the further increase of the heat treatment time of ethylene glycol, the dissolution of Bi_2Te_3 film in the longitudinal direction is intensified, which makes the surface roughness of the film larger. The change of the film surface roughness means that the electron and phonon transmission methods on the film surface will also change, and the scattering intensity will increase. The corresponding changes will also have a significant impact on the thermoelectric properties of the film, which is analyzed later.

In order to facilitate the comparison and analysis later, the basic parameters of the samples have been arranged, see Table 7.1.

XRD was used to identify the structure of the Bi_2Te_3 nanosheet film before and after the ethylene glycol heat treatment, as shown in Fig. 7.3a. The peak spectra in the film can be indexed to the Bi_2Te_3 phase (JCPDS card number 72-2036) and the Te phase (JCPDS card number 85-0555). Among them, the (006) peak of the Bi_2Te_3 phase is the sharpest and has the highest relative intensity. Due to the combination of the intrinsic hexagonal structure of Bi_2Te_3 crystal and the inter-layer van der Waals force, when Bi_2Te_3 grows freely, it tends to the (00l) direction, and its morphology is a hexagonal nanosheet. According to the (006) peak of the Bi_2Te_3 in the XRD

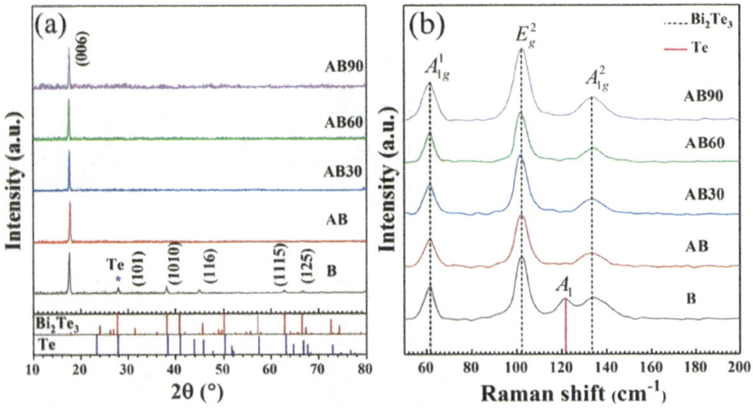

Fig. 7.3 XRD and Raman spectra of samples: **a** XRD; **b** Raman

pattern, the corresponding grain size of each sample was obtained, and the grain sizes were 26.3 nm, 28.6 nm, 28.9 nm, and 24.8 nm, respectively. The grain size did not change much, and the grain size of the sample treated with ethylene glycol for 90 min was the smallest. In addition, a peak of Te (101) appears in the XRD pattern of each sample, which is consistent with the excessive Te component in the energy spectrum detection result. The peak-to-average differences of the XRD patterns are not significant, indicating that the heat treatment of ethylene glycol has a small effect on the structure of the film.

The Raman spectrum was used to detect the phonon vibration mode of the Bi_2Te_3 nanosheet film before and after ethylene glycol heat treatment, as shown in Fig. 7.3b. All samples have three distinct peaks with peak shifts of approximately 62, 103, and 120 cm^{-1}, corresponding to the Raman active phonon vibration modes A_{1g}^1, E_g^2 of Bi_2Te_3, and A_1 of Te, respectively [12, 13]. The Raman shift of the fourth peak of samples B, B30, and B60 is about 135 cm^{-1}, which corresponds to the A_{1g}^2 of Bi_2Te_3 mode. The fourth peak in the B90 sample is about 140 cm^{-1} to the right. There are two possibilities: one is the Bi_2Te_3 mode due to the longer heat treatment time in ethylene glycol, the reaction time is longer Bi_2Te_3 is here. The phonon vibration direction is subject to a certain stress, which makes the corresponding Raman shift to the right. Another possibility is the E' of Te model. Compared with the peak spectra of single crystal Bi_2Te_3 and Te, the Raman peak obtained by the test is slightly shifted and may be subject to slight stress, which is in a normal range.

Sample B was selected for transmission electron microscope analysis to further observe the fine structure of the nanosheets. The structure of the Bi_2Te_3 nanosheets is shown in Fig. 7.4. When preparing TEM samples, the blade method was used which has less damage to the film. Due to the dispersion of anhydrous ethanol,

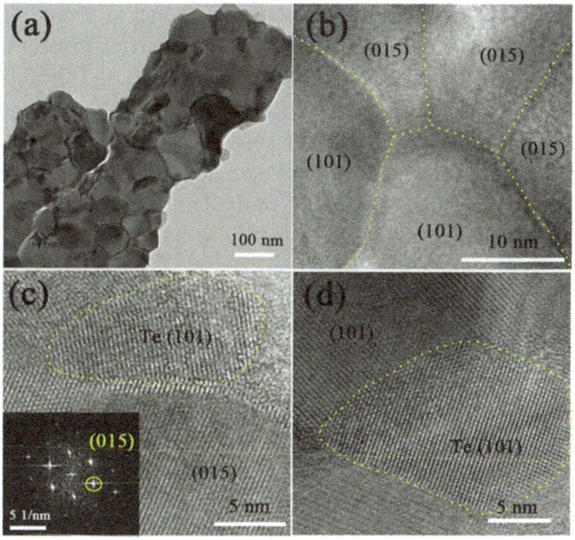

Fig. 7.4 TEM image of sample B: **a** Low-resolution TEM image; **b** Bi_2Te_3 grain interface with different orientations; **c** Bi_2Te_3 (015) and Te (101) interface, inset is Fourier in the Bi_2Te_3 region transformation diagram; **d** Bi_2Te_3 (101) and Te (101) interface

multiple nanoplates are continuously tiled into a large area, as shown in Fig. 7.4a. The interface in the Bi_2Te_3 film is mainly the interface between the nanosheets. As shown in Fig. 7.4b, the interface between multiple Bi_2Te_3 nanosheets with different orientations can be observed. At the same time, in addition to the homojunction interface between Bi_2Te_3 nanosheets, a heterojunction nano interface between Bi_2Te_3 and Te can also be observed, as shown in Fig. 7.4c–d. Observation of the surface by TEM shows that there are multiple interfaces in Bi_2Te_3 nanosheet films. Different interfaces contribute differently to carrier and phonon scattering, which affects the thermoelectric properties of the material.

7.1.3 Thermoelectric Properties of Crossed Nanosheet Films

The thermoelectric properties of Bi_2Te_3 nanosheets at room temperature were tested, as shown in Table 7.2. As the ethylene glycol heat treatment time increased, the carrier concentrations in samples B, B30, B60, and B90 increased sequentially. After the ethylene glycol heat treatment, the dissolution of Bi_2Te_3 led to Te vacancy defects dominating, and the n-type carrier time increased sequentially with the increase of heat treatment time. After the heat treatment of ethylene glycol, the carrier mobility increases in order, but decreases relative to that before the treatment, as shown in Fig. 7.5a. The change trend is consistent with the surface roughness of the film. Due to the dissolution of different sites on Bi_2Te_3 nanosheets during the heat treatment of ethylene glycol, the roughness of the surface changes. The surface roughness is mainly caused by the change of the height of the nanosheets. The nanometer interface on the surface will scatter carriers to make the mobility change. According to the conductivity expression and the Mott formula of the Seebeck coefficient, the conductivity increases as the carrier concentration and mobility increase, while the Seebeck coefficient decreases as the carrier concentration and mobility increase. The change trend of carrier concentration and mobility makes the conductivity of sample B decrease and then increase after ethylene glycol heat treatment, which is consistent with the trend of surface roughness of the film. The change of Seebeck coefficient first increases and then decreases, which is closely related to the change of the roughness of the surface of the nanosheet film detected in the foregoing. Finally, the power factor obtained from the conductivity and Seebeck coefficient shows a

Table 7.2 Thermoelectric properties of samples at 300 K

Sample	n (cm^{-3})	μ ($cm^2\ V^{-1}\ s^{-1}$)	σ ($S\ cm^{-1}$)	α ($\mu V\ K^{-1}$)	PF ($\mu W\ cm^{-1}\ K^{-2}$)
B	13.9E19	26.2	582.9	−148.8	12.9
B30	15.4E19	16.5	406.3	−154.0	9.6
B60	16.2E19	18.0	468.0	−165.2	12.8
B90	22.2E19	21.4	760.0	−132.8	13.4

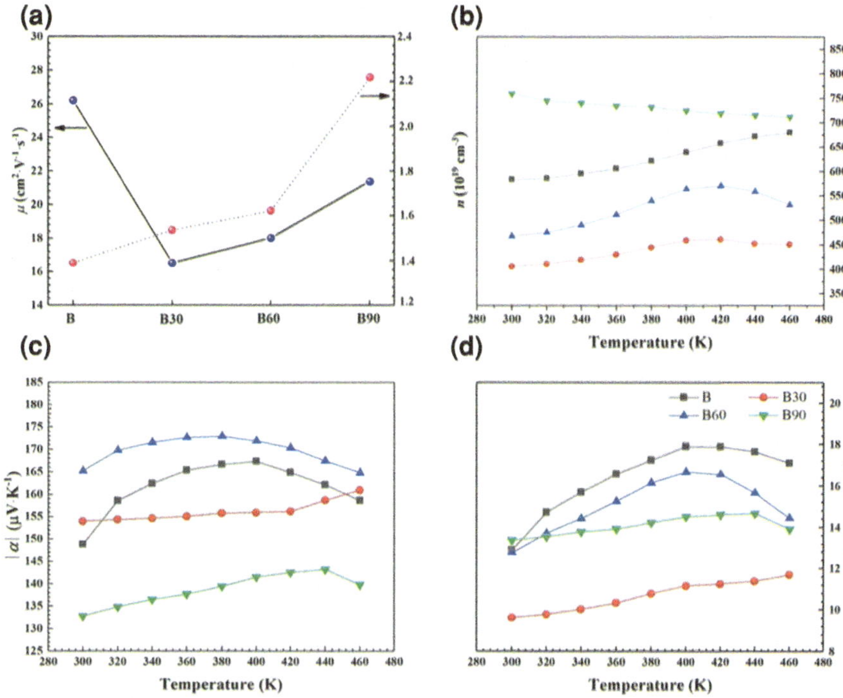

Fig. 7.5 Dependence of thermoelectric properties on temperature: **a** carrier and mobility trends of samples at 300 K; **b–d** are conductivity and Seebeck coefficients, respectively, and the relationship between power factor and temperature

trend of decreasing first and then increasing. After the ethylene glycol heat treatment for 90 min, the power factor was highest at 300 K reaching 13.4 $\mu W\ cm^{-1}\ K^{-2}$.

In order to explore the optimal temperature range for Bi_2Te_3 nanosheet films, the thermoelectric properties at different temperatures were tested, as shown in Fig. 7.5b–d. The electrical conductivity of samples B and B30 increases with increasing temperature. The electrical conductivity of sample B60 increases first and then decreases with increasing temperature. The electrical conductivity of sample B90 continues to decrease with increasing temperature. When the temperature is low, the Seebeck coefficient of all samples gradually increases with the increase of temperature, and it decreases after higher temperature. The Seebeck coefficient of sample B30 changes most slowly with temperature, and increases slightly at high temperatures. This may be that during the heat treatment of ethylene glycol, the Bi_2Te_3 nanosheets have just begun to dissolve, and a narrow amorphous region exists at the edge of the dissolution site [11]. Making Seebeck changes with temperature changes. The power factor of all samples increased roughly with the increase of temperature. The power factor of Sample B at 400 K reached a maximum of 18 $\mu W\ cm^{-1}\ K^{-2}$.

7.2 Effect of Tile Bi_2Te_3 Nanosheet Film and EG Heat Treatment on Thermoelectric Properties

Buha et al. [14] studied the thermal stability and anisotropic sublimation of Bi_2Te_3 at a certain annealing temperature. Zhu et al. [15] prepared a $Bi_{0.5}Sb_{1.5}Te_3$ thin film in the (015) crystal orientation by magnetron sputtering, and achieved the transition to the (00l) direction by annealing. This article refers to the related literature, and based on the research in the previous section, annealed and combined with ethylene glycol heat treatment, the behavior and thermoelectric properties of the nanosheets are studied.

7.2.1 Annealing and Glycol Heat Treatment

This section is based on the cross-type Bi_2Te_3 nanosheet films prepared in the previous section. Because Te is volatile at higher temperatures, the composition of the film will shift. In this section, the sample B is annealed at a certain temperature. The sample B is placed in the annealing furnace, and the vacuum in the annealing furnace is evacuated to less than 0.1 Pa, and then filled with high purity N_2 (purity 99.999%) to a vacuum degree of 9.9×10^4. Pa, and repeat the above steps 3 times, try to exclude the air in the cavity. The annealing temperature was then set to 573 K, the annealing time was maintained at 2 h, and the temperature was raised and lowered at a rate of 0.5 K/s. After cooling to room temperature, the sample was taken out and the sample was named AB. Place the sample AB in a hydrothermal reactor (capacity 100 mL) filled with ethylene glycol (purity 99.99%), then put the hydrothermal reactor in an oven and heat it to 343 K for 30, 60, and 90 min, the samples were named AB30, AB60, and AB90, respectively. The Bi_2Te_3 film was then characterized and tested for thermoelectric properties. For comparison, Sample B and the annealed AB series samples are listed in the analysis below.

7.2.2 Morphology and Structure Characterization of Tiled Nanosheet Films

The surface morphology of samples B and AB series is shown in Fig. 7.6. Sample B was in the form of intersecting nanosheets. After annealing, sample AB became tiled nanosheets with pores [15]. The crossed nanosheets are dumped during the heat treatment process, and the crossed nanosheets are irregular. Therefore, the nanosheets after dumping cannot be arranged seamlessly, which leads to the appearance of nanopores. In addition, the evaporation of Te during the annealing process will also make the nanosheets irregular and easier to tile, and nanopores may appear [14]. In addition, the annealing process also results in changes in the size of the nanosheets,

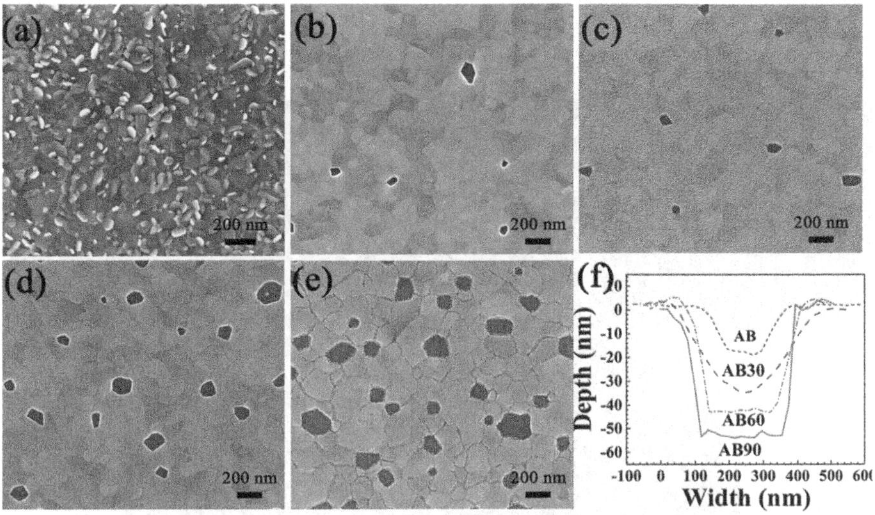

Fig. 7.6 SEM images of Bi_2Te_3 surface morphology: **a** B, **b** AB, **c** AB30, **d** AB60 and **e** AB90; **f** typical nanopore depth and size in AB series samples detected by AFM

which will be analyzed in more detail in the XRD analysis later. After heat treatment with ethylene glycol, more holes appeared in the tiled nanosheet film. With the increase of ethylene glycol heat treatment time, the total area of nanopores tends to increase, which is due to the intensification of the dissolution reaction of Bi_2Te_3 in ethylene glycol with time. The size and depth of typical nanopores in AB, AB30, AB60, AB90 were detected using AFM. With the increase of the heat treatment time of ethylene glycol, the depth of typical nanopores in AB, AB30, AB60, and AB90 increased sequentially from 18, 36, 46 to 52 nm, as shown in Fig. 7.6f. The above results indicate that the dissolution reaction of Bi_2Te_3 in the ethylene glycol heat treatment includes a direction parallel to the film and a direction perpendicular to the film.

The surface morphology of the film was characterized by AFM, and the microstructure was further probed. The observation area was 5 $\mu m \times 5$ μm, as shown in Fig. 7.7. The range of the height color bars of samples B, B30, B60, and B90 increased in turn, which was due to the fact that the dissolution of Bi_2Te_3 in the direction perpendicular to the film became stronger with the increase of the reaction time. Compared with the surface roughness change of the B series samples, the surface change of the AB series samples is mainly reflected in the increase of the hole depth. After the annealing process, the intersecting nanosheets become tiled and the longitudinal protrusions are reduced. During the heat treatment of ethylene glycol, the dissolution reaction of Bi_2Te_3 mainly starts from the exposed surface of the tiled nanosheets and expands in-plane and in the depth direction. The appearance of nanopores will greatly change the way of electron and phonon transmission on the surface of the film. The corresponding changes will also have a significant impact

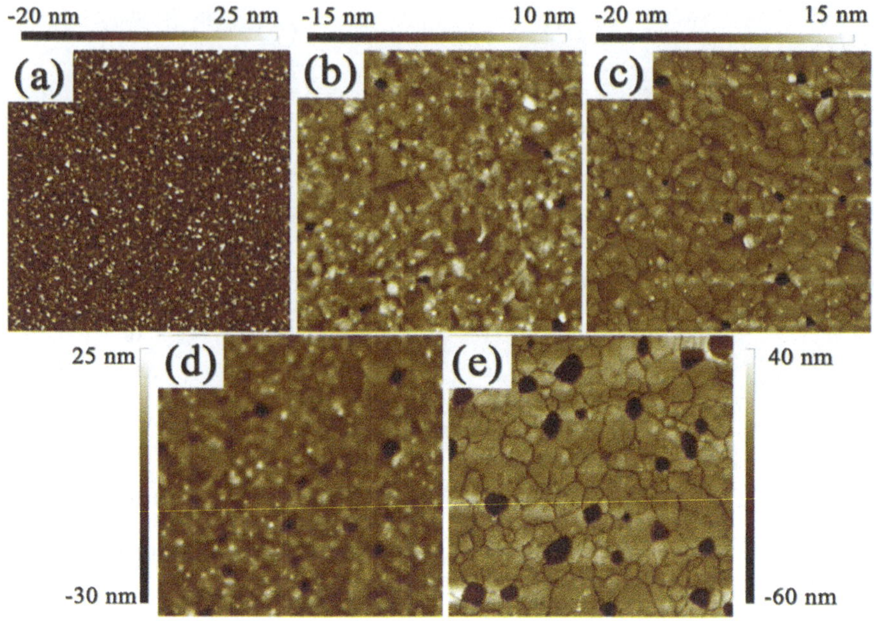

Fig. 7.7 AFM images of the sample surface morphology represent the samples: **a** B, **b** AB, **c** AB30, **d** AB60, and **e** AB90

on the thermoelectric properties of the film. The relationship between thermoelectric properties and pores is analyzed later.

As the morphology of the nanosheets of the thin film changed greatly after the annealing treatment and the ethylene glycol heat treatment, it is necessary to detect the change of its elemental composition (Table 7.3). The Te/Bi ratios in the samples AB, AB30, AB60, and AB90 are in the range of 1.5–1.6. Compared with the before annealing, the Te content in the film is reduced. This is due to the sublimation of some Te during the 573 K annealing treatment [14]. In addition, the morphology of Bi_2Te_3 tiled nanosheets and cross nanosheets are different, resulting in different starting sites and violent degrees of the dissolution reaction in the heat treatment of

Table 7.3 Basic parameters of Bi_2Te_3 film after annealing

Sample	Atomic percentage (Te/Bi)	EG heat treatment time (min)	Thickness (nm)	Hole depth (nm)
B	63.8/36.2	0	~60	–
AB	61.6/38.4	0	~50	18
AB30	61.06/38.94	30		36
AB60	60.97/39.03	60		46
AB90	60.85/39.15	90		52

ethylene glycol, but it does not affect that they are still in a certain proportion (Bi/Te = 2:3) dissolved.

XRD was used to identify the structure of Bi_2Te_3 nanosheet film before and after annealing, and before and after ethylene glycol heat treatment, as shown in Fig. 7.8a. The peak spectra in the annealed films can be indexed to the Bi_2Te_3 phase (JCPDS card number 72-2036). Compared with that before annealing, only the (006) peak of the Bi_2Te_3 phase in the AB series samples, the peak of Te disappeared. It is shown that the annealed Bi_2Te_3 nanosheets are single wafers parallel to the substrate height. According to the (006) peak of Bi_2Te_3 in the XRD spectrum, the corresponding grain size of each sample of the AB series was obtained, and the grain size was between 36–40 nm. The ethylene glycol heat treatment had little effect on the grain size, and the annealing treatment made the crystal grain size of the film becomes larger.

The Raman spectroscopy was used to detect the phonon vibration mode of the Bi_2Te_3 nanosheet film before and after ethylene glycol heat treatment, as shown in Fig. 7.8b. The annealed AB series samples had three distinct peaks with peak shifts of approximately 62, 103, and 135 cm^{-1}, corresponding to the Bi_2Te_3, and Raman active phonon vibration modes, respectively. Compared with sample B, the Te mode in the AB series samples disappeared, leaving only the phonon vibration mode of Bi_2Te_3, indicating that the Te content decreased due to annealing, which is consistent with the results of XRD and EDS detection. During the heat treatment of ethylene glycol, the Raman shift of each peak in the Raman spectrum of the AB series thin

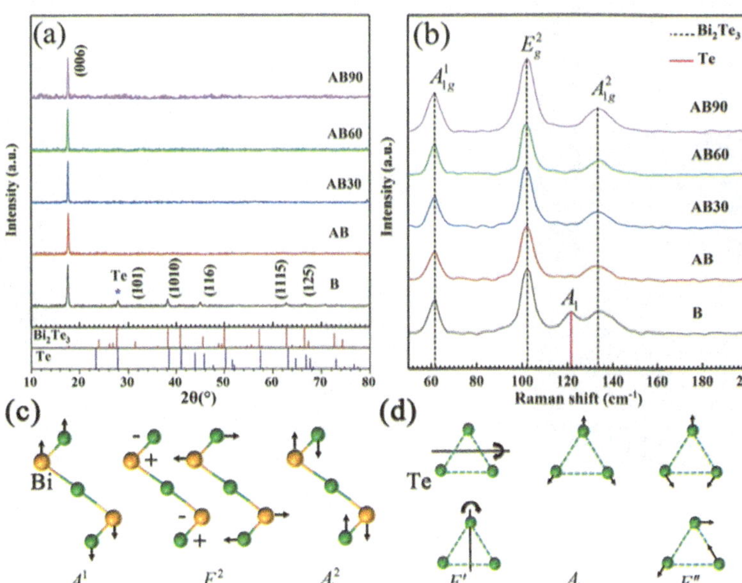

Fig. 7.8 XRD patterns and Raman patterns of samples: **a** XRD patterns of samples; **b** Raman patterns of samples; **c–d** Raman active phonon vibration modes of Bi_2Te_3 and Te, respectively

film did not change significantly with the increase of the processing time. Based on the behavior of the intersecting nanosheets in the ethylene glycol solution, it is shown that under the same composition, the Raman phonon vibration mode of the nanosheets is not affected by the heat treatment of ethylene glycol, regardless of the shape of the nanosheets being crossed or tiled.

Sample AB was selected as a representative for TEM analysis to further observe the fine structure of nanosheets and nanopores. In order not to damage the structure of the film and nanopores, the FIB method was used to prepare TEM observation samples. The structure of multiple Bi_2Te_3 nanosheets with nanopores is shown in Fig. 7.9a, and the element distribution is shown in Fig. 7.9b–c. Multiple nanosheets were selected for electron diffraction analysis. The diffraction pattern was a polycrystalline ring, indicating that it was polycrystalline, as shown in Fig. 7.9d. Since the selected area contains multiple Bi_2Te_3 nanoplates, different nanoplates contact at different angles, which is the reason for the appearance of polycrystalline rings.

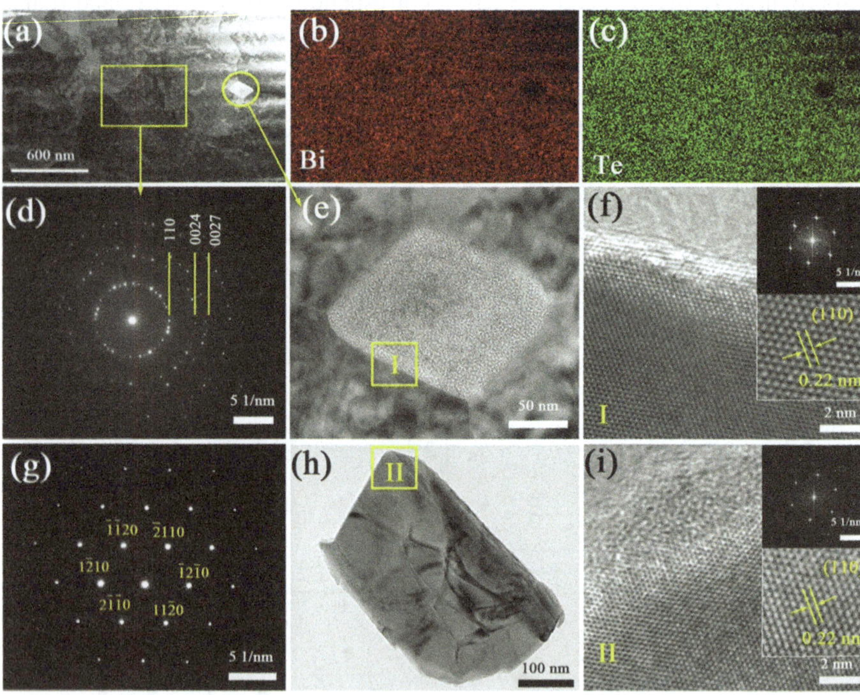

Fig. 7.9 TEM image of sample AB: **a**–**c** morphology and elemental distribution of the AB sample under low-resolution TEM. The sample was prepared using the FIB method; **d** shows the graph in **a** Selected electron diffraction patterns of selected areas; **e** and **f** represent nanopores and corresponding boundaries in figure (**a**). The inset is the Fourier transform and high-resolution images of Bi_2Te_3; **g**, **h** and **i** respectively represent the selected electron diffraction pattern of a single Bi_2Te_3 nanosheet, the corresponding nanosheet and the nanosheet boundary, and the inset is the Fourier transform map and high-resolution map of Bi_2Te_3

In the observation area, the nanopores formed between multiple nanosheets can be observed, as shown in Fig. 7.9e. The analysis of the nanosheet lattice and Fourier transform around the nanosheet shows that a single nanosheet is a single crystal with a crystal plane of (110), as shown in Fig. 7.9f. In order to observe the structure of a single nanosheet more closely, the ultrasonic oscillation method is used for sample preparation. The structure of a single nanosheet is shown in Fig. 7.9h, and the selected area electron diffraction is shown in Fig. 7.9g, indicating The nanosheets are single crystals. At the same time, the corresponding high-resolution graph and Fourier transform also prove this, as shown in Fig. 7.9i. Therefore, the tiled Bi_2Te_3 nanosheets are (00l) oriented single crystals, which is consistent with the XRD pattern detection results. The main reason for the formation of single wafers is due to the tiling and recrystallization of nanosheets in the annealing process.

7.2.3 Characteristics and Formation Mechanism of Nanopores in Tiled Nanosheet Films

In order to explore the distribution of pores in each sample after annealing and ethylene glycol heat treatment, when the magnification is 10,000 times under the scanning electron microscope, 10 regions are randomly selected for each sample for statistical distribution of pores, as shown in Fig. 7.10. The average pore diameter of the sample is obtained from the Gaussian distribution. When calculating the porosity, the total area of all pores is divided by the total area of the statistical area.

The pores in the sample AB are pores between the nanosheets. In order to distinguish the pores generated by the heat treatment of ethylene glycol, they are classified as type I. The average pore size of Type I pores in sample AB was 122 nm, and the number of pores was 28 per 100 μm^2. After 30 min of heat treatment with ethylene glycol, with the dissolution of Bi_2Te_3, the pore size increased to 154 nm, and the number increased to 69 per 100 μm^2. The increase in the average pore size may be due to the dissolution of the nanosheets around the type I pores, leading to further pore expansion as a result, the increase in the number is mainly due to the generation of new holes in Bi_2Te_3. When the ethylene glycol heat treatment time was extended to 60 min, the average pore size decreased compared to the sample AB30 to 138 nm, but the number increased sharply to 263 per 100 μm^2, because a large number of new pores were generated and gradually grew, making the average pore size slightly reduced. When the ethylene glycol heat treatment time was extended to 90 min, the average pore size increased to 251 nm, while the number was 104 per 100 μm^2. With the increase of processing time, the pores in the sample AB60 grew, and some of the pores expanded due to the dissolution of the nanosheets. The last two small pores merged into a larger pore, and the pore diameter increased rapidly, while the number of pores decreased slightly. When the heat treatment time of sample AB in ethylene glycol was increased from 30, 60 to 90 min, its porosity increased from 0.6% to 1.7%, 5.8% and 6.4%, respectively. The dissolution mechanism of the Bi_2Te_3 nanosheets

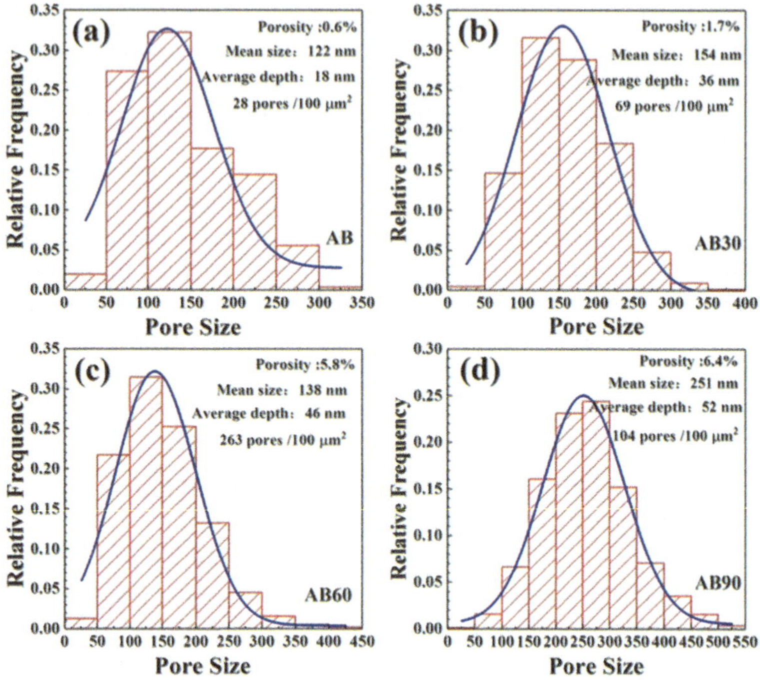

Fig. 7.10 Number and size distribution of 20 nanopores: **a** AB, **b** AB30, **c** AB60 and **d** AB90

in the heat treatment of ethylene glycol can be preliminarily reflected through the distribution trend of the average pore size and number. It also shows that the porosity, pore size, and number of pores of the film can be controlled by appropriate ethylene glycol heat treatment.

In order to further explore the dissolution process of tiled Bi_2Te_3 nanosheet films over time during the heat treatment of ethylene glycol, a large number of areas were selected on the surface of the sample using a scanning electron microscope to observe, and a representative pore was used to analyze the dissolution mechanism, such as Fig. 7.11. First, the crossed nanoplates become tiled nanoplates after annealing. The pores are mainly the pores between the nanoplates, that is the type I nanopores in the sample AB. The edges and corners of the nanoplates are clear. In order to analyze the dissolution process of Bi_2Te_3 during the heat treatment of ethylene glycol, the crystal structure of Bi_2Te_3 was first analyzed, as shown in Fig. 7.11a. The hexagonal Bi_2Te_3 nanosheets are stacked with multiple five-layer atoms in the c-axis direction, and Te and Te between each five-layer atom and the five-layer atom are combined by van der Waals force. At room temperature, Bi_2Te_3 is stable in ethylene glycol. When the ethylene glycol was heat-treated, due to the relatively weak van der Waals' bonding force between the layers, the layers fractured, leading to the dissolution of Bi_2Te_3. Relevant literature [11] compared the different surface energies of Bi_2Te_3 and obtained the disassembly rates of different crystal planes.

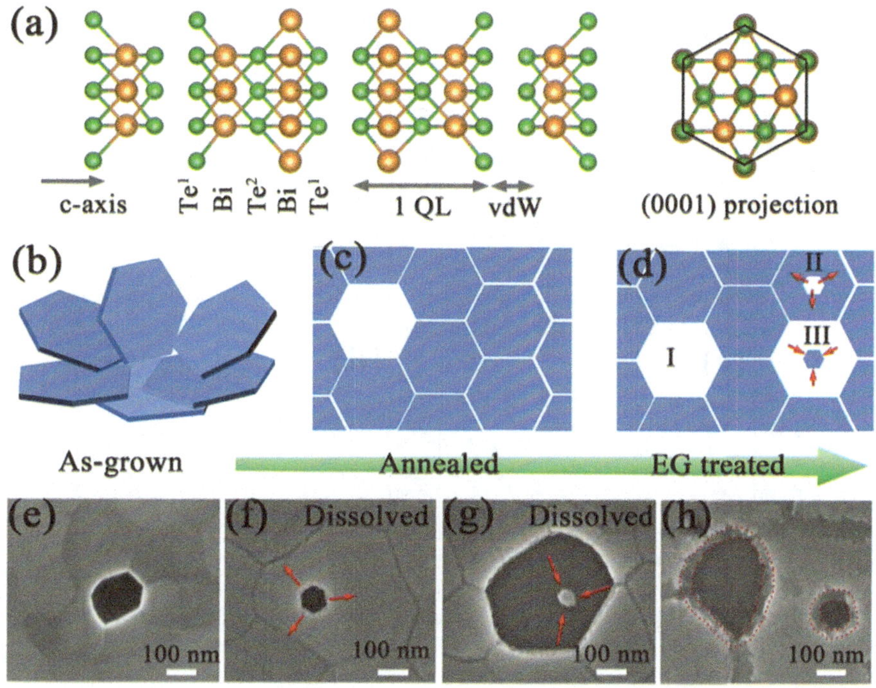

Fig. 7.11 The formation of nanopores in Bi_2Te_3: **a** the crystal structure of Bi_2Te_3; **b** the nanoplates in sample B; **c** the nanopores between the grains (Type I); Pores (Types II and III); **e**–**h** represent typical nanopores in samples AB, AB30, AB60 and AB90, respectively

The Bi_2Te_3 nanosheets synthesized by the solution method were subjected to ethylene glycol heat treatment, and hexagonal nanopores appeared in the center. Here, hexagonal nanopores also appeared in the center of the tiled nanosheets in sample AB30. This hole is divided into type II, as shown in Fig. 7.11f. The edges of type II nanopores are relatively rough, indicating that the dissolution of Bi_2Te_3 extends from the center to the surroundings. As the ethylene glycol heat treatment time increases, the dissolution of Bi_2Te_3 intensifies, and the nanopores will gradually grow. In addition, there is another type of hole's generation. The dissolution of Bi_2Te_3 proceeds along the boundary between the tiled nanosheets toward the center of the nanosheets, and this hole is divided into type III, as shown in Fig. 7.11g. It can be observed from the figure that there are still some undissolved Bi_2Te_3 islands in the center of the nanopore. The distribution of the above types of nanopores is summarized. Type I nanopores are the nanopores in sample AB, type II nanopores are mainly present in sample AB30, and type III nanopores are mainly present in sample AB60. After the sample is annealed and treated with ethylene glycol, the generation of different nanopores is shown in Fig. 7.11b–d. When the ethylene glycol was heat-treated for 90 min, the dissolution of Bi_2Te_3 intensified, and the edges of the nanopores became rougher, as shown in Fig. 7.11h.

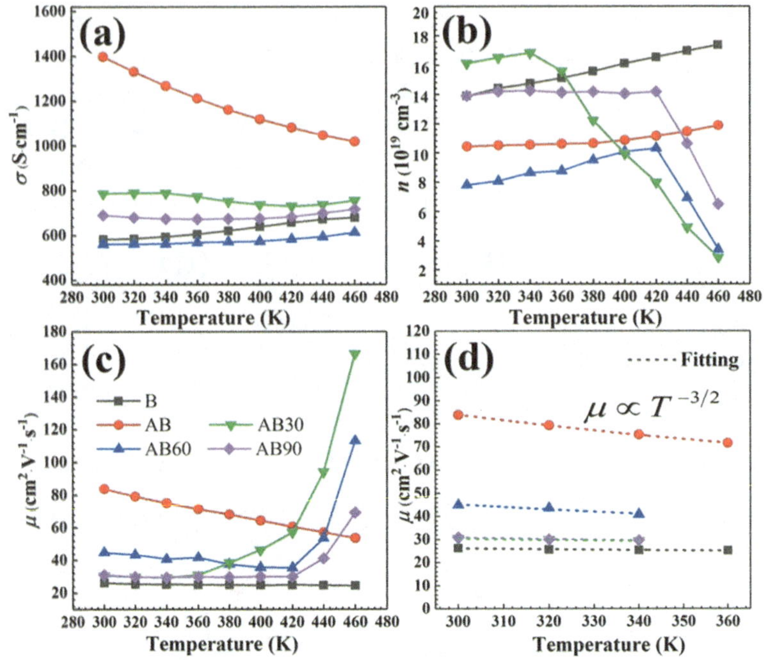

Fig. 7.12 Relationship between electrical properties and temperature: **a** conductivity; **b** carrier concentration; **c** and **d** carrier mobility

7.2.4 Thermoelectric Properties of Tiled Nanosheet Films

In order to explore the effect of nanopores on the electron transport in the tiled Bi_2Te_3 nanosheet film, the electrical properties of the film at different temperatures were tested as shown in Fig. 7.12.

After annealing, the conductivity of sample AB decreases with increasing temperature, which is opposite to the change trend of sample B, as shown in Fig. 7.12a. However, the conductivity of the AB series samples is greatly reduced after ethylene glycol heat treatment. This is due to the strong scattering of carriers transported along the film surface due to the presence of pores. After increasing the heat treatment time of ethylene glycol, the porosity of the film increased in turn, and the conductivity showed a decreasing trend. However, the conductivity of sample AB60 was lower than AB90, which was related to the abnormal number of nanopores. The above results indicate that, relative to the porosity, the number of pores has a greater effect on the conductivity. The change of carrier concentration with temperature is shown in Fig. 7.12b.

Before the ethylene glycol heat treatment, the carrier concentration in the film increases with the temperature, regardless of whether the sample is annealed or not. After the heat treatment of ethylene glycol, the carrier concentration in the film increased slowly with the increase of temperature, and then decreased rapidly. From the previous analysis, it is known that the thermal dissolution of Bi_2Te_3 in ethylene glycol results in rough interfaces at the edges of nanopores, which may lead to a very narrow amorphous transition region, resulting in a large difference in carrier concentration with temperature as compared with what was before the treatment [11]. The nanopores distributed along the surface of the nanosheet film will have a strong scattering effect on the electron transport on the surface of the film, and the mobility of the carriers is shown in Fig. 7.12c. Before the ethylene glycol heat treatment, the carrier mobility of sample B changed little with temperature, and the carrier mobility of sample AB gradually decreased with increasing temperature. After ethylene glycol heat treatment, the carrier mobility decreases slowly with increasing temperature and then increases. The interval in which the carrier mobility increases with increasing temperature may be related to the narrow amorphous region existing at the edges of the nanopores generated after the ethylene glycol treatment. The carrier scattering mechanism has changed with temperature. At this time, impurity scattering is the main scattering mechanism. In the lower temperature range, the fitted relationship of the carrier mobility with temperature is proportional to $T^{-3/2}$, as shown in Fig. 7.12d. It shows that this temperature range is dominated by acoustic scattering with a scattering factor of $-1/2$.

The change of Seebeck coefficient of each sample in Bi_2Te_3 nanosheet film with temperature is shown in Fig. 7.13a. In Sample B before annealing, the Seebeck coefficient first increased and then decreased with increasing temperature. For the AB series samples after annealing treatment, the Seebeck coefficient of the nanosheet film samples decreases with the increase of temperature, regardless of whether or not they have been subjected to ethylene glycol heat treatment. According to Mott's

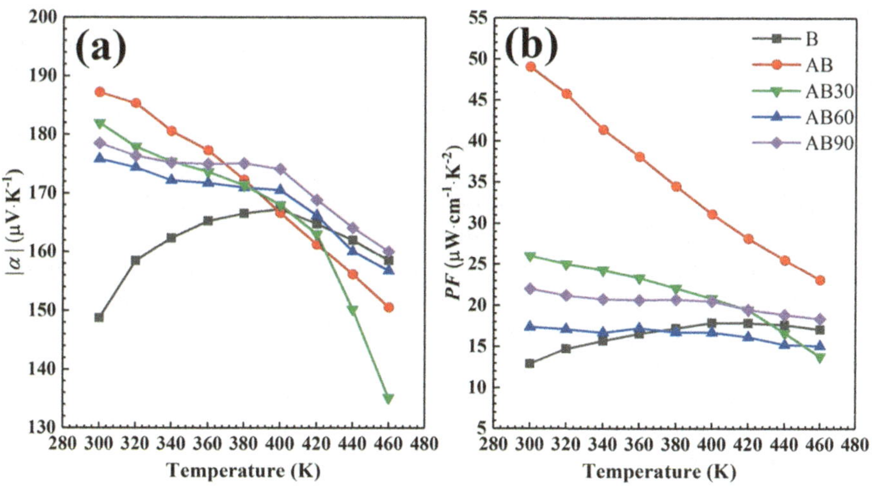

Fig. 7.13 Relationship between thermoelectric performance and temperature: **a** seebeck coefficient; **b** power factor

Table 7.4 Thermoelectric properties of samples at 300 K

Sample	n (cm^{-3})	l (nm)	L (V^{-2} K^{-2})	m^*/m_0	η	σ (S cm^{-1})	α (μV K^{-1})	PF (μW cm^{-1} K^{-2})
B	13.9 E19	2.8	1.73E–8	2.27	1.05	582.9	−148.8	12.9
AB	10.4 E19	8.0	1.64E–8	2.66	0.29	1398.6	−187.3	49.1
AB30	16.1 E19	3.9	1.65E–8	3.39	0.39	787.4	−182.0	26.1
AB60	7.8 E19	3.4	1.66E–8	1.98	0.50	562.6	−175.9	17.4
AB90	13.9 E19	3.3	1.66E–8	2.98	0.45	690.9	−178.5	22.0

formula, higher carrier concentration and mobility will lead to a decrease in Seebeck coefficient. The trend of Seebeck coefficient with temperature changes reflects the comprehensive result of changes in carrier concentration and mobility with temperature. The variation of the power factor of each sample in the Bi_2Te_3 nanosheet film with temperature is shown in Fig. 7.13b. The power factor of the AB sample decreases with increasing temperature and changes significantly. After the sample AB was treated with ethylene glycol, the power factor decreased, but it was still higher than that of the sample B. With the increase of ethylene glycol heat treatment time, the trend of the power factor of AB series samples becomes more and more gentle with temperature.

In order to investigate the effect of nanopores on the thermoelectric properties of Bi_2Te_3 nanosheet films, the thermoelectric parameters at room temperature of the films are listed, as shown in Table 7.4. After annealing, the power factor of AB series samples at 300 K was higher than that of sample B. The average free path of the carriers in sample AB was the largest, which resulted in a conductance of 1398.6 S cm^{-1} and a maximum room temperature power factor of 49.1 μW cm^{-1} K^{-2}. As the ethylene glycol heat treatment time increases, the porosity in the sample increases, and the power factor decreases in turn. The power factor of the sample AB60 is abnormal. Unlike this trend, its porosity is smaller than that of the sample AB90, but its power factor is less than AB90. Due to the abnormal number of nanopores in sample AB60, the variation trend of its electrical conductivity also appears abnormal. At the same time, the corresponding Seebeck coefficient trend has also changed accordingly. After annealing, the Seebeck coefficient of the nanosheet film was higher than that before annealing.

The relationship between Seebeck coefficient and carrier concentration (Pisarenko) is shown in Fig. 7.14a. As can be seen from the figure, except for sample AB60, the effective carrier mass of other AB series samples has been increased compared to B. Through the size and number of nanopores, the effective mass of the carriers can be adjusted, and then the Seebeck coefficient of the film can be adjusted. The relationship between the thermoelectric properties of nanosheet films and nanopores is shown in Fig. 7.14b. To sum up, the appropriate Seebeck coefficient and conductivity in AB make it obtain a higher power factor.

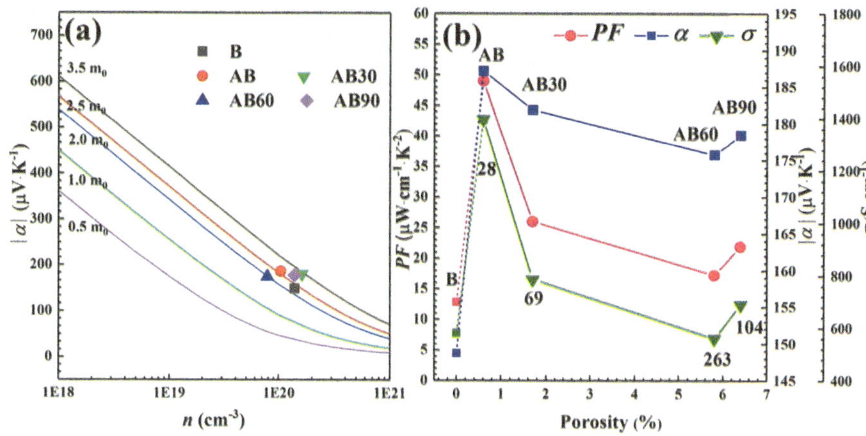

Fig. 7.14 Pisarenko diagram and thermoelectric performance: **a** the relationship between Seebeck coefficient and carrier concentration at 300 K (Pisarenko plot); **b** The relationship between nanopores and thermoelectric properties. The figure in the figure is the number of nanopores per 100 μm^2 area

After the cross-type nanosheet film becomes a tiled nanosheet film, the power factor has been greatly improved. In order to further compare the changes in thermoelectric properties, its thermal conductivity was tested. Since the thin film is deposited on a silicon substrate, and the thickness of the thin film is too thin, there is great difficulty in measuring the in-plane thermal conductivity. In this paper, the TDTR method is used to conduct longitudinal (out-of-plane) thermal conductivity measurement of the film, and the in-plane thermal conductivity can be estimated from the out-of-plane thermal conductivity of the nanosheet film by referring to related literature [16, 17]. Because too much porosity of the AB series samples after ethylene glycol heat treatment will affect the longitudinal thermal conductivity test, only the out-of-plane thermal conductivity of samples B and AB at room temperature were tested. The TDTR test data and fitting the curves are shown in Fig. 7.15.

Sample B was a cross-type nanosheet film, and its out-of-plane thermal conductivity ($\kappa_{\perp}$) was measured to be 1.271 W m^{-1} K^{-1}. According to the XRD pattern detection results, Bi_2Te_3 has no obvious preferred orientation, so it can be roughly considered to be isotropic, and the in-plane thermal conductivity ($\kappa_{\parallel}$) is equal to the out-of-plane thermal conductivity. According to its in-plane power factor, the in-plane ZT value at 300 K was calculated to be 0.3. The sample AB is a tiled nanosheet film. The XRD pattern results show that it is a (00 l) oriented nanosheet. The direction outside the nanosheet film is the c-axis direction of the Bi_2Te_3 crystal. According to the Wiedemann-Franz law and the Lorentz constant calculated in Table 4.4, the in-plane electronic thermal conductivity ($\kappa_{\parallel,e}$) is calculated to be 0.69 W m^{-1} K^{-1}. Based on the anisotropy factor in bulk Bi_2Te_3 single crystals with similar carrier concentrations, the thermal conductivity of the tiled nanosheet film can be estimated

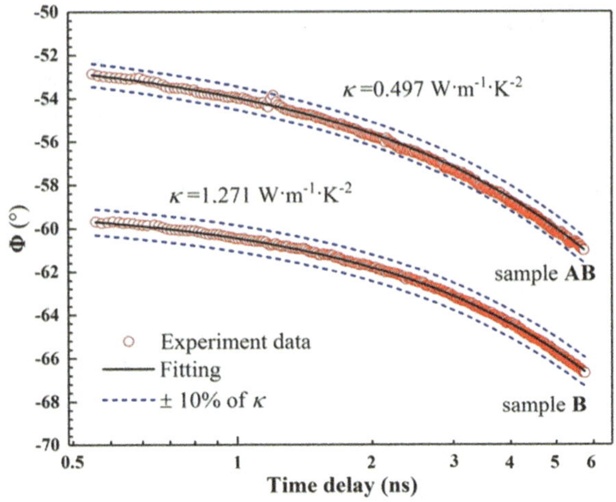

Fig. 7.15 Thermal conductivity data for the TDTR test

Table 7.5 Thermal conductivity of sample AB and related materials [19]

Sample structure		W m^{-1} K^{-1}			$\kappa_{\parallel,e}/\kappa_{\perp,e}$	$\kappa_{\parallel,L}/\kappa_{\perp,L}$
		κ	κ_e	κ_L		
AB	In-plane	1.22	0.69	0.53	6.67	1.34
	Out-plane	0.50	0.10	0.40		
Bulk crystal	In-plane	3.24	1.89	1.35		
	Out-plane	1.29	0.28	1.01		

initially. The relationship between the anisotropy of the electronic thermal conductivity is and the relationship between the anisotropy of the lattice thermal conductivity is $\kappa_{\perp,L}/\kappa_{\parallel,L} = 1.34$ [18]. The related calculation parameters and results of the bulk Bi_2Te_3 single crystal and the tiled nanosheet film in this section are shown in Table 7.5.

The out-of-plane thermal conductivity of the AB sample is 0.497 W m^{-1} K^{-1}. Without considering the existence of nanopores in the plane, the out-of-plane electrons are obtained based on the in-plane electronic thermal conductivity and anisotropic factor. The thermal conductivity is 0.10 W m^{-1} K^{-1}, and the out-of-plane lattice thermal conductivity is obtained based on the measured out-of-plane thermal conductivity 40 W m^{-1} K^{-1}. The outer lattice thermal conductivity and anisotropy factor obtained in-plane lattice thermal conductivity was 0.53 W m^{-1} K^{-1}. Finally, the estimated in-plane thermal conductivity is 1.22 W m^{-1} K^{-1}, which is 2.44 times the out-of-plane thermal conductivity. The in-plane thermal conductivity thus obtained can obtain an in-plane ZT value of the film of 1.2. This method is used to estimate the in-plane thermal conductivity without considering the nanopores. Due to the large number of nanopores in the plane, the in-plane phonons will be strongly scattered, making the actual thermal conductivity in the plane lower than the estimated value, so its actual ZT value will be greater than 1.2.

7.3 Chapter Summary

This chapter mainly explores the experimental process of preparing bismuth telluride material by molecular beam epitaxy:

1. Cross-shaped Bi_2Te_3 nanosheet films were prepared using MBE, and the films were heat treated with ethylene glycol for 0, 30, 60, and 90 min. During heat treatment, Bi_2Te_3 is dissolved in ethylene glycol at a certain ratio (Bi/Te = 2/3). Due to the difference in dissolution reaction time, the roughness of the nanosheet film surface changes, resulting in corresponding changes in the electron-phonon transmission on the surface of the Bi_2Te_3 nanosheet film. At 300 K, the maximum power factor of the film after heat treatment of ethylene glycol for 90 min was 13.4 μW cm^{-1} K^{-2}. Bi_2Te_3 film without ethylene glycol obtained a maximum power factor of 18 μW cm^{-1} K^{-2} at 400 K.
2. Through the annealing process, the Bi_2Te_3 cross-nanosheet thin film becomes a tiled single crystal nanosheet, and there are holes between the multiple nanosheets. The tiled nanosheet films were heat treated with ethylene glycol for 0, 30, 60, and 90 min. During heat treatment, Bi_2Te_3 was dissolved in ethylene glycol at a certain ratio (Bi/Te = 2/3), and new pores appeared in the tiled nanosheet film. With the increase of the heat treatment time of ethylene glycol, the number and size of pores increased roughly. Due to the difference in pore generation mechanism, the number of pores increased sharply and the average pore size decreased slightly during the 60 min heat treatment. On the whole, as the heat treatment time of ethylene glycol increases, the porosity increases in turn. Relative to porosity and pore size, the number of pores has a more significant effect on the power factor of the film. As the number of holes per unit area of the film increases, the Seebeck coefficient, electrical conductivity, and power factor gradually decrease. The tiled nanosheet film without ethylene glycol heat treatment has the largest power factor, up to 49.1 μW cm^{-1} K^{-2} at 300 K, and the corresponding porosity, pore size, and number are 0.6%, 122 nm, and 28 pcs/100 μm^2. At 300 K, the out-of-plane thermal conductivity of the cross-nanosheet and tile-type nanosheet films without ethylene glycol heat treatment are 1.27 W m^{-1} K^{-1} and 0.50 W m^{-1} K^{-1}, respectively. Through the analysis of the anisotropy of the film, the in-plane thermal conductivity was estimated to be 1.27 W m^{-1} K^{-1} and 1.22 W m^{-1} K^{-1}, respectively. Finally, the room temperature ZT values of the obtained cross-type and flat-type nanosheet films were 0.3 and 1.2, respectively. Because there are a large number of nanopores in the plane of the tiled nanosheet film, its actual ZT value will be greater than 1.2.

References

1. Wright DA (1958) Thermoelectric properties of bismuth telluride and its alloys. Nature 181(4612):834
2. Hicks LD, Dresselhaus MS (1993) Effect of quantum-well structures on the thermoelectric figure of merit. Phys Rev B 47(19):12727–12731
3. Plucinski L, Mussler G, Krumrain J, Herdt A, Suga S, Grutzmacher D, Schneider CM (2011) Robust surface electronic properties of topological insulators: Bi[sub 2]Te[sub 3] films grown by molecular beam epitaxy. Appl Phys Lett 98(22):222503
4. Wang G (2011) MBE-grown binary chalcogenide one-dimensional nanostructures: synthesis, structural analysis and formation mechanisms. PhD, Hong Kong University of Science and Technology (Hong Kong), Hong Kong
5. Shahil KMF, Hossain MZ, Teweldebrhan D, Balandin AA (2010) Crystal symmetry breaking in few-quintuple Bi_2Te_3 films: applications in nanometrology of topological insulators. Appl Phys Lett 96(15):153103
6. Li H, Cao J, Zheng W, Chen Y, Wu D, Dang W, Wang K, Peng H, Liu Z (2012) Controlled synthesis of topological insulator nanoplate arrays on mica. J Am Chem Soc 134(14):6132–6135. https://doi.org/10.1021/ja3021395
7. Teweldebrhan D, Goyal V, Balandin AA (2010) Exfoliation and characterization of bismuth telluride atomic quintuples and quasi-two-dimensional crystals. Nano Lett 10(4):1209–1218. https://doi.org/10.1021/nl903590b
8. Zhang Y, Mehta RJ, Belley M, Han L, Ramanath G, Borca-Tasciuc T (2012) Lattice thermal conductivity diminution and high thermoelectric power factor retention in nanoporous macroassemblies of sulfur-doped bismuth telluride nanocrystals. Appl Phys Lett 100(19):193113
9. Zhao K, Duan H, Raghavendra N, Qiu P, Zeng Y, Zhang W, Yang J, Shi X, Chen L (2017) Solid-state explosive reaction for nanoporous bulk thermoelectric materials. Adv Mater 29(42):1701148. https://doi.org/10.1002/adma.201701148
10. Shi X, Wu A, Liu W, Moshwan R, Wang Y, Chen ZG, Zou J (2018) Polycrystalline SnSe with extraordinary thermoelectric property via nanoporous design. ACS Nano 12(11):11417–11425. https://doi.org/10.1021/acsnano.8b06387
11. Dun C, Hewitt CA, Jiang Q, Guo Y, Xu J, Li Y, Li Q, Wang H, Carroll DL (2018) Bi_2Te_3 plates with single nanopore: the formation of surface defects and self-repair growth. Chem Mater 30(6):1965–1970. https://doi.org/10.1021/acs.chemmater.7b04985
12. Richter W, Becker CR (1977) A Raman and far-infrared investigation of phonons in the rhombohedral V_2–VI_3 compounds Bi_2Te_3, Bi_2Se_3, Sb_2Te_3 and $Bi_2(Te_{1-x}Se_x)_3$ $(0 < x < 1)$, $(Bi_{1-y}Sb_y)_2Te_3$ $(0 < y < 1)$. Phys Status Solidi (b) 84(2):619–628
13. Pine AS, Dresselhaus G (1971) Raman spectra and lattice dynamics of tellurium. Phys Rev B 4(2):356–371. https://doi.org/10.1103/PhysRevB.4.356
14. Buha J, Gaspari R, Del Rio Castillo AE, Bonaccorso F, Manna L (2016) Thermal stability and anisotropic sublimation of two-dimensional colloidal Bi_2Te_3 and Bi_2Se_3 nanocrystals. Nano Lett 16(7):4217–4223
15. Zhu W, Deng Y, Wang Y, Luo B, Cao L (2014) Preferential growth transformation of $Bi_{0.5}Sb_{1.5}Te_3$ films induced by facile post-annealing process: enhanced thermoelectric performance with layered structure. Thin Solid Films 556:270–276. https://doi.org/10.1016/j.tsf.2014.02.041
16. Takashiri M, Tanaka S, Miyazaki K (2010) Improved thermoelectric performance of highly-oriented nanocrystalline bismuth antimony telluride thin films. Thin Solid Films 519(2):619–624. https://doi.org/10.1016/j.tsf.2010.08.013

17. Sun T, Samani MK, Khosravian N, Ang KM, Yan Q, Tay BK, Hng HH (2014) Enhanced thermoelectric properties of n-type $Bi_2Te_{2.7}Se_{0.3}$ thin films through the introduction of Pt nanoinclusions by pulsed laser deposition. Nano Energy 8:223–230
18. Rowe IToECToECM (1995) CRC handbook of thermoelectrics. CRC Press
19. Saito K, Benenti G, Casati G (2010) A microscopic mechanism for increasing thermoelectric efficiency. Chem Phys 375(2–3):508–513

Chapter 8
Summary and Outlook

Thermoelectric materials can be designed by controlling at lower dimensions. Low-dimensional thermoelectric materials have excellent thermoelectric properties. After the thermoelectric materials are reduced in dimension, the state density near the Fermi level will increase, and the effective mass of the carriers and the absolute value of the Seebeck coefficient will increase. On the other hand, the quantum confinement effect of phonons reduces thermal conductivity. In addition, effects such as quantum constraints can increase carrier mobility. Compared with the more mature bulk thermoelectric materials and devices, thermoelectric thin films are easily combined with modern micro-nano processing technology to make micro-devices, which is suitable for a wider field. This book aims to build nano-scale thermoelectric thin films, regulates the factors affecting the performance of thermoelectric thin films, and provides reference for the low-dimensional application of thermoelectric thin films.

Magnetron sputtering is one of the most common thin film deposition methods in micro-nano processing technology. Firstly, multilayer thermoelectric thin films were constructed by magnetron sputtering. Si materials were selected and Si/($Si_{0.75}Ge_{0.25}$, Cr, Ti), Si/$Si_{0.75}Ge_{0.25}$/Au, and Si/Au multilayer films were constructed. Sb_2Te_3 materials were selected to construct Sb_2Te_3/(Au, Ag, Cu, Pt, Cr, Mo, W, Ta) multilayer films. By controlling different interfaces, periods, and thicknesses, and by constructing interfaces to scatter phonons, the thermal conductivity of the film is reduced. In addition, some multilayer films were selected for annealing treatment, thermal conductivity and power factor were adjusted, and the relationship between interface evolution and thermoelectric performance was studied.

Sb_2Te_3 and Bi_2Te_3 based films were deposited by magnetron sputtering, and the effects of sputtering methods, sputtering power, annealing temperature, thickness and other factors on the structure and thermoelectric properties of the films were controlled. In addition, Sb_2Te_3 and Bi_2Te_3 based films were deposited by molecular beam epitaxy, and the structure and thermoelectric properties of the films were investigated by precisely controlling the doping, growth mode, nanopore and other factors.

Z. Hu and Z. Wu, *Nanostructured Thermoelectric Films*,
https://doi.org/10.1007/978-981-15-6518-2_8

In the past few years, taking advantage of the microelectronics industry, more and more components, equipment and instruments have been miniaturized. In addition, thermoelectric materials have evolved from bulk thermoelectric materials to low-dimensional thermoelectric materials such as nanowires, superlattices, and multilayer films. In addition, miniaturization can increase the integration density of thermoelectric modules, thereby increasing the power output, and the smaller the feature size of the thermoelectric device, the greater the power density. Therefore, the miniaturization of thermoelectric devices is receiving increasing attention. Miniaturized thermoelectric devices can provide microwatts or milliwatts of energy to some micro-electromechanical systems (MEMS) device equipment. According to the structural characteristics of MEMS thermoelectric devices, MEMS thermoelectric devices can be divided into planar structure MEMS thermoelectric devices and vertical structure MEMS thermoelectric devices. Among them, the structure characteristic of the vertical structure thermoelectric device is that the thermoelectric column is perpendicular to the substrate, that is, the directions of heat flow and current flowing through the thermoelectric material are perpendicular to the substrate. Traditional thermoelectric devices are mainly vertical structures. Compared with the planar structure, the processing of the vertical structure is relatively simple, so most of the miniature thermoelectric devices are also vertical structures. In the existing research status at home and abroad, vertical structure MEMS thermoelectric devices are processed in order from the bottom electrode, thermoelectric material, and top electrode, and there are also sub-module processing and then bonding and assembly. The P-type and N-type thermoelectric units are connected in series in a certain order through the electrode material and sandwiched between two substrates with good thermal conductivity. When heat passes from the substrate, the thermoelectric units form a thermal parallel connection, generating electrical energy. Thermoelectric material processing methods include electron beam evaporation, electrochemical deposition, screen printing, molecular beam epitaxy, and magnetron sputtering. In summary, MEMS technology has been widely used in the processing of micro thermoelectric devices, and the application of MEMS technology has greatly promoted the development of the structure and performance of micro thermoelectric devices.

Radiation cooling is a passive cooling method developed in recent years. The heat of objects on the earth can be transmitted to outer space through the "atmosphere window" in the form of thermal radiation without using electrical energy and other energy sources, thereby reducing the temperature of the objects. The temperature in outer space is about 4 K, which is a huge source of cold for objects on Earth. However, the existence of water vapor, ozone, carbon dioxide and other gases in the earth's atmosphere hinders the heat exchange between earth objects and outer space through thermal radiation. However, for certain bands of electronic waves, the atmosphere can pass through them unhindered or with little obstruction. For electromagnetic waves in these bands, the atmosphere is transparent, so these can make the atmosphere Transparent bands are called "atmospheric windows". Atmospheric windows are mainly 0.3–2.5, 3.2–4.8, 8–13 μm, etc. Through these windows, objects on the earth can transfer heat to the outer space (approximately zero degrees) through thermal radiation. According to Wien's displacement law, the wavelength of the heat radiation wave of objects on the earth (temperature is in the range of 20–50 °C) corresponds to the atmospheric window of 8-13 μm. Therefore, people are most interested in the atmospheric window of 8-13 μm, and use this window to achieve radiative cooling. After decades of research, great progress has been made in improving radiator performance. In order to obtain good infrared emitters, many researchers have studied natural compounds, polymer films, pigment coatings, and gases. On the other hand, in recent years, more and more researchers have begun to focus on radiative cooling during the day and have made breakthrough progress. Achieve radiant cooling 24 h a day. This research group innovatively put forward and preliminary verified the research of using radiant cooling technology and micro/nano thermoelectric chip to realize power generation.

MEMS thermoelectric chip systems can convert small temperature differences (such as indoor and outdoor temperature differences, seawater temperature differences at different depths in oceans, cave temperature differences, infrared radiation cooling temperature differences, etc.) that are ubiquitous, timeless, and long neglected in nature. Solve the current problem of long-term stable supply of passive energy. The problem of complex maintenance and insufficient service life of the energy supply system has greatly improved its capabilities in outer space exploration. Aiming at the tiny temperature difference power generation chip based on radiant cooling, it can realize unattended power generation in all weather. In polar, island, mountain, desert and other harsh and unattended places, such as automatic unmanned weather stations, buoys and lighthouses, earthquake observation stations, aircraft navigation, microwave communication relay stations, etc. can use maintenance-free, long-life micro-temperature difference power generation chip. In medicine, the long-life micro-temperature difference power generation chip can be widely used in various implantable sensors. Although small in size, it can be maintenance-free and replaceable, eliminating the pain of patients replacing batteries. The small temperature difference power generation chip is also expected to be applied in the field of mobile phone manufacturing, which can make full use of the waste heat generated by the operation of the mobile phone and improve the battery life. In the natural world and in the production and life of human beings, there are various conditions of temperature differences. For example, the temperature difference between indoor and outdoor, the temperature difference between the exhaust pipe of the automobile engine, the temperature difference between the human body and the environment, the temperature difference between the sun-illuminated surface and the back, the temperature difference between the earth and outer space, the temperature difference between industrial waste gas and the environment, the temperature difference between the back of the solar panel and the environment, etc. All above-mentioned small temperature differences that have been ignored for a long time can be achieved by the small temperature difference power generation chip.

In general, the physical vapor deposition method is used to deposit thermoelectric thin films, and the adjustment and optimization of different parameters of thermoelectric materials are conducive to the preparation of micro-devices in combination with micro-processing technology, which can expand a wider range of application scenarios. The research work on thermoelectric thin films is still continuing, and at the same time, the work of applying them to micro-devices is also unremitting. We hope this book will be useful for workers studying thermoelectric thin films and their applications.

Appendix A
Thermal Conductivity of Si and Sb_2Te_3-Based Multilayer Films

Multilayer			Thickness (nm)	Number of layers	κ (W m^{-1} K^{-1})
Material A	Material B	Material C	A/B/C	A + B + C	
Amorphous Si	–	–	–	–	1.44
–	Amorphous $Si_{0.75}Ge_{0.25}$	–	–	–	0.76
Si	$Si_{0.75}Ge_{0.25}$	–	1.5/1	200 + 200	0.98
			3/2	100 + 100	1.09
			6/4	50 + 50	1.09
			12/8	25 + 50	0.85
			30/20	10 + 50	1.12
			12/10/0	10 + 10	0.94
		Au	12/10/10	10 + 8 + 2	0.97
			12/10/10	10 + 5 + 5	1.02
			13.1/8.5	10 + 10	1.01
	–	Cr	14.2/9.4		0.89
		Ti	12.0/7.9		0.44
		Au	12/1		0.67
			12/3		0.60
			12/5		0.62
			12/10		1.31
			12/20		1.55
			12/40		2.28

(continued)

Z. Hu and Z. Wu, *Nanostructured Thermoelectric Films*,
https://doi.org/10.1007/978-981-15-6518-2

(continued)

Multilayer			Thickness (nm)	Number of layers	κ (W m^{-1} K^{-1})
Material A	Material B	Material C	A/B/C	A + B + C	
Sb_2Te_3	–				1
		Au	13/1	10 + 10	0.85
			13/3		0.5
			13/5		0.45
			13/10		0.55
			13/20		0.72
			15/5		0.43
			15/10		0.60
			15/15		0.77
		Ag	15/5		0.16
			15/10		0.33
			15/15		0.50
		Cu	15/5		0.68
			15/10		1.39
			15/15		1.93
		Pt	15/5		1.44
			15/10		2.07
			15/15		2.52
		Cr	15/5		0.31
			15/10		0.43
			15/15		0.55
		Mo	15/5		0.21
			15/10		0.41
			15/15		0.53
		W	15/5		0.26
			15/10		0.37
			15/15		0.42
		Ta	15/5		0.22
			15/10		0.30
			15/15		0.37

Note The preparation method is magnetron sputtering, and the out-plane thermal conductivity obtained is at the room temperature

Appendix B
Thermoelectric Properties of Sb_2Te_3-Based Multilayer Thin Films

Z. Hu and Z. Wu, *Nanostructured Thermoelectric Films*,
https://doi.org/10.1007/978-981-15-6518-2

Multilayer		Thickness (nm)	Number of layers	Substrate temperature	Annealing temperature/time	Temperature range	σ (S cm^{-1})	α (μV K^{-1})	*PF* (μW cm^{-1} K^{-2})	κ (W m^{-1} K^{-1})
Material A	Material B	A/B	A + B							
Sb_2Te_3	–	200	1	Room temperature	423 K/6 h	300 K	200	90	1.62	–
					473 K/6 h		300	92	2.40	
	Cu	20/0.1	10 + 10		423 K/6 h		800	65	3.40	
					473 K/6 h		1587	43	2.90	
		20/0.3			423 K/6 h		1120	66	4.80	
					473 K/6 h		1250	58	4.20	
	–	200	1		423 K/6 h	210 K–425 K	127–174	100–138	1.28–3.25	
	Ag	20/1	10 + 10				150–197	138–162	2.85–5.10	
		20/2					60–129	160–197	1.61–4.84	
		20/4					342–589	88–121	4.56–4.89	
	Te	250	1	423 K	–	300 K	456.96	121	6.69	0.42
		5/1	50 + 50				425.54	147.5	9.26	0.31
		5/2					262.09	171	7.66	0.35
		5/4					99.74	195	3.79	0.4

Note The molecular beam epitaxy method is used to prepare Sb_2Te_3 by co-evaporation. The power factor is in-plane and the thermal conductivity is out-of-plane

Appendix C
Thermoelectric Properties of Sb_2Te_3/ Bi_2Te_3 Based Thin Film

Z. Hu and Z. Wu, *Nanostructured Thermoelectric Films*,
https://doi.org/10.1007/978-981-15-6518-2

Film	Deposition method	Substrate	Control parameters		Thickness (nm)	σ (S cm^{-1})	α (μV K^{-1})	*PF* (μW cm^{-1} K^{-2})	κ (W·m^{-1} K^{-1})
Sb_2Te_3	Sb_2Te_3 (RF) sputtering	Glass	Control sputtering power	15 W	~240	82.37	136.43	1.53	
				20 W		115.47	147.24	2.50	
				25 W		125.47	140.26	2.47	
			Sputtering power 20 W, annealing for 6 h, control annealing temperature	373 K	243	277.78	119.60	3.97	
				423 K	240	907.44	116.63	12.34	
				473 K	238	1043.84	119.01	14.78	
				523 K	236	1176.47	124.00	18.09	
				573 K	180	4.23	330.7	0.46	
			Sputtering power 20 W, control thickness	–	90	36.70	149.48	0.82	
					140	1	150.44	1.34	
					190	93.02	146.63	2.00	
					230	115.47	150.34	2.61	
					270	119.33	147.61	2.60	
				Annealing at 523 K for 6 h	90	1111.11	121.77	16.48	
					140	1190.48	120.93	17.41	
					190	1234.57	119.31	17.57	
					230	1298.70	117.95	18.07	
					270	1449.28	116.14	19.55	

(continued)

(continued)

Film	Deposition method	Substrate	Control parameters		Thickness (nm)	σ (S cm^{-1})	α (μV K^{-1})	*PF* (μW cm^{-1} K^{-2})	κ ($W{\cdot}m^{-1}$ K^{-1})
Sb_2Te_3–Te	Sb_2Te_3 (RF 150 W) + Te (DC 10 W) co-sputtering Te/Sb atomic ratios is1.8	SiO_2/Si	Annealing at 423 K, control annealing time	0	85	5.33	335.4	0.6	0.41
				30 min	80	524.4	145.9	11.2	0.55
				180 min	75	557.02	112.9	7.1	0.69
Sb_2Te_3	Sb (DC, 15 W) + Te (RF,10 W) co-sputtering	SiO_2/Si	Annealing for 2 h, control annealing temperature	473 K	915	75	104.2	0.8	–
				523 K	924	112	120.6	1.6	
				573 K	951	121	125.0	1.9	
				623 K	981	126	127.5	2.1	
				673 K	1200	101	121.5	1.5	
Bi_2Te_3	Bi (DC,13 W) + Te (RF,10 W) co-sputtering			473 K	687	73	−108.8	0.9	
				523 K	699	74	−126.4	1.2	
				573 K	673	82	−129.7	1.4	
				623 K	571	116	−134.4	2.1	
				673 K	552	98	−127.9	1.6	
Bi–Sb–Te	Sb_2Te_3 (RF, 20 W) + Bi	Glass	Control sputtering power of Bi	3 W	~240	32.82	66.1	0.14	
				4 W		28.89	64.07	0.12	
				5 W		23.80	58.73	0.08	

(continued)

(continued)

Film	Deposition method	Substrate	Control parameters		Thickness (nm)	σ (S cm^{-1})	α (μV K^{-1})	*PF* (μW cm^{-1} K^{-2})	κ (W·m^{-1} K^{-1})
	Sb_2Te_3 (RF, 20 W) + Bi (DC, 3 W		Annealing for 6 h, control annealing temperature	373 K		66.72	115.98	0.90	
				423 K		104.53	124.77	1.63	
				473 K		203.22	127.41	3.30	
				523 K		714.29	154.64	17.08	
				573 K		620.45	190.6	22.54	
				623 K		68.86	19.9	0.03	
			Control thickness	–	90	16.28	93.9	0.14	
					140	21.34	86.93	0.16	
					180	24.14	77.95	0.15	
					240	32.71	66.34	0.14	
					280	34.65	66.17	0.14	
				Annealing at 673 K for 6 h	90	589.57	167.19	16.48	
					140	603.25	171.95	17.84	
					180	613.21	180.92	20.07	
					240	623.50	190.96	22.74	

(continued)

(continued)

Film	Deposition method	Substrate	Control parameters		Thickness (nm)	σ (S cm^{-1})	α (μV K^{-1})	*PF* (μW cm^{-1} K^{-2})	κ (W·m^{-1} K^{-1})
					280	636.94	203.82	26.41	
			The substrate is at room temperature, annealed for 6 h, control annealing temperature	Room temperature		24.24	78.45	0.15	
				423 K		60.31	125.15	0.94	
				473 K		141.67	130.70	2.42	
				523 K		689.37	153.63	16.27	
				573 K		613.31	181.16	20.13	
				623 K		35.19	58.40	0.12	
			The substrate is at 423 K, annealed for 6 h, control annealing temperature	Room temperature		117.67	112.72	1.49	
				423 K		173.25	121.06	2.54	
				473 K		251.85	133.77	4.51	
				523 K		605.84	161.81	15.86	
				573 K		694.33	190.99	25.32	
				623 K		34.87	86.00	0.26	
			The substrate is at 473 K, annealed for 6 h, control annealing temperature	Room temperature		102.22	109.09	1.22	
				423 K		159.06	124.13	2.45	
				473 K		200	131.72	3.47	
				523 K		536.13	168.45	15.21	
				573 K		624.53	190.87	22.75	
				623 K		44.44	93.16	0.39	

(continued)

(continued)

Film	Deposition method	Substrate	Control parameters		Thickness (nm)	σ (S cm^{-1})	α (μV K^{-1})	PF (μW cm^{-1} K^{-2})	κ (W·m^{-1} K^{-1})
			The substrate is at 523 K, annealed for 6 h, control annealing temperature	Room temperature		73.92	103.87	0.80	
				423 K		152.81	120.55	2.22	
				473 K		181.33	128.65	3.00	
				523 K		526.49	155.16	12.68	
				573 K		533.51	195.64	20.42	
				623 K		68.86	102.88	0.73	

Note The above table shows the thermoelectric performance at room temperature, the power factor is in-plane, and the thermal conductivity is out-of-plane

Appendix D
Thermoelectric Properties of Sb_2Te_3/Bi_2Te_3 Thin Films Prepared by Molecular Beam Epitaxy

Z. Hu and Z. Wu, *Nanostructured Thermoelectric Films*,
https://doi.org/10.1007/978-981-15-6518-2

Film	Substrate	Control parameters			σ (S m^{-1})	α (µV K^{-1})	*PF* (µW cm^{-1} K^{-2})	κ (W m^{-1} K^{-1})	ZT
Sb_2Te_3–Sb	Quartz	Control Sb (at%) percentage		10%	97.58	121	1.43	0.13	0.33
				40%	144.71	90	1.17	0.17	0.21
				63%	7.32	536	2.10	0.10	0.63
Sb_2Te_3	Si (111)	Control film thickness		28 nm	425.713	–			
				67 nm	623.83				
				98 nm	898.473				
				121 nm	1036				
		No-doping		Sb_2Te_3	64.3	159.86	1.64	–	–
		Bi-doping		$Bi_{0.4}Sb_{1.6}Te_3$	801	134.71	14.5		
Bi_2Te_3	SiO_2/Si	–	Heat treated by ethylene glycol 343 K, control time	0	582.9	−148.8	12.9	1.27	
				30 min	406.3	−154.0	9.6	–	
				60 min	468.0	−165.2	12.8		
				90 min	760.0	−132.8	13.4		
		Annealing at 573 K for 2 h		0	1398.6	−187.3	49.1	0.50	
				30 min	787.4	−182.0	26.1	–	
				60 min	562.6	−175.9	17.4		
				90 min	690.9	−178.5	22.0		

Note The film is prepared by molecular beam epitaxy co-evaporation. The above table shows the thermoelectric performance at room temperature, the power factor is in-plane, and the thermal conductivity is out-of-plane

Bibilography

1. Scherrer P (1918) Determination of the size and internal structure of colloidal particles using X-rays. Nachr Ges Wiss Göttingen 2:98–100
2. Pal K, Anand S, Waghmare UV (2015) Thermoelectric properties of materials with nontrivial electronic topology. J Mater Chem C 3(46):12130–12139
3. Shekar NVC, Polvani DA, Meng JF, Badding JV (2005) Improved thermoelectric properties due to electronic topological transition under high pressure. Phys B 358(1–4):14–18
4. Zhang H, Liu CX, Qi XL, Dai X, Fang Z, Zhang SC (2009) Topological insulators in Bi_2Se_3, Bi_2Te_3 and Sb_2Te_3 with a single Dirac cone on the surface. Nat Phys 5(6):438–442
5. Zhao K (2017) First principle study of electronic topological transition in Bi_2Te_3 and Sb_2Te_3 under pressure Doctor, Harbin Institute of Technology
6. Gomis O, Vilaplana R, Manjón FJ, Rodríguez-Hernández P, Pérez-González E, Muñoz A, Kucek V, Drasar C (2011) Lattice dynamics of Sb_2Te_3 at high pressures. Phys Rev B 84(17):174305
7. Zhao K, Wang Y, Xin C, Sui Y, Wang X, Wang Y, Liu Z, Li B (2016) Pressure-induced anomalies in structure, charge density and transport properties of Bi_2Te_3: a first principles study. J Alloys Compd 661:428–434
8. Bera A, Pal K, Muthu DVS, Sen S, Guptasarma P, Waghmare UV, Sood AK (2013) Sharp Raman anomalies and broken adiabaticity at a pressure induced transition from band to topological insulator in Sb_2Se_3. Phys Rev Lett 110(10):107401
9. Yu Z, Wang L, Hu Q, Zhao J, Yan S, Yang K, Sinogeikin S, Gu G, Hk Mao (2015) Structural phase transitions in Bi_2Se_3 under high pressure. Sci Rep 5:15939
10. Zhang J, Hu T, Yan J, Ke F, Wang J, Cui X, Li X, Ma Y, Yang J, Gao C (2017) Pressure driven semi-metallic phase transition of Sb_2Te_3. Mater Lett 209:78–81
11. Meng JF, Shekar NVC, Badding JV, Chung DY, Kanatzidis MG (2001) Multifold enhancement of the thermoelectric figure of merit in p-type $BaBiTe_3$ by pressure tuning. J Appl Phys 90(6):2836–2839
12. Polvani DA, Meng JF, Chandra Shekar NV, Sharp J, Badding JV (2001) Large improvement in thermoelectric properties in pressure-tuned p-type $Sb_{1.5}Bi_{0.5}Te_3$. Chem Mater 13 (6):2068–2071
13. Secor J, Harris MA, Zhao L, Deng H, Raoux S, Krusin-Elbaum L (2014) Phonon renormalization and Raman spectral evolution through amorphous to crystalline transitions in Sb_2Te_3 thin films. Appl Phys Lett 104(22):221908
14. Richter W, Renucci JB, Cardona M (1973) Hydrostatic pressure dependence of first-order Raman frequencies in Se and Te. Phys Status Solidi (b) 56(1):223–229
15. Wu Y, Chen Z, Nan P, Xiong F, Lin S, Zhang X, Chen Y, Chen L, Ge B, Pei Y (2019) Lattice Strain Adv Thermoelectr. Joule 3:1276–1288
16. Klemens PG (1958) Thermal conductivity and lattice vibrational modes. In: Solid state physics, vol 7. Elsevier, pp 1–98

Z. Hu and Z. Wu, *Nanostructured Thermoelectric Films*,
https://doi.org/10.1007/978-981-15-6518-2

17. Carruthers P (1959) Scattering of phonons by elastic strain fields and the thermal resistance of dislocations. Phys Rev 114(4):995–1001
18. Liu E, Zhu B, Luo J (2008) Semiconductor physics, 7th edn. Electron Industry Press, Beijing
19. Zhu T, Hu L, Zhao X, He J (2016) New insights into intrinsic point defects in V_2VI_3 thermoelectric materials. Adv Sci 3(7):1600004
20. Noguchi S, Sakata H (1980) Electrical properties of undoped In_2O_3 films prepared by reactive evaporation. J Phys D: Appl Phys 13(6):1129–1133
21. Li YY, Qin XY, Li D, Zhang J, Li C, Liu YF, Song CJ, Xin HX, Guo HF (2016) Enhanced thermoelectric performance of $Cu_2Se/Bi_{0.4}Sb_{1.6}Te_3$ nanocomposites at elevated temperatures. Appl Phys Lett 108 (6):062104
22. Shen S, Zhu W, Deng Y, Zhao H, Peng Y, Wang C (2017) Enhancing thermoelectric properties of Sb_2Te_3 flexible thin film through microstructure control and crystal preferential orientation engineering. Appl Surf Sci 414:197–204
23. Saleemi M, Ruditskiy A, Toprak MS, Stingaciu M, Johnsson M, Kretzschmar I, Jacquot A, Jagle M, Muhammed M (2014) Evaluation of the structure and transport properties of nanostructured antimony telluride (Sb_2Te_3). J Electron Mater 43:1927–1932. https://doi.org/10.1007/s11664-013-2911-6
24. Van Thiet D, Van Quang N, Hai NTM, Huong NT, Cho S, Tuan DA, Dung DD, Van Tam T (2018) Optimizing the carrier density and thermoelectric properties of Sb_2Te_3 films by using the growth temperature. J Korean Phys Soc 72(8):915–919
25. Schumacher C, Reinsberg KG, Akinsinde L, Zastrow S, Heiderich S, Toellner W, Rampelberg G, Detavernier C, Broekaert JA, Nielsch K (2012) Optimization of electrodeposited p-doped Sb_2Te_3 thermoelectric films by millisecond potentiostatic pulses. Adv Energy Mater 2(3):345–352
26. Shi W, Yu S, Liu P, Fan W (2013) Hydrothermal synthesis and thermoelectric transport properties of Sb_2Te_3–Te heterogeneous nanostructures. CrystEngCommun 15(15):2978–2985
27. Sun Y, Zhang E, Johnsen S, Sillassen M, Sun P, Steglich F, Bøttiger J, Iversen BB (2011) Growth of FeSb2 thin films by magnetron sputtering. Thin Solid Films 519(16):5397–5402
28. Nuthongkum P, Sakdanuphab R, Horprathum M, Sakulkalavek A (2017) [Bi]:[Te] control, structural and thermoelectric properties of flexible BixTey thin films prepared by rf magnetron sputtering at different sputtering pressures. J Electron Mater 46(11):6444–6450
29. Kim D-H, Byon E, Lee G-H, Cho S (2006) Effect of deposition temperature on the structural and thermoelectric properties of bismuth telluride thin films grown by co-sputtering. Thin Solid Films 510(1–2):148–153

The manufacturer's authorised representative in the EU is Springer Nature Customer Service Centre GmbH, Europaplatz 3, 69115 Heidelberg, Germany. If you have any concerns regarding our products, please contact ProductSafety@springernature.com

Printed and bound by CPI Group (UK) Ltd, Croydon, CR0 4YY
30/09/2026
02225331-0002